Theories of Scientific Method

Classics in the History and Philosophy of Science

Series Editor
Roger Hahn, University of California, Berkeley

Volume 1
Historical and Philosophical Perspectives of Science
Edited by Roger Stuewer

Volume 2
Theories of Scientific Method: The Renaissance through the Nineteenth Century
by Ralph M. Blake, Curt J. Ducasse and Edward H. Madden
Edited by Edward H. Madden

Volume 3
Science and Technology in the Industrial Revolution
A. E. Musson and Eric Robinson

Volume 4
Identity and Reality
Emile Meyerson

Volume 5
Lavoisier: The Crucial Year—1772
Henry Guerlac

Volume 6
Jean D'Alembert - Science and the Enlightenment
Thomas L. Hankins

Volume 7
The Newtonians and The English Revolution 1689–1720
Margaret C. Jacob

Theories of Scientific Method
The Renaissance through the Nineteenth Century

by

Ralph M. Blake, Curt J. Ducasse
and **Edward H. Madden**

Edited by

Edward H. Madden

Gordon and Breach

New York Philadelphia London Paris Montreux Tokyo Melbourne

Gordon and Breach Science Publishers

Post Office Box 786
Cooper Station
New York, New York 10276
United States of America

5301 Tacony Street, Slot 330
Philadelphia, Pennsylvania 19137
United States of America

Post Office Box 197
London WC2E 9PX
United Kingdom

58, rue Lhomond
75005 Paris
France

Post Office Box 161
1820 Montreux 2
Switzerland

3-14-9, Okubo
Shinjuku-ku, Tokyo 169
Japan

Private Bag 8
Camberwell, Victoria 3124
Australia

Library of Congress Cataloging-in-Publication Data

Blake, Ralph M.
Theories of scientific method : the Renaissance through the nineteenth century / by Ralph M. Blake, Curt J. Ducasse, and Edward H. Madden ; edited by Edward H. Madden.
p. cm. — (Classics in the history and philosophy of science ; 2)
Reprint. Originally published: Seattle : University of Washington Press, 1960.
Bibliography: p.
Includes index.
ISBN 2-88124-351-7 (France)
1. Science—Methodology. 2. Science—Philosophy. 3. Science—History. I. Ducasse, Curt John, 1881–1969. II. Madden, Edward H. III. Title. IV. Series.
Q175.B58 1989
500—dc20

89-76:3
CIP

Contents

Preface

THESE THIRTEEN studies are historical and philosophical investigations of the theories of scientific method held by some leading thinkers from the Renaissance through the nineteenth century. The studies primarily concern the philosophy of science, that is, the logic and assumptions of science, and only incidentally, for background purposes, the history of science.

We chose our particular authors because we think they are important historically and still have something significant to say philosophically, although we do not mean to suggest, of course, that we think the ones we omitted are historically or intrinsically unimportant. Our authors are primarily concerned with these persistently recurring philosophical topics and problems: theory construction, hypothesis, causality, lawfulness, probability, the experimental methods, the uniformity of nature and the rule of succession (the problem of induction), the universality of causality, and the problem of discriminating the a priori and the empirical elements of science. We hope that our book will take its place as a helpful reference volume in which one can find detailed statements of what these authors said, correlations and developments among their thoughts, and insight into the contemporary discussions of these fundamental and fascinating problems.

Although the chapters are interrelated and contain explicit comparisons, each chapter is a complete study in itself and can be read as a separate unit.

Professor Ducasse is responsible for chapters 3, 7, 8, 9, and 10; Professor Blake for chapters 1, 2, 4, and 6; and I am responsible for chapters 5, 11, 12, and 13. I have used the usual editorial devices. The quotations are from original sources, and the translations frequently are our own; where this is not the case the translator is noted.

Several chapters have appeared in print previously and are reprinted, with minor changes, by the permission of the editors and publishers. The following appeared in *The Philosophical Review:* Chapter 4, XXXVIII (1929), 125–43, 201–18; Chapter 6, XLII (1933), 453–86; Chapter 9, LX (1951), 56–59, 213–34. Chapter 3 appeared originally in *Structure, Method, and Meaning: Essays in Honor of Henry M. Sheffer,* ed. Paul Henle, Horace M. Kallen, and Susanne K. Langer (New York: Liberal Arts Press, 1951). Chapter 8 is reprinted from *Studies in the History of Cultures: The Disciplines of the Humanities* (Menasha, Wis.: George Banta, published for the Conference of Secretaries of the American Council of Learned Societies, 1942).

I am indebted to Timothy Duggan, Paul Richards, and Roberta Smith for bibliographical aid; to Delvin Covey, Leonard Dean, and Joseph Palermo for advice on translations; to Betty Seaver for typing and proofreading; and to Philip P. Wiener for reading the entire manuscript with care and insight and making innumerable valuable additions and corrections. To Marian Madden I am indebted deeply for assistance of every sort, great and small, at every stage of preparation of the manuscript.

E. H. M.

Theories of Scientific Method: The Renaissance through the Nineteenth Century

CHAPTER ONE

Natural Science in the Renaissance

I. NATURAL SCIENCE AND HUMANISM

AT THE CLOSE of the fourteenth century the intellectual life of Europe was manifesting itself in three principal tendencies, each of which continued to play an important part also throughout the fifteenth century.

There were first of all the followers of Thomas Aquinas and of Duns Scotus who interpreted Aristotelian metaphysics and natural science in ways designed to bring it into harmony with the revealed dogmas of Christianity. These later followers of the great thirteenth-century scholastics were, however, by no means the equals of their masters. They increasingly appear rather as the representatives and defenders of a past tradition than as living thinkers, and they exhibit little but indifference or hostility to the newer scientific ideas of the time.

In the second place there were the dogmatic defenders of Latin Averroism, active chiefly in the north of Italy, who cared more for the letter of the Aristotelian teaching, as interpreted by Averroes, whether in logic, metaphysics, or natural science, than for either orthodox theology or progressive science. Indeed it was they more than any others who stood out most irreconcilably for the unshakable truth of even the details of Aristotelian science and met all the newer scientific developments, however well buttressed by empirical observation or fortified by experiment, with nothing but an unsympathetic and dogged resistance.

Finally there were the Occamists, particularly at Paris, the keenest and most progressive thinkers of the time, busy with their acutely critical philosophy and with the prosecution of their forward-looking scientific investigations. With this latter aspect of their activity the famous Cardinal Nicolas of Cusa (1401–64), whose own philosophy was inspired chiefly by Platonic and Augustinian influences, was in full sympathy, welcoming especially the increasing cultivation of mathematical studies and the utilization of mathematical methods in the physical sciences.[1]

The importance of this school of scientific studies at Paris is difficult to overestimate. The notion that modern mechanical science was completely created by Galileo and his emulators and disciples, Baliani, Torricelli, Descartes, Beeckman, and Gassendi, dies slowly. The truth of the matter is, as Pierre Duhem writes,[2] that Galileo and his contemporaries used the mathematical skill they had obtained in their studies of the ancient geometers to explicate and make more precise a mechanical science of which the Christian Middle Ages, in one of its aspects, had provided the most fundamental concepts and principles. The physicists who taught at Paris in the fourteenth century, taking observation as their guide, had clearly reached the conception of "modern mechanics," substituting it for the dynamics of Aristotle, which, they had come to believe, was powerless to "save the phenomena." Of the importance of the Parisians, Duhem says:

> . . . from the field cultivated by the philosophers and the theologians of Paris there was destined to arise the most marvellous harvest that Science had ever gathered in. The *Calculationes* of Suiseth, the discussions on infinite divisibility, on the intensity of forms, on uniform movement, or uniformly varied movement, were so many seeds that were to spring up in the following century and produce analytic geometry, the infinitesimal calculus, kinematics and dynamics. Such men as Gregory of Rimini, John Buridan, Albert of Saxony and Nicolas of Oresme . . . were the precursors of Galileo and Descartes, of Cavalieri and Torricelli, of Fermat and of Pascal.[3]

Their prefigurement, indeed, is impressive—although, it must be remembered, they retained some of the teleological elements of the concepts of Aristotelian physics. Buridan developed a concept of inertia and conceived of gravity as uniformly accelerated motion. Albert of Saxony formulated with some precision the laws of falling bodies and of mass. Nicholas of Oresme did three important things: "He invented the idea of analytic geometry; he discovered the formula for uniformly accelerated motion; and he argued cogently for the rotation of the earth, a widely known hypothesis."[4] By 1375, then, "modern" mechanics was beginning.

In Italy, however, in the fifteenth century there had developed against this new scientific movement a notable opposition. By the time the fifteenth century was well under way, the enthusiastic new literary and artistic movements characteristic of the Renaissance seem to have been absorbing most of the best energies of the time. Impatient of close and patient philosophical analysis, which they despised as logic-chopping, and of sober scientific investigation of nature, many of the leading spirits were inspired by the "new learning," and especially by the newly discovered writings of Plato and his Neoplatonic interpreters, to an impressionistic and eclectic philosophical dilettantism, frequently combined with an enthusiastic revival of occultism, magic, astrology, and all the fantastic extravagances of the "hermetic" sciences. Of the "Humanists," as these leading spirits came to be known, Duhem writes:

> Excited by poetry and eloquence, delicate admirers of Roman or Attic elegance, the Humanists felt no desire to take part in the discussions which were rife at the Sorbonne, in the noisy Rue de Fouarre or at the College of Montaigu; the subjects of these discussions seemed to them too abstract; the methods by which they were conducted appeared to them too subtile; and above all their refined Latinism could not endure the "style of Paris," the rude technical language which these arguments knew not how to dispense with. Hermaleo Barbaro, for example "pursues his insults of these barbaric philosophers; they are commonly accounted, says he, dirty, rude and uncultivated; while they live they are scarcely alive, and after their death they live no more; or if they live it is in misery and opprobrium." The humility in which these monks and masters of arts had buried their laborious existence offended to the point of disgust the Italians of the Renaissance, thirsty for fame as they were.[5]

In fact the Humanists continued to display, well on into the sixteenth century, a characteristic hostility toward empirical scientific research itself. Petrarch had already expressed his scorn for the study of natural science. "Even if these things were true, they would be of no assistance in securing us a happy life. For what would be the advantage of knowing the nature of animals, birds, fishes and reptiles, while remaining ignorant of the nature of man, and neither knowing nor caring whence he comes nor whither he goes?"[6] Erasmus, too, constantly ridicules the pursuit of science. He catalogues among the "fools" those plodding virtuosos who plunder the inward recesses of nature for a new invention and rake over sea and land trying to turn up some latent mystery.

A honeyed hope cajoles them, so that they begrudge no pains or costs, but with marvellous ingenuity contrive that by which they may deceive themselves; and they go on in this pleasant imposture till, having gone through their possessions, there is not enough left to build a new furnace. Even then they do not leave off dreaming their pleasant dreams, but with all their strength they urge others to seek the same happiness.[7]

And more to the same effect. In a later passage he ridicules the students of natural science:

The scientists, reverenced for their beards and the fur on their gowns . . . teach that they alone are wise while the rest of mortal men flit about as shadows. How pleasantly they dote, indeed, while they construct their numberless worlds, and measure the sun, moon, stars, and spheres as with thumb and line. They assign causes for lightning, winds, eclipses, and other inexplicable things, never hesitating a whit, as if they were privy to the secrets of nature, artificer of things, or as if they visited us fresh from the council of the gods. Yet all the while nature is laughing grandly at them and their conjectures. For to prove that they have good intelligence of nothing, this is a sufficient argument: they can never explain why they disagree with each other on every subject. Thus knowing nothing in general, they profess to know all things in particular. . . . When they especially disdain the vulgar crowd is when they bring out their triangles, quadrangles, circles, and mathematical pictures of the sort, lay one upon the other, intertwine them into a maze, then deploy some letters as if in line of battle, and presently do it over in reverse order—and all to involve the uninitiated in darkness. Their fraternity does not lack those who predict future events by consulting the stars, and promise wonders even more magical; and these lucky scientists find people to believe them.[8]

And we find in Rabelais a characteristically violent echo of the same spirit:

And I wonder much at a rabble of foolish philosophers and physicians, who spend their time in disputing whence the heat of the said waters cometh, whether it be by reason of borax, or sulphur, or alum, or saltpetre that is within the mine. For they do nothing but dote, and better were it for them to rub their arse against a thistle, than to waste away their time thus in disputing of that whereof they know not the original; for the resolution is easy, neither need we to inquire any further than that the said baths came by a hot piss of the good Pantagruel.[9]

Such Renaissance "Platonists" as Marsilio Ficino and Francesco Pico della Mirandola also paint a dim picture of the significance, and even the possibility, of science, maintaining that a genuine knowledge of the natural world solely through empirical science is altogether impossible. For Ficino, Ernst Cassirer writes, the characteristic superiority of Plato consists in his devotion, from the very beginning, to an investigation of the Divine, while other philosophers lost themselves in the study of nature, of which one can have only incomplete and dreamlike knowledge. "In thus relegating the corporeal world to a lower sphere of existence and of knowledge," Cassirer says, "Ficino is sharply distinguished from the truly modern form of Platonism, which grew up on the basis of exact science." [10]

Francesco Pico's position is very similar; like Ficino he attacks the empirical epistemology of the Aristotelians. In Aristotle's writings "we discover that sense is the sole foundation of the whole structure, and the beginning of the whole demonstration: for he fortifies that demonstration with universal propositions, derives the universal propositions from particulars by induction, and obtains the particulars by sense." [11] But this foundation is altogether insecure. Aristotle discovered only a few ways in which the judgment of sense may err and thought that apart from these the senses could not go astray.

> But we hold that there are as many modes or species of sense as there are human temperaments, and in the fifth book of this work we shall prove this more precisely. So that it is impossible to derive any certain rule for judging of a sensible thing. And in that same fifth book we shall also show, by means of its own very rules, that the judgment of sense is in great part uncertain.[12]

In fact "the very nature of sense is various, not only because of variations in the object of sense, but because of variations in the human constitution, which is also various in its own nature." [13] And Pico proceeds to make the most of all the devices of the Greek skeptics in order to show at length the deceitfulness of sense experience and the impossibility of any secure criterion by which judgments of sense may be corrected.

Lodovico Vives' attack on the Aristotelian theory of science, on the other hand, is directed chiefly against the possibility of demonstrative knowledge. Aristotle "will have demonstration to be from primary and necessary principles, from proximate principles and from causes." [14] But how are such principles to be certified? Such principles come from induction. But induction can certify nothing. "For all universal principles are derived by us from particulars. Since these particulars are infinite in number we cannot enumerate them all: but if one individual is lacking, the universal is not established." Moreover, "there have been universal propositions attested by long experience, concerning the heavens or the elements, which lapse of time or change of place have shown cannot be asserted universally." Are they then to be certified by some sort of rational self-evidence? No stable basis for science can be found in this direction. What is evident to some men will by no means be so to others. "You take us men as standards for these immediate propositions, concerning which there shall be no need that any should be taught. But if you are teaching men, you will not thus obtain any single and constant demonstration: for some principles will be immediately evident and primary to some men, others to others," and so forth. But if there is no solid foundation for demonstrative science the whole tradition concerning demonstration is empty and useless. Here Vives appeals to direct experience of nature as itself giving us the only valid demonstration.[15] "All arts are thus discovered, from particular experiments, which the senses show forth, the rule of the art being derived by the work of the mind." [16] "But you seem never to have so much as cast your eyes upon nature herself, for you take us men as standards for these immediate propositions," and so forth. "Tell me, is your mind set on us, and on capturing us, or on the nature of things?" In truth "demonstration does not submit itself to the nature of our understanding, but keeps to the straight way to the truth; nor does it wait upon our ability, but applies itself solely to the nature of the thing itself, that it investigates, that it explores, even if it be to us unknown and invisible." [17]

However, as to the method of such an empirical science as he desiderated Vives is extremely vague. But his scorn for the actual achievements of the "*Moderni*" at Paris is not vague!

What science can there be of things so far removed and separated from God, from the senses, and from the whole mind? Yet from these things there arises, as it were from some foundation in the void, a vast edifice of assertions and discordant opinions: concerning increase or decrease of intensity, concerning the dense and the rare, concerning uniform movement, non-uniform, uniformly varied, non-uniformly varied. What of the fact that there are many who with unbridled license discuss cases that never occur, nor even can occur in nature: as the infinitely dense or rare, an hour divided into proportional parts in this or that ratio, so that in each of these there shall be a certain rate of motion or of alteration of motion, or of rarefaction? They leave aside as it were already explored and sufficiently known the secrets of nature; they have done with things which are, will be or have been, and give their attention to things which can scarcely happen at all: they are ignorant of what lies under their feet, and spy out things which never exist. . . .[18]

So the Humanists and Platonists pursued the Parisians with cries of mockery; or worse yet they, and others, simply ignored them. The future, however, lay with the logicians of Paris: "What would they not have given to have plumbed the future, and what comfort they would have found there! The centuries to come were destined to tire very soon of Humanism; the Latin elegances of an Erasmus or a Vives were scarcely adapted long to retain the favor of men of taste, once the modern languages were in condition to produce their more beautiful masterpieces." [19] On the other hand, those very notions of the Parisians which Vives scorned most were the seeds that proliferated into the fruits of modern mechanical science.

The Averroistic Aristotelians of the fifteenth century were, if anything, even more hostile to the Parisians than the Humanists and Platonists. Averroistic Aristotelians were particularly strong at Padua, where they worked in close contact with the medical school; so their denial of the dynamics of the Parisians in favor of the inadmissible dynamics of Aristotle was particularly powerful in Italy. Nevertheless, in spite of this stiff resistance, the Parisian tradition found supporters in fifteenth-century Italy, both in and out of the universities. However, during this period the masters and savants, even those favorable toward the terminalistic doctrines of Paris, mainly limited themselves to reproducing, in an abbreviated and sometimes hesitating fashion, the essential theses of this mechanics and were far from making it produce any of the fruits of which it was the flower. Leonardo da Vinci, however, as we shall see, was an exception: "He was not content to admit the general principles of the Dynamics of the Impetus; he meditated upon these principles constantly, turned them in every direction, constraining them, as it were, to yield the consequences they implicitly contained." [20]

Early in the sixteenth century there was a renewal of interest in the dynamics of the Masters of Paris at the Sorbonne, which issued in the printing of their works. The Parisians continued their work in spite of the derision of the Humanists, Platonists, and Averroists until finally in 1545 Domenico de Soto brought together the work of Albert of Saxony and Nicholas of Oresme to form the exact law of falling bodies.[21] And during this century the minority of radical scientists in Padua and elsewhere in Italy increased their activity. John Herman Randall, Jr., writes that "Tartaglia, Cardano, Scaliger, and Baldi developed [their ideas on the theory of projectiles and ballistics] in successive stages of criticism, until finally in 1585 Benedetti formulated the main principles of Galilean dynamics."[22] This achievement was made possible by the mathematical interpretation of nature that was facilitated by Tartaglia when he published the first Latin edition of Archimedes' mathematical methods of analysis and synthesis in 1543. Randall concludes: "The science of dynamics as it reached Galileo was thus the result of the careful reconstruction of the Aristotelian physics in the light of the mathematical interpretation of nature."[23] And Duhem, on the same theme, writes:

It is the tradition of Paris of which Galileo and his emulators were the heirs. When we see the science of a Galileo triumphing over the stubborn Aristotelianism of a Cremonini [died 1631], we may think, in our ignorance of the history of human thought, that we are witnessing the victory of the youth of modern thought over the philosophy of the Middle Ages grown obstinate in its Psittacism; in truth we are viewing the triumph, long prepared, of the science that was born at Paris in the XIVth Century, over the doctrines of Aristotle and of Averroes, restored to honor by the Italian Renaissance.[24]

2. LEONARDO DA VINCI ON THE METHOD OF SCIENCE

Leonardo da Vinci (1452–1519), it is clear, was influenced by the Scholastics of Paris in his thinking on mechanics. In his day, as we have seen, the Averroists in Italy were opposing vigorously the mechanics of the "*Moderni*"; and even the masters and savants who were inclined favorably toward the Parisian doctrine limited themselves mainly to abbreviated and hesitating résumés. Leonardo, on the contrary, went into detail and followed out the consequences of the new doctrine.

It is doubtless true that Leonardo did not always recognize the full richness of the treasure heaped up by the Scholasticism of Paris; he neglects some features the borrowing of which would have contributed very much to the perfection of his mechanical doctrines. None the less the complexion of his Physics as a whole puts him among the number of those whom the Italians of his time called Parisians. This title, moreover, was justly conferred upon him; he had in fact derived the principles of his physics from the assiduous reading of Albert of Saxony, probably also from his meditations upon the writings of Nicolas of Cusa; and Nicolas of Cusa was himself also an adept of the Mechanics of Paris. Thus Leonardo rightly takes his place among the Parisian precursors of Galileo.[25]

It is precisely in Leonardo's writing also that we find any noteworthy consideration, in the fifteenth century, of the subject of scientific method. Yet not even here have we any systematic discussion of the subject. Like most active scientific workers Leonardo exhibits in practice a familiarity with principles of method of which he is scarcely reflectively conscious. Edmondo Solmi observes:

Leonardo da Vinci, like Galileo and Kepler, did not develop a theory of method, but his constant effort after truth leads him in practice to so clear a conception of methods of procedure and proof that his work deserves an important place in the development of scientific methodology. Presupposed and often not expressed, followed and often not openly stated, the rules of observation, of hypothesis, of experiment, of description and comparison come out clearly in his investigations as a whole. His work, moreover, leaves continuous and evident traces of the road the investigator has followed in his effort after truth—a road sometimes tortuous, but always interesting from the methodological point of view.[26]

Indeed it is only in isolated statements, gathered here and there from among the scattered and fragmentary remains of his notebooks, unpublished during his lifetime (and buried in the Biblioteca Ambrosiana of Milan for over three centuries), that we find any expressly formulated account of the nature and methods of science as these appeared to the keenest mind of his age. But this account in itself, interesting as it is, will be found to contain little with which our study of the past has not made us abundantly familiar. Leonardo emphasizes the necessity of experience and of experiment, the indispensability of mathematics, the cooperation of reason with experience, and the like. But he scarcely gives us any more definite statement of the methods and aims of science than had been given before, and a large measure of vagueness and confusion still hangs about his treatment of the matter. As we shall see, he is much less of a pure empiricist than were the Occamists. He represents rather, in the fifteenth century, a position very similar to that advocated in the thirteenth century by Roger Bacon. And indeed, whatever the superiority of his understanding and practice, his actual account of the subject scarcely takes us beyond that of Bacon.

Leonardo is always most eloquent in his denunciation of those who, like the contemporary Humanists, prefer to study authorities rather than nature herself. Both art and science suffer when imitative of the works of others rather than of those of nature. Thus

> . . . painting declines and deteriorates from age to age, when painters have no other standard than painting already done. Hence the painter will produce pictures of small merit if he takes for his standard the pictures of others, but if he will study from natural objects he will bear good fruit. As was seen in the painters after the Romans who always imitated each other and so their art constantly declined from age to age . . . those who take for their standard any one but nature—the mistress of all masters—weary themselves in vain.[27]

Just as the work of a painter suffers if he takes as a model the painting of others rather than natural objects themselves, so does that of a scientist who takes as a guide the writings of others rather than the observed facts of nature.

> Those who only study the authorities and not the works of nature are descendants but not sons of nature, the mistress of all good authors. Oh! how great is the folly of those who blame those who learn from nature, setting aside the authorities who were disciples of nature.[28]

> Anyone who in discussion relies upon authority uses, not his understanding, but rather his memory.[29]

The literary adornments and learned citations so much affected by the Humanists Leonardo regards as worthless superfluities in comparison with the direct report of first-hand experience; and he denounces with withering scorn those who are content to parade in these borrowed garments.

> I am fully conscious that, not being a literary man, certain presumptuous persons will think that they may reasonably blame me; alleging that I am not a man of letters. Foolish folks! do they not know that I might retort as Marius did to the Roman Patricians by saying: That they, who deck themselves out in the labours of others will not allow me my own. They will say that I, having no literary skill, cannot properly express that which I desire to treat of; but they do not know that my subjects are to be dealt with by experience rather than by words; and [ex-

perience] has been the mistress of those who wrote well. And so, as mistress, I will cite her in all cases.[30]

In the same vein he writes:

Though I may not, like them, be able to quote other authors, I shall rely on that which is much greater and more worthy: on experience, the mistress of their Masters. They go about puffed up and pompous, dressed and decorated with [the fruits], not of their own labours, but of those of others. And they will not allow me my own. They will scorn me as an inventor; but how much more might they —who are not inventors but vaunters and declaimers of the works of others—be blamed.[31]

Leonardo is next most eloquent in his praise of experience, although he eventually qualifies his enthusiasm. All true knowledge, he believes, begins with experience.

All our knowledge has its origin in our perceptions.[32]

Wisdom is the daughter of experience.[33]

Good judgment is born of clear understanding, and a clear understanding comes of reasons derived from sound rules, and sound rules are the issue of sound experience—the common mother of all the sciences and arts.[34]

Experience never errs; it is only your judgments that err by promising themselves effects such as are not caused by your experiments.[35]

Leonardo makes constant appeal to experience for confirmation of his own theories. Such expressions as the following abound: "The above proposition is plainly shown and proved by experiment. . . ." [36] "Experience shows us. . . ." [37] "I remind you . . . to demonstrate the things above mentioned by examples and not by affirmations, which would be too simple. And say thus: experience." [38]

Moreover, a true science of nature presupposes an observation of natural phenomena as *full and complete* as possible:

If you wish to have a sound knowledge of the forms of objects, begin with details of them, and do not go on to the second till you have the first well fixed in memory and in practice.[39]

Abbreviators do harm to knowledge. . . . Certainty is born of a complete knowledge of all the parts, which, when combined, compose the totality of the thing. . . . Of what use then is he who abridges the details of those matters of which he professes to give thorough information, while he leaves behind the chief part of the things of which the whole is composed? It is true that impatience, the mother of stupidity, praises brevity, as if such persons had not life long enough to serve them to acquire a complete knowledge of one single subject, such as the human body; and then they want to comprehend the mind of God. . . .[40]

However, the unaided senses are weak and imperfect. Their use must be supplemented, extended, and rendered more precise by the help of instruments. Leonardo, realizing this need, spent much time and ingenuity in devising mechanical aids to increase the scope and accuracy of scientific observation. These include devices for exact quantitative measurement, especially of space and time, of velocity, weight, and force or energy.[41] Among these aids to accurate observation drawing and painting, in Leonardo's view, occupy places of high importance.[42] A merely verbal description of an object is insufficient. Graphic depiction gives a more faithful and a more complete image of the reality. The pictures supplement direct observation, they sum up its results, and they give details that might easily be missed.

If you despise painting, which is the sole imitator of all visible works of nature, you will certainly despise a subtle invention which brings philosophy and subtle speculation to the consideration of the nature of all forms—the sky and the land, plants, animals, grass and flowers—which are surrounded by shade and light. And truly this is a science and the legitimate issue of nature. . . .[43]

O painter, if you do not know how to manage your figures, [you] are like an orator who knows not how to use his words.[44]

Though you may be able to tell or write the exact description of forms, the painter can so depict them that they will appear alive, with the shadow and light which show the expression of a face; which you cannot accomplish with the pen though it can be achieved by the brush.[45]

And you, who say that it would be better to watch an anatomist at work than to see these drawings, you would be right, if it were possible to observe all the things which are demonstrated in such drawings in a single figure. . . .[46]

If you keep the details of the spots of the moon under observation you will often find great variation in them, and this I myself have proved by drawing them.[47]

This pupil in Man dilates and contracts according to the brightness or darkness of (surrounding) objects; and since it takes some time to dilate and contract, it cannot see immediately on going out of the light and into the shade, nor, in the same way, out of the shade into the light, and this very thing has already deceived me in painting an eye, and from that I learnt it.[48]

As in the medieval writers Leonardo's "*esperienza*" still hovers in meaning between *experience* and *experiment*. He is aware, however, of the insufficiency of mere observation of nature, and is conscious of the necessity also of active experimentation. He himself made and in his notebooks describes many such experiments.[49] Moreover, he knows the value of repeating experiments. "Before making of this case a general rule, make trial of it two or three times, observing whether the experiments produce each time the same effects." [50] Such repetition guards the experimenter against confusing accidental with essential circumstances. "And let this experiment be repeated, in order that no accident impede or falsify the proof." [51] Repetition of the experiment under *varied* conditions is especially useful in eliminating factors irrelevant to the production of the given effect, and revealing the essential.[52] Leonardo is evidently familiar with the principle of Mill's "method of agreement." [53] He also makes use in practice of the methods of "difference" and "concomitant variations." [54] But he gives no theoretical account of these methods.

He naturally makes frequent use of hypotheses; and these are often framed by way of analogy with other natural phenomena.[55] And he understands that hypotheses are to be taken merely as suggestions until confirmed by further investigation.[56] If experience fails to confirm the hypothesis it must be abandoned; [57] and apart from positive experimental confirmation it has no value. "Therefore ye observers, trust not those authors who with their imagination alone have desired to set themselves up as interpreters between nature and man." [58]

While Leonardo agrees that reason may guide experience,[59] he nevertheless insists that the ideal of a purely rational, a priori science is vain and empty.

Flee the precepts of those speculators whose reasonings are not confirmed by experience.[60]

They say that that knowledge is mechanical which issues from experience, and that that is scientific which is born and ends in the mind. . . . But it seems to me that those sciences are vain and full of errors, which are not born of experience, mother of all certitude, and which do not terminate in definite experience, that is whose origin or middle or end does not come through one of the five senses.[61] And if thou sayest that the sciences that begin and end in the mind are the true sciences, that cannot be admitted, but we deny it for many reasons, and in the first place, because into such mental discourses experience does not enter, without which there exists no certitude.[62]

But if Leonardo thus sets great store by experience, his high estimate of the value of mathematics, consisting in its certainty, and his insistence upon its necessity for the investigation of nature, are expressed with equal emphasis.

Among the great features of Mathematics the certainty of its demonstrations is what predominantly [tends to] elevate the mind of the investigator.[63]

No human inquiry can be called true science, unless it proceeds through mathematical demonstrations.[64]

There exists no certitude where some branch of the mathematical sciences cannot be applied, or which is not bound up with the said science.[65]

Mathematical relationships are found in all nature.

Proportions are found not only in numbers and measures, but also in sounds, weights, times and places, and in every force.[66]

The supreme certitude of mathematical truth alone can put to an end the eternal quibblings of contentious "science."

We will not argue that twice three makes more or less than six, nor that a triangle has the sum of its angles less than two right angles. In an eternal silence all quibbling is abolished and the devotees of these sciences have ended it in peace.[67]

He who despises the sovereign certitude of mathematics nourishes himself on confusion and can never silence the contradictions of the sophistical sciences which lead to an eternal clamor.[68]

Let no man who is not a Mathematician read the elements of my work.[69]

Therefore O students study mathematics and do not build without foundations.[70]

In fact Leonardo seems to have believed that the most perfect science would only be attained when to the original inductive derivation of truths from experience would be added a rigorously mathematical demonstration in which their necessity should be shown by deduction from first principles—as in the Aristotelian ideal of science.

But the true sciences are those which experience has caused to enter through the senses, silencing the tongue of the disputants, and not feeding its investigators upon dreams; but proceeds always upon true and known first principles, step by step, and with true conclusions to the end; as is indicated in the elementary mathematical sciences, that is number and measure, called arithmetic and geometry, which treat with highest truth of discontinuous and continuous quantity.[71]

I will make an experiment before I proceed, because my intention is first to appeal to experience, and then by reason to demonstrate why such experience is constrained to work in such fashion. And this is the true rule to be followed by the investigators of natural phenomena; while nature begins with the cause and ends in experience, we must follow a contrary procedure—that is, begin . . . with experience and with that seek for the cause.[72]

Reason will thus show the necessity of the facts—"reason constrains them in this form." [73] A fully developed science "is proved by reason and confirmed by experience." [74]

Consequently, the scientist may proceed either inductively or deductively, "according to the nature of the subject," sometimes "deducing the causes from the effects" in "natural demonstrations," sometimes, on the contrary, "deducing the effects from the causes," by "mathematical demonstrations." [75] Our knowledge of nature may even be extended by purely deductive reasoning from previously established principles. "One must not blame those who in the course of the systematic development of a science appeal to general rules drawn from a conclusion already established." [76] And conclusions so derived may be so certain as to require no experimental confirmation. "There is no result in nature without a cause; understand the cause, and you will have no need of experience." [77] At times Leonardo is even tempted to define the science of perspective in purely rationalistic terms. "Perspective is nothing more than a rational demonstration applied to the consideration of how objects in front of the eye transmit their image to it, by means of a pyramid of lines. . . ." Or, again, "Perspective is a rational demonstration, by which we may practically and clearly understand how objects transmit their own image, by lines forming a pyramid (centered) in the eye." [78] But in the end he recognizes that this science is in truth, like all the natural sciences, a blend of reason and experience. "Perspective is a rational demonstration by which experience confirms that every object sends its image to the eye by a pyramid of lines. . . ." [79]

It is evident that Leonardo is no Occamist. He believes, as we have seen, that there is a "necessity" in nature.

Nature is constrained by reason of her law which lives infused in her.[80]

Necessity is mistress and governess of nature . . . her bridle and eternal rule.[81]

Experience the interpreter between resourceful nature and the human species teaches that that which this nature works out among mortals constrained by necessity cannot operate in any other way than that in which reason which is its rudder teaches it to work.[82]

Again, cause and effect are connected by a necessary bond.

Where anything that is cause of another produces by its motion any effect, it is necessary that the motion of the effect follow the motion of the cause.[83]

A principle being given, it is necessary that its consequences flow from it, if it has not been hindered; and if there is some hindrance, the effect, which should follow according to the given principle, depends more or less upon the said hindrance, according as the latter is more or less powerful than the given principle.[84]

Moreover, there is no action at a distance. "Necessity wills that the corporeal agent be in contact with that which employs it." [85] It is further characteristic of this necessary order of nature that a law of simplicity or economy rules over all natural processes. "Every action done by nature is done in the shortest way." [86] "Given the cause nature produces the effect in the briefest manner that it can employ." [87] "Human ingenuity . . . will never devise any invention more beautiful, nor more simple, nor more to the purpose than Nature does; because in her inventions nothing is wanting and nothing is superfluous. . . ." [88]

Sometimes, indeed, Leonardo seems for the moment to follow a radical empiricism. He contrasts "those things which can at any time be clearly known and proved by experience" with such intrinsically insoluble problems as chiefly concerned "the ancients, who tried to define what the soul and life are—which are beyond proof. . . ." [89] Again he seems at times to forbid investigators any attempt to penetrate below the surface of observable phenomena. "What is an element? It is not in the power of man to define the 'quiddity' of any element, but a great part of their effects is known to us." [90] But this phenomenalism is no more than momentary. "Nature is full of infinite causes which were never set forth in experience." [91] In fact nature is even animated by "desires." "Naturally all things desire to maintain themselves in their natural state." [92] "Every heavy body desires to fall to the centre by the shortest way." [93] And Leonardo has much to say of the activity in nature of an invisible and immaterial "force," "impulse," or "power."

Force I define as a spiritual power, incorporeal and invisible, which with brief life is produced in those bodies which as the result of accidental violence are brought out of their natural state and condition.

I have said spiritual because in this force there is an active, incorporeal life; and I call it invisible because the body in which it is created does not increase either in weight or in size; and of brief duration because it desires perpetually to subdue its cause. . . .[94]

For any fuller treatment of the problems of scientific method superior, or even equal, to that of Leonardo we must wait until the time of Galileo. But in the meantime Leonardo clearly sounds for us a note that is to be increasingly dominant through the whole of the sixteenth and seventeenth centuries. For Leonardo's ideal of a genuine science of nature is throughout that of a system of truths possessing no mere probability, of whatever degree, but rather complete certitude and finality; while for the foundation of that certainty which is to put an end to all disputes he looks sometimes to "experience" and sometimes to "reason"—particularly in the form of rational, or even mathematical, demonstration.

Now it is altogether characteristic of the attitude of the champions of the new science to manifest an almost boundless confidence in man's power, if only the right *method* be employed, of attaining the absolute and final truth about nature. And as to the nature of this method itself they are so far agreed, that it is to be some combination of experience and reason. In the seventeenth century, as we shall see, there are no pure "rationalists" and no pure "empiricists." Some, like Descartes and Hobbes (and these represent, on the whole, the prevailing tendency), press to the limit the prerogatives of rational and mathematical demonstration. Others, like Francis Bacon, lay the chief emphasis on experience and induction based upon experience. But either party is in the end forced, however grudgingly, to admit in large part the indispensability of the factor emphasized by the other. Yet most of all they are agreed, whatever their difference of emphasis as to the correct formulation of the principles of scientific method, that in any case the methods of modern science are such as to insure that the conclusions it yields shall have about them nothing merely tentative, hypothetical, or problematic. They are to possess full and irrefragable certitude. Francis Bacon expects no less than this from his "empirical" method of induction. He is as confident as is Descartes himself that the use of a right method confers upon the results thereby attained a final and assured certainty.

Our attitude in this matter has now changed. We are now prevailingly agreed that no method, however excellent, can confer upon the conclusions of natural science a status greater than that of some degree of probability. How then can we explain the contrary conviction of the greatest minds of the sixteenth and seventeenth centuries? The answer will perhaps be plain if we consider the circumstances of the time. The advocates of the new science were rightly confident that by means of their new combination of experimental and mathematical procedure they were making real and startling progress in the understanding of nature. If they had agreed that their results were, after all, "merely probable," they would then, in the first place, seem to be claiming for them no higher prerogative than the sort of loose and informally grounded "probability" to which Aristotle and his followers had long since accustomed the world—the probability of a merely "plausible opinion" or "persuasive speculation"—such as many Aristotelians of the day thought to be the utmost attainable in the study of the imperfect realm of "matter." The advocates of the new science felt strongly that they had succeeded in attaining more than this—that their conclusions were immeasurably better grounded and immeasurably more securely certified than those of their Aristotelian opponents. In this they were clearly right. They were doubtless mistaken in supposing that this difference could not be adequately expressed in terms of "probability," adequately interpreted. But their error was not wholly unfounded, and not without its historical reason.

Again, too many contemporary Aristotelians, fascinated by the authority of the master, and secure in the support of tradition, were accustomed to claim for the very details of Aristotelian science a complete and demonstrative certitude to which neither Aristotle himself nor his more judicious followers had ever pretended. If the advocates of the newer views were to claim for their doctrines no greater authority than that of probable opinion, their position against such adversaries would seem but weak indeed. Hence the search for certitude. And it was not predominantly to experience, however much its indispensability might be recognized, that men turned to guarantee the conclusions of natural science and to raise them from mere probability to the point of certainty. It was rather to demonstrative reasoning, and especially to the conclusive force of mathematical proofs, that they were inclined to look. In Galileo, as we shall soon see, this confidence in the power of mathematics is already strongly to the fore, although for him its excesses are restrained by pronounced experimental sympathies. But the time was to come when such men as Descartes and Hobbes could in their devotion to the ideal of rational certitude make the attempt of reducing to a rather grudging minimum the office of experience, of which in the end they are inevitably obliged, nevertheless, fully to confess the indispensability.

CHAPTER TWO

Theory of Hypothesis among Renaissance Astronomers

I. ANCIENT AND MEDIEVAL BACKGROUND

THE USE OF THE method of hypothesis in connection with planetary theory goes back at least to the Pythagoreans, but it was Plato who set the main problem for astronomers for many centuries to come. He set his pupils in the Academy the task of working out a system of geometrical hypotheses which, by substituting uniform and circular movements for the apparently irregular movements of the heavenly bodies, would make it possible to explain the latter in terms of the former—in his own famous phrase to "save the phenomena." Ancient astronomers devised three different mathematical solutions of this problem. Eudoxus of Cnidus solved it by a geocentric system of homocentric spheres. Heraclides of Pontus proposed a partially heliocentric system, taking the sun as the center for the orbits of Mercury and Venus and the earth as the center for the motion of the sun itself and the other planets. Aristarchus of Samos (ca. 280 B.C.) "saved the phenomena" by means of a completely heliocentric system.

Aristotle adopted the system of Eudoxus with some modifications in detail. But instead of being satisfied with the purely mathematical combinations of Eudoxus he gave the system a realistic interpretation. In his view the various heavenly bodies were carried around by a machinery of solid, though hollow, corporeal spheres. He made his astronomy an integral part of a physics in which the central position and immobility of the earth and the homocentricity of all the celestial spheres were essential factors.

It soon became evident that with homocentric spheres alone it is impossible to "save the phenomena." To overcome this difficulty astronomers introduced additional eccentric and epicyclic spheres. Ptolemy (second century A.D.) systematized geocentric astronomy on this basis. He worked it out with a complete and detailed mathematics. He "saved the phenomena" better than anyone had ever done before; but unfortunately his hypotheses were in flat contradiction with the principles of Aristotelian physics. The men who accepted Aristotle's physics were confronted with a dilemma. Several ancient thinkers resolved the difficulty by resorting to a nonrealistic interpretation of astronomical hypotheses. Ptolemy himself had already proposed such an interpretation in the *Syntaxis* (Almagest). We are not to suppose that there is really any physical system of spheres in the heavens. Astronomy is simply a mathematical device for calculating the apparent motions. It makes no claim that the motions it supposes in its hypotheses are the "real" motions. Among alternative systems that equally enable us to "save the phenomena," that is to be preferred which is the simplest. That is the only test to be imposed. Conformity with physical fact is not necessary.

This interpretation was revived in the fifth century by Proclus. Epicyclic and eccentric spheres are physically impossible, and no system of hypotheses that "saves the phenomena" can be reconciled with the principles of physics. But there is no need for astronomical hypotheses to conform with physical fact. The motions imagined by astronomers are merely mathematical fictions that enable us to calculate the apparent motions. In the sixth century Simplicius and his Christian contemporary John Philoponus adopted similar views.

In the early Middle Ages the partially heliocentric system of Heraclides of Pontus, some knowledge of which (or some geometrical equivalent) had been transmitted through Chalcidius, Martianus Capella, and Macrobius, was that followed by most writers. It was only during the course of the twelfth century that Western Europe gradually became familiar with the physics and astronomy of Aristotle and with the system of Ptolemy. The enormous superiority of Ptolemy's astronomy, which alone made possible detailed mathematical calculations and definite predictions of observable celestial phenomena, was immediately evident. Yet the physics of Aristotle seemed equally convincing, and Ptolemy's eccentric and epicyclic spheres were inconsistent with its fundamental principles. Strict Aristotelians therefore followed the lead of the Arabian commentator Averroes in rejecting Ptolemy's astronomy in favor of Aristotle's system of homocentric spheres, though this could not be shown to "save the phenomena." Moses ben Maimon (1135–1204) and Thomas Aquinas (1225?–74?), however, accepted the Ptolemaic system for astronomical purposes and reconciled this attitude with their adhesion to Aristotelian physics by reverting to nonrealistic theories of astronomical hypotheses similar to those proposed by Simplicius (sixth century) and John Philoponus (fifth century). Views of this type had a wide following during the thirteenth century and seem on the whole to have prevailed. Except among Averroists, active chiefly at Padua, who remained dogmatically attached to the strict homocentric theory interpreted in a literally physical sense, similar views continued to prevail until toward the end of the sixteenth century. Among followers of Ptolemy adherents of a realistic interpretation seem to have been few. I have met with definite references to only two. About 1300 Bernard of Verdun argued as follows:

> By means of this system one avoids all the difficulties which have just been enumerated, and saves all the phenomena which have been formulated in the preceding chapter; by taking it as a principle one can advance to everything that it is useful to know with regard to the celestial movements, the size and distances of the heavenly bodies; and, up to the present, everything of the sort that has been discovered has been found to be verified; this would certainly not have happened if the point of departure of these deductions had been false, for, in any subject, a small error at the beginning entails a large error in the end.[1]

Similarly Francesco Capuano of Manfredonia in 1495 defended the Ptolemaic system by arguing that, by showing an exact accord between the data of experience and the results deduced from a set of astronomical hypotheses, one demonstrates with certainty the truth of these hypotheses.[2]

A most interesting development of the nonrealistic type of theory was its extension by certain thinkers of the fifteenth and early sixteenth centuries from astronomy to physics. Certain nonrealists of later antiquity, including Proclus (fifth century) and Simplicius, were accustomed to reason as follows. Physics can discover the true nature and causes of the objects with which it deals. Astronomy, however, can never succeed in penetrating to the true nature and causes of the movements of the heavenly bodies. It must content itself with plausible suppositions which will at any rate "save the appearances" and will be simple enough to be followed and used by our finite human intelligences. Only the Divine Intelligence can follow and comprehend the true movements of the celestial bodies.

Nicolas of Cusa in the fifteenth century extended this line of thought to physics. He no longer believed in the Aristotelian distinction between celestial and terrestrial matter. For him, the nature and constitution of the heavenly bodies are not fundamentally different from those of terrestrial bodies. The physical universe is of one stuff throughout. But he still maintains a distinction between the sort of knowledge of the physical world that is accessible to the human mind and another sort that can be attained only by God himself. None but the Divine Intelligence is capable of grasping the true essential nature and the veritable causes, not only of the celestial movements, but of any part of the physical universe whatsoever. Human science is doomed to everlasting ignorance of the true causes and the true essences of things. It must content itself perforce with fictitious and hypothetical conceptions and with fictitious and hypothetical causes. By these means it may approximate indefinitely to the absolute truth, but it can never attain this goal. And in the writings of Luiz Coronel (1511) we find again the application of a nonrealistic theory to physics as well as to astronomy. Physics no more than astronomy is a deductive science, the propositions of which are derived from principles self-evident a priori; it is a science the origin of which is to be found in experience, and the principles of physics, like those of astronomy, are nothing but hypotheses invented in order to "save the phenomena" known to us by experience.

2. THE SIXTEENTH CENTURY

A. *Copernicus and his disciples*

Such was the situation at the moment of the publication of Copernicus' *De revolutionibus orbium caelestium* (1543), and such it remained, practically unchanged, for some forty years. For the advent of the new system, revolutionary as it was, brought with it no considerable alteration in the status of the various theories concerning the nature of astronomical hypotheses.

Copernicus himself, together with his enthusiastic disciple Joachim Rheticus, adopted a realistic view. In fact their theory of the nature of the hypotheses of astronomy differs in no essential respect from that of Ptolemaists of the type of Capuano. With Capuano and the Averroists they believe in the physical reality of their astronomical system; with Capuano (and on this point differing from men of all other schools) they are convinced that, if only we can invent a set of astronomical hypotheses that shall correspond with sufficient accuracy and faithfulness to the observed phenomena, we are warranted in concluding that we have discovered the only theory that can conceivably "save the phenomena," and have thus given a cogent demonstration of its truth.

Copernicus at times seems willing to content himself with a more moderate interpretation of the scope and bearing of astronomical theories. In his prefatory letter to Pope Paul III, for example, he represents himself as at first experimenting with the apparently "absurd opinion" of the motion of the earth merely as a mathematical fiction, comparable to those which others had long been accustomed freely to imagine, simply in the hope, by its means, of more efficiently "saving the phenomena." [3] Likewise in the *Commentariolus* he introduces the immobility of the sun and the movement of the earth merely as postulates which he asks to be granted—"if certain postulates are granted us." [4]

But a very different attitude is revealed by the *De revolutionibus* as a whole. Even in the dedicatory epistle itself Copernicus makes it a reproach against the current form of the theory of eccentric spheres that, despite its utility as a device for calculating the celestial phenomena, it conflicts with fundamental physical principles and cannot therefore represent the actual system of the world.

> Surely the ones who devise [a theory of] eccentricity, though to a large extent they may seem to have freed certain motions with consistent numbers by means of the eccentricity idea, still they permit other [things] which seem to disagree with the prime principles about the equality of motion.[5]

He evidently expects more from an astronomical system than explanation of the observed facts. In his view it will be acceptable only if it is capable also of physical reality. And he goes on to confess a disappointment that, despite all the labors of the philosophers, there is not some "more certain account of the motions of the machine of the world." Moreover, Copernicus found his trial of the "absurd opinion" of the motion of the earth extraordinarily successful in "saving the phenomena." The new hypotheses, and the calculations based upon them, explained all the data of observation with great exactitude, whereas if any feature of the system was altered in the slightest degree the whole would be thrown out of joint.

And so in respect to the residing motions, which I assign to the earth's labors, I have finally discovered with much long observation that if the movements of the other wandering stars are compared to the movements of the earth and are computed by the revolution of each star, not only will their phenomena follow, but even the ranks and sizes of the stars and of the whole orb and the sky itself will be so connected that nothing is able to be transposed in any part without confusion in the other parts and in the whole universe.[6]

From the fact that any modification of his mathematical system would upset the whole scheme and render it inapplicable to the phenomena, he thereupon seems to have drawn the conclusion that no other system could account for them—that his own was therefore not merely *a* possible, but the *only* possible, and therefore demonstrably and certainly the *true* system: "And so we have found an amazing symmetry with this mathematical system of the universe and a certain tying together of the harmony of movement and of the size of the orbs, such as can be found in no other way." [7]

It was thus at any rate that Rheticus, writing in 1540, understood the mind of his master:

Aristotle says *that is most true which is cause of the truth of its consequences.* So when my master determined that he ought to assume hypotheses as should contain causes that would confirm the observations of previous ages—such also it was to be hoped, as should be causes, for the future, of the establishment as true of all astronomical predictions of phenomena.[8]

Rheticus does not trouble to draw the conclusion, but it is evident. In terms of Aristotle's own dictum the hypotheses of Copernicus are "most true"—*verissimae.*[9]

It was thus, too, that Bishop Tiedemann Giese of Culm understood his old friend's attitude when in a letter to Rheticus (June 20, 1543) he bitterly assailed the publisher and the writer (Osiander) of the anonymous preface to the *De revolutionibus* for their betrayal of trust in misleading the public as to the author's real convictions.[10]

And it was thus, finally, that the Holy Congregation of the Index understood Copernicus when, in the decree of 1620, it censured his work precisely on the ground that, instead of treating the motion of the earth as a mathematical hypothesis, he had represented it as a physical reality.[11]

The anonymous preface with which the *De revolutionibus* made its first appearance before the public, and to which the Bishop of Culm referred, was the composition of one Andreas Osiander (*né* Hossmann), added by him without the knowledge of Copernicus and his other friends, with the aim of making the work more acceptable to Peripatetics and theologians.[12] In this preface Osiander presented the new doctrine in accordance with the nonrealistic theory of hypotheses which he himself had long held.[13] According to Osiander, no astronomer can ever be justified in pretending to have discovered the true causes of the celestial movements. He can only invent mathematical fictions—hypotheses—which will enable him to calculate the observed motions. It is not necessary that these hypotheses be true, or even probable. It is enough if they give rise to calculations that fit the observations. The new hypotheses of Copernicus thus have as much, and as little, claim to truth as any that fit the facts; and we should be willing to use them freely, as mathematical and empirical exigencies may require.

For it is the part of the astronomer to record a picture of celestial motion with diligence and scientific observation. And then since he can in no way perceive the true causes of these movements, it becomes his part to contrive and to invent some sort of hypotheses by which these movements can be calculated by geometric principles, i.e., for future as past motions. . . . For it is not obligatory that these hypotheses be true, nor probable even, but this one thing suffices: that [the hypotheses] show a consistent conclusion from the observations. . . . For it is evident enough that astronomy ignores completely the causes of the apparently unequal movements. And if by inventing [the causes] one is fabricating, by no means does he do this to persuade anyone that it is true but only that others will draw a conclusion correctly. . . . Wherefore let us permit these new hypotheses to take their place among the old, no longer true hypotheses, particularly since they are both admirable and facile and carry a great treasury of learned observations. And no one should expect anything to be made sure by astronomy which depends upon hypotheses for even astronomy is unable to offer any such proof except a person accept as the truth a proof invented for another purpose and leave the study more stupid than he came.[14]

In adopting this theory of the matter Osiander was by no means alone. It was rather the realistic view of Copernicus and Rheticus that was exceptional, nor does their advocacy seem to have brought it into any greater favor; while the theory which Osiander had adopted (though in no wise invented) continued to prevail, whether among those who employed the Copernican or those who retained the Ptolemaic hypotheses.[15]

The state of mind of the majority of astronomers, during the thirty or forty years that followed the publication of the book of Copernicus, is clearly apparent. The work of the Astronomer of Thorn attracts their lively attention, because it seems to them very well adapted for the construction of exact astronomical tables, and because the combinations it proposes in order to "save the phenomena" seem to them preferable to those which Ptolemy had imagined. As for the hypotheses from which Copernicus derived his kinematical combinations, are they true, probable, or purely fictitious? They leave to the physicist, to the Philosopher of Nature, the business of discussing this question, and they show no concern over the answer that will be given. They act with regard to these hypotheses as Osiander recommended them to do. It is not, however, that their attitude with regard to the Copernican assumptions is in the least imposed upon them by the anonymous preface of the treatise *De revolutionibus;* this attitude they had long since been accustomed to hold; it is the attitude which, from the time of Greek antiquity, throughout the Middle Ages, and down to the beginning of the Renaissance, had allowed the partisans of the system of Ptolemy to make astronomy progress despite the Peripatetics and the Averroists, and to disregard the constantly repeated and always fruitless attempts to restore the system of homocentric spheres.

B. *The Wittenberg School*

It is difficult to see how even theologians could object to the teaching and utilization of the Copernican hypotheses when presented in this light, and indeed there seems for some time to have been relatively little opposition. Luther and Melanchthon, however, were strongly against the theory from the start. Luther is reported to have said of Copernicus: "The fool wants to overturn the whole science of astronomy. But as Holy Writ informs us, Joshua bade the sun stand still, and not the earth."[16] And Melanchthon, following the lead of his master, argued against the Copernican system, utilizing both considerations derived from the Peripatetic physics and texts drawn from the Holy Scriptures. In his view the best way to deal with the new theory is to ignore it. If left alone it will die a natural death. But "wise governors ought to repress the excesses of clever wits."[17]

In view of statements that Melanchthon makes elsewhere it is difficult to see how he can have adopted such an attitude of uncompromising hostility. For Melanchthon was well aware that the Ptolemaic hypotheses of which he himself made use were by no means in strict accordance with the principles of the Peripatetic physics. Yet when the Ptolemaic assumptions were ridiculed by the Averroists, on the ground of physical impossibility, he denounced their perversity in no measured terms. Let students not be deceived by these calumnies. Astronomers know well that their hypothetical constructions represent no physical realities; but that destroys neither their value for setting forth the laws of the celestial motions, nor the necessity of giving them careful study. Either let the Averroists cease from troubling, or let them produce a system better adapted for calculating the observed motions than that of Ptolemy.[18]

It is obvious that a partisan of the Copernican hypotheses might with perfect justice have attacked Melanchthon's own position on exactly these same lines. In fact, Melanchthon himself confesses that "the system of lunar orbits recently invented by Copernicus is very well arranged." He refuses indeed to utilize the Copernican assumptions even here:

> I follow the existing system handed down by Ptolemy, which many others have followed thus far, and, though the recent system of lunar orbits invented by Copernicus is well arranged, still let us refuse [to use] it so that we may entice the studious ones to the common training received in the schools. . . .[19]

He is satisfied to represent the astronomical hypotheses that he himself accepts in terms of a nonrealistic theory, and indeed even insists that they be understood in this way:

> Wherefore again the hearer has to be reminded that he realizes that such a system of orbs and epicycles was invented by geometricians, and that however much laws

and times of motions may be indicated it is not because there are such great missiles in the sky, although it is agreed there are certain orbs. And the sagacity of such schemers is to be praised, namely, the ones who sculpt as it were another image of that amazing orb.[20]

However, Melanchthon is unwilling even to consider the possibility of admitting the Copernican alternatives on the same basis. Men who introduce the motion of the earth and the other Copernican hypotheses into the discussion cannot be taken seriously. They are only amusing themselves in clever play. They make such assumptions, he tells us, merely "from the love of novelty," or "in order to display their ingenuity," or "to exercise their wits." But to assert such absurdities is altogether unseemly and "sets a bad example." A sober and honest man will not abandon the Ptolemaic assumptions, which have been received by common consent for so many centuries, for such clever imaginations. He will rather reverently accept the truth as revealed by God and set forth in the Scriptures.[21]

Melanchthon's opposition did not, however, prevent his colleague Reinhold from adopting a more consistent attitude. Sharing to the full Melanchthon's nonrealistic theory of the nature of astronomical hypotheses, Reinhold nevertheless welcomed the system of Copernicus, so interpreted, with open arms—simply on the ground of its immense superiority for purposes of calculation. In 1551 he published his *Prutenicae tabulae* [22]—a set of new astronomical tables based upon the Copernican hypotheses. And in introducing the work to the public he took occasion to sing the praises of Copernicus in no uncertain terms.

All posterity [he writes] will celebrate with gratitude the name of Copernicus; the science of the celestial movements was almost in ruins; the studies and works of this author have restored it; the goodness of God has lighted in him a great light, so that he has discovered and explained a multitude of things which, up to our time, had remained unknown or obscure. . . .[23]

3. THE TRIUMPH OF REALISM

A. *Anti-Copernicans*

The half century that preceded the condemnation of Galileo (in 1633) witnessed the growth on all sides of a decidedly realistic temper of mind, and a consequent demand that astronomical hypotheses be brought into agreement with the doctrines of physics and with the texts of Scripture.

Such, for example, is the teaching of Christopher Clavius, whose interesting defense of the Ptolemaic system first appeared in 1581.[24] Against those who regard the Ptolemaic assumptions as mere mathematical fictions he argues at some length. Their position he first summarizes as follows:

They concede that on the assumption of eccentric and epicyclic spheres it is possible to explain all the phenomena; but hold that it does not thence follow that the said spheres are to be found in nature: first, because it may be that all the appearances can be accounted for in some yet more convenient way as yet unknown to us; second, because it may well be that the phenomena can be explained by the said spheres even though they be purely fictitious [*omnino fictiti*], and in no wise the true cause [*vera causa*] of those phenomena: just as a true conclusion may be inferred from a false premise, as is established in Aristotle's Dialectic.[25]

To this argument Clavius replies as follows. It must be admitted that it is not altogether *certain* that the Ptolemaic spheres actually exist, and that it is *possible* that the phenomena might be accounted for in some other way unknown to us. In fact the success of the Copernican arrangement is sufficient to show that "it is not altogether certain that the true disposition of the Eccentrics and Epicycles is exactly that which Ptolemy supposed: seeing that many of the phenomena can be accounted for in another way." [26]

But Copernicus himself does not pretend to be able to account for the phenomena without eccentric and epicyclic spheres. He merely gives them a different arrangement.[27] We know that by means of the assumption that such spheres exist we can give a very exact and convenient explanation of the celestial motions. No known theory that dispenses with them can do this.[28] If those who oppose them "have some more convenient way [sc. of dealing with the appearances], let them show it to us and we shall be content, and will give them very hearty thanks." [29] But if this challenge cannot be met, if the assumption of some system of epicycles and eccentrics is the only known way of accounting satisfactorily for the phenomena, then we have every right to conclude that it is at least "*highly probable* [*credibile valde*] that the celestial spheres are of this sort." [30] If such a conclusion is not legitimate, then we have no right to believe that *any natural science whatever* tells us anything of the true causes of natural phenomena. For the conclusions of all these sciences are established in precisely the same manner. If the explanations of astronomers are "fictions," then those of physics are equally so.

If they cannot show us some better way [sc. of accounting for the phenomena], they certainly ought to accept this way, inferred as it is from so wide a variety of phenomena: unless in fact they wish to destroy not only the Natural Philosophy that is studied in the schools but also to bar the way to all other arts that investigate causes by inferring them from their effects. For as often as anyone inferred a certain cause from its observable effects, I might say to him precisely what they say to us—that forsooth it may be possible to explain those effects by some cause as yet unknown to us. . . . Wherefore, if one is not justified in inferring on the basis of the phenomena that there are Eccentrics and Epicycles in the heavens, just because a true conclusion may be inferred from a false premise, then the whole of Natural Philosophy will be destroyed. For in precisely the same manner, when anyone infers from a known effect that this or that is its cause, I can say, that this is not true, because a true conclusion can be inferred from a false premise, and thus all the principles of natural science that have been discovered by philosophers will be destroyed. Since, however, this is absurd, the force and validity of our argument seems not really weakened by our adversaries.[31]

Furthermore the consideration that a true conclusion can be inferred from a false premise is not really *ad rem*. It is true that when we know a certain proposition to be true we can arrange false premises in such a way that this true proposition will follow necessarily by the structure of the argument. For example, since I know that every animal has sensations I can construct the following syllogism:

Every plant has sensations;
Every animal is a plant;
Therefore every animal has sensations.

But, Clavius continues, if I am in doubt of the truth of a certain conclusion I shall never acquire certitude of its truth from false premises, even if it validly follows according to the rules of the syllogism; for otherwise I might easily in this fashion reach any conclusion you please.

Thus if I am in doubt whether every star is round, it is true that by the force of the following syllogism,

Every stone is round;
Every star is a stone;
Therefore every star is round,

I infer this conclusion validly from false premises, nevertheless I shall never thus become assured of the truth of the aforesaid conclusion of which I was doubtful. But by the assumption of Eccentric and Epicyclic spheres not only are all the appearances already known accounted for, but also future phenomena are predicted, the time of which is altogether unknown: thus, if I am in doubt whether, for example, the full moon will be eclipsed in September, 1583, I shall be assured, by calculation from the motions of the Eccentric and Epicyclic spheres, that the eclipse will occur, so that I shall doubt no further. I even know, from the same motions, at what hour that eclipse will begin, and how much of the moon will be in shadow.[32]

The point is important. If a given hypothesis not only provides explanation of the phenomena already observed, but also yields precise and numerically detailed predictions of hitherto unobserved phenomena; if, moreover, these latter are actually found to occur as predicted, this is surely enough to establish the truth of the hypothesis, at least with a high degree of probability. "For it is incredible that we force the heavens (but we seem to force them, if the Eccentrics and Epiycles are figments, as our adversaries will have it) to obey the figments of our own minds, and to move as we will, or in accordance with our principles." [33]

Clavius rejects the system of homocentric spheres on two grounds. In the first place it leaves many of the observed phenomena wholly unexplained.[34] In the second place it is less *simple* than the Ptolemaic system. It requires the assumption of an enormously greater number of spheres—and it is a well-known maxim of philosophy that "it is useless for something to be done by more that can be done equally well by fewer." [35]

But what of the Copernican system? Clavius concedes that the Copernican hypotheses account for the known facts as well as, and even at some points more accurately than, those of Ptolemy. But this is not sufficient to prove their truth.

His sole aim was to correct the periods of the planetary motions, which he found were already defective. For it is extremely difficult to determine the periods of their motions so accurately that they shall not be a trifle too great or too small; and this, after an interval of many years, will bring about an observable error. . . .

Thus it is not strange that Copernicus could account for the phenomena in another way. For, since by calculation from the motions of the Eccentrics and Epicycles he knew the time, the quantity, and quality of both future and past phenomena, he was able (for he was a man of great genius) to think out a new way of accounting for those phenomena more conveniently (as he thought), and to correct somewhat the periods of the planetary motions, which he had already noticed were defective—and this, as we have said, seems to have been his chief aim: just as we can infer any already known conclusion by means of many different syllogisms, even if they have false premises.[36]

Still it must be admitted that if there were no further criteria by which to judge between them the Copernican system would have as great a probability as the Ptolemaic. "If the position of Copernicus involved nothing false and nothing absurd it would indeed be doubtful which opinion, that of Ptolemy or that of Copernicus, ought to be accepted." [37] This, however, is not the case. In the first place "many things absurd and erroneous are contained in the Copernican theory . . . all of which are contradictory to the common doctrine of Philosophers and Astronomers." [38]

It is not sufficient that an astronomical hypothesis should be consistent with astronomical phenomena alone. It must also be consistent with the rest of what we know about the physical world. Clavius is careful to argue that the Ptolemaic assumptions, despite their inconsistency with the Aristotelian physics, really satisfy this requirement. "They are in no respect repugnant to Natural Philosophy, as is established by our reply to the arguments of Averroes and his followers . . . nothing follows from them that is absurd or inconvenient in Natural Philosophy." [39] Aristotle himself would have modified the principles of his physics if he had been familiar with the evidence upon which modern astronomy is based.[40] Copernicus' theory, however, fails to satisfy this requirement. And, moreover, it contains many things "that seem to contradict what Sacred Scripture teaches in many passages. . . . Consequently the opinion of Ptolemy seems preferable to the invention of this Copernicus." [41]

Tycho Brahe (1546–1601) manifests a very similar attitude. He is content with no mere mathematical fictions but insists on the contrary that legitimate astronomical hypotheses must state the real physical disposition and movements of the heavenly bodies.[42] After rejecting both the Ptolemaic and the Copernican systems on grounds that include both physical and theological considerations, he seeks to work out "another system of hypotheses that should stand at all points in accord with the principles both of mathematics and of physics, and should not make use of subterfuges [43] in order to escape the censures of theology, and at the same time should fully satisfy the celestial appearances." [44] He no longer believes in the physical reality of the solid spheres but regards them as mere pedagogical expedients to assist the imagination and to aid the understanding of scholars. And it is precisely because of the physical difficulties involved that he rejects them. The motion of the comets is such as to take them right through the supposedly solid spheres. Thus Tycho writes to Kepler:

In my opinion the reality of all the spheres, however they may be conceived, is to be excluded from the heavens; this I have learnt from all the comets which have appeared since the new star of 1572, and which are in truth heavenly phenomena —for they followed completely the laws of none of the spheres, but rather acted in contradiction thereto.[45]

And again he writes:

It will be clearly proved by the motions of the comets that the machine of the heaven itself is no hard and impenetrable body composed of various real spheres, as has hitherto been believed by most, but very fluid and free, open in all directions and opposing absolutely no obstacle to the free course of the planets, which is regulated without machinery or the turning of any real spheres, according to the legislative wisdom of the Deity. . . . And thus no real and inconsistent penetration of the spheres (for these do not really exist in the heavens, but are set forth only for the sake of teaching and learning) is admitted.[46]

Georg Horst, writing in 1604, had taken a further and final step. The hypotheses of astronomy, in his view, can be demonstrated as physically true, not merely, as Clavius had held, with probability, but with absolute certainty. Astronomy "establishes most of its conclusions by means of apodictic principles, and that too in a manner so certain and infallible that Pliny . . . says with reason: It is shameful that anyone can fail to believe in them." [47] And he proceeds to establish the truth of the Ptolemaic hypotheses by proofs drawn partly from the facts of observation and from Peripatetic physics, and partly from texts of Scripture.

The increasing tendency to realistic conceptions brought with it an increasing hostility to the Copernican hypotheses on the part of those who advocated alternative theories. Thus Petrus Ramus (1515–72) wrote:

And would that Copernicus had rather pondered the problem of founding an astronomy without hypotheses. For it would have been much easier to draw up an astronomy corresponding to the true nature of the stars themselves, rather than as it were by some giant's labor to set the earth to moving, in order that in relation to the motion of the earth we might view the stars as at rest.[48]

According to Francis Bacon, also, the assumptions of Copernicus are merely "the speculations of one who cares not what fictions he introduces into nature, provided his calculations answer." And he writes: "The opinion of Copernicus touching the rotation of the earth (which has now become prevalent) cannot be refuted by astronomical principles because it is not repugnant to any of the phenomena; yet the principles of natural philosophy rightly laid down may correct it [*revinci*]."[49]

B. *Kepler and Galileo*

The increasing opposition was far from causing the followers of Copernicus to conciliate opinion by any recourse to the nonrealistic views so widely prevalent a few decades before. They insisted quite as strongly as any of their adversaries that the only legitimate astronomical hypotheses are those which are physically true and demonstrable with complete certainty. The rising enthusiasm and self-confidence of the energetic champions of the new theory did not allow them to rest content with any cautious refinements of interpretation. They felt strongly that Copernicus had for the first time discovered the objectively real order and system of the heavens, and that the method he had employed was thus yielding truths about the universe that were capable of cogent demonstration.[50] To hold that astronomical theories are indifferent to considerations of physical reality, or that they represent at best a merely possible rather than a demonstrably actual disposition of things, seemed to them to be plain treason to the advancing cause of scientific reform. The minimizing views of Osiander particularly aroused the ire of Giordano Bruno (1548–1600). He could see in them merely time-serving subterfuges.[51] In his opinion astronomical hypotheses evidently can be taken seriously only if viewed as demonstrable propositions about physical realities.[52]

Kepler's language is more temperate, but his opinion of Osiander is fundamentally the same. Like Clavius and Coronel he insists strongly upon the homogeneity of physics and astronomy, and upon the identity of their methods; and he believes that these methods lead to conclusions which, alike in physics and in astronomy, are demonstrably true. If the empirical method can give the physician knowledge of the true causes of disease, then precisely similar methods of the astronomer can likewise give him knowledge of the true causes of celestial phenomena. If reasoning from effect to cause leads to valid conclusions in the one sphere, how can we deny that it may be similarly efficacious in the other? To cast doubt upon conclusions thus established is to indulge in a groundless skepticism.[53] In his earliest work he regards the principles of the Copernican system, by virtue of their consistent explanation of so many phenomena, as demonstrated with certainty.[54] In this treatise, indeed, Kepler goes even further. He asserts that whatsoever Copernicus has proved a posteriori can also be proved a priori and with such cogency that Aristotle himself, were he still living, could not fail to be content with the demonstration.[55] This a priori demonstration was to rest on metaphysical and physical grounds, and the *Prodromus* offers such proofs.[56]

It was probably due to the influence of Tycho Brahe that Kepler later somewhat modified his attitude. Tycho, on December 9, 1599, addressed to Kepler a letter in which he points out that in astronomy the a posteriori is the true method of proof, upon which its exact results are founded.[57] Accordingly, when Kepler came to write his *Epitome astronomiae Copernicanae* (1618), although he was still concerned to justify astronomical hypotheses on physical and metaphysical grounds—for, as he says, "astronomy is a part of physics"[58]—he nevertheless no longer claims for this line of reasoning more than that it should be able to give "probable grounds" in physics and metaphysics for these hypotheses.[59] In fact he even regards it as permissible, on the basis of results attained in astronomy, to modify the doctrines of the "higher sciences" of physics and metaphysics themselves.[60]

Kepler here seems in effect to be refusing to physics and metaphysics any certainty ultimately superior to or differing in kind from that of astronomy, and to be insisting upon the solidarity of all the natural sciences. What is true for the astronomer must be true also, and in the same sense, for the physicist or for the metaphysician, and vice versa. As Cassirer remarks, Kepler demanded that the same laws that apply to terrestrial physics must be applicable also to celestial physics—that the latter must not be developed from independent hypotheses.

In order that a hypothesis should be "true" it is not enough that it should succeed in expressing in a brief formula simply and solely astronomical phenomena, which indeed constitute only a limited section of our total experience: it must represent these in such a way as corresponds to our understanding of all concrete processes of nature in general. The foundations of astronomy can be laid down only in connection with those of scientific physics. The truth of a given assumption, accordingly, is not only attested by the immediate confirmation that it finds in particular facts of perception, but requires verification and control by reference to a system of principles of mathematical physics. Only by finding its place in this universal interconnection of things is a phenomenon truly accredited and "saved." [61]

When Nicolas Reymer Baer in 1597 publicly defended Osiander's nonrealistic theory of astronomical hypotheses,[62] Kepler was moved to reply. About 1600 or 1601 he began to compose to this end his *Apologia Tychonis contra Nicolaum Ursum*—a work which, however, he never completed or published.[63]

Reymer had apparently expounded the nonrealistic theory in a highly exaggerated form. Not only were astronomical hypotheses merely "fictitious descriptions of an imaginary form of the system of the world, and not the true and veritable form of this system," but also "hypotheses would not be hypotheses if they were true," and "it is the proper function of hypotheses to seek to derive true conclusions from false premises." Kepler's reply is correspondingly heated, but it affords us, nevertheless, an admirable insight into his theory of the matter.[64] It constitutes, as Prantl remarks, "a formal treatise concerning the essence and the significance of hypotheses." [65]

A hypothesis, in the generic sense, Kepler defines as "whatsoever for the purpose of any demonstration is taken for certain and demonstrated." [66] Hypotheses in this sense are to be found in geometry, in logic, and in astronomy. The use of such assumptions in geometry is justified as follows:

For just as in architecture it is enough if the builder constructs the foundations for the future mass of the building in the earth, and does not trouble himself to ask whether the earth further down falls away or is solid down to the centre, so in geometry the first writers were not so foolish as to wish to bring all things into doubt, as the followers of Pyrrho later did, or to lay hold on nothing upon which, as upon a foundation certain and recognized by all, to raise the rest of the structure. They called these assumptions hypotheses or suppositions with respect to that which they intended to demonstrate. For they were assumed as known, and upon them things less known were built up, and were shown to the learner.[67]

These hypotheses of geometry may be divided into three classes: (1) axioms, i.e., principles that are certain and acknowledged by all men; [68] (2) postulates, i.e., principles not universally admitted, but sufficiently known to the author himself, and which he asks the learner to concede as the basis of a given demonstration; [69] (3) a third kind of principle, such, namely, as "either cannot be, or are not, true; but are assumed in order that it may be shown by demonstration what would follow if they were true. This kind of principle has a place also when, leaving geometry behind, we carry over the method of demonstration into cognate sciences." [70]

In astronomy the term hypothesis is used in two senses: (1) it may refer simply to the observation of the phenomena themselves, since this is after all the foundation of all our demonstrations; [71] (2) but in a more specific sense the term refers to the whole system of conceptions employed by an author as premises from which he calculates deductively to the observed motions of the stars—whether the propositions utilized be such as he has previously derived inductively from observation of astronomical phenomena or such as he has simply taken over from physics or geometry.[72] A further distinction, however, is necessary—that between *astronomical* hypotheses strictly speaking and *geometrical* hypotheses employed by astronomers, a distinction between the physically real and the hypothetical that Osiander has willfully ignored.[73]

If an astronomer says that the path of the moon is an ellipse, it is an astronomical hypothesis; when he shows by what combinations of circular movements such an elliptical orbit may be brought about, he is using geometrical hypotheses. . . . In sum there are three things in astronomy: geometrical hypotheses, astronomical hypotheses, and the apparent motions of the stars themselves; and, consequently, the astronomer has two distinct functions, the first, truly astronomical, to set up such astronomical hypotheses as will yield as consequences the apparent motions; the second, geometrical, to set up geometrical hypotheses of whatsoever form (for in geometry there may often be many) such that from them the former astronomical hypotheses, that is, the true motions of the planets, uncorrupted by the variability of the appearances, both follow and can be calculated.[74]

Now at least the strictly astronomical hypotheses used in astronomy must be regarded as *true*. Nor can this contention be impugned by adducing the example of a true conclusion drawn from false premises. That a true conclusion should follow from false premises is due to chance, and what is false in nature, as soon as it is applied also in some other cognate matter, betrays itself: "unless you freely allow the arguer to assume an infinite number of further false propositions, and never to be consistent with himself in his arguings up and down." However,

. . . the matter is very different with him who assigns to the sun a central position. For command him to demonstrate anything you will that really appears in the heavens from that hypothesis once for all laid down, to go backward or forward, to infer one thing from another, or to do anything that the truth of things permits: and he will not hesitate to demonstrate anything that is genuinely so, and even after the most intricate windings of his demonstration will always return upon himself with perfect consistency.[75]

If we start with a false assumption errors are bound to show themselves in the long run.

So at last it comes about, through the interweaving of syllogisms in demonstrations, that, given one proposition not agreeing with the truth, an infinite number of such will follow.[76]

It can scarcely ever occur, and indeed I can think of no example of it, that from the assumption of an unsound hypothesis there should follow a conclusion in all respects sound and conformable to the celestial motions, or such as one wishes to demonstrate.[77]

But an objection must here be met. Is it not the case that there is often more than one hypothesis that agrees perfectly with all the observed phenomena? And in this case must not one of them be false and nevertheless lead to true conclusions? How then can we be sure that any given hypothesis may not be such a one? Kepler replies that it is not often the case—even if one considers only strictly astronomical phenomena—that such "equipollent" hypotheses, even again if one considers only their agreement with strictly astronomical phenomena, can be constructed. "Different hypotheses have not the same consequence as often as the less skilled suppose."[78] The Ptolemaic hypotheses, for example, did not have altogether the same consequences as the Copernican. They did not account, as does the latter, for some of the most important phenomena—e.g., they did not give the causes of the number, magnitude, and time of the retrograde movement of the planets.[79] Further, "even if diverse hypotheses give altogether the same result as far as astronomy is concerned . . . nevertheless there is often a difference in their conclusions because of some physical consideration."[80] For example, the hypotheses of Tycho Brahe, although they give the same astronomical predictions as those of Copernicus, do not require that the fixed stars be of such enormous size as they must be according to the Copernican assumptions.

If one considers everything in accordance with this rule, I know not indeed whether any hypothesis will ever occur to him, whether simple or complex, which will not exhibit some conclusion peculiar to itself and distinct and different from all others. For if in their geometrical conclusions two hypotheses coincide, nevertheless in physics each will have its own peculiar additional consequence. . . . Authors do not always consider this variety in the physical consequence, but more often confine their thoughts within the boundaries of geometry or astronomy, and within some one science raise the question of the equipollence of hypotheses, neglecting the diverse consequences which in respect of the neighboring sciences lessen and destroy that reputed equipollence.[81]

Moreover if two hypotheses have the same consequences in astronomy it is because they contain a common part. Astronomically speaking, then, they are not two hypotheses, but one. The astronomical consequences follow

. . . in so far as they are of one kind [*sub uno genere*], not in so far as they differ from each other. Since, therefore, as a cause of the demonstration they are one, as a cause of the demonstration they will not be contradictory. . . . Nor can it happen in astronomy that the conclusion derived from a false hypothesis should be true in all respects: and accordingly it is characteristic of hypotheses (if we have in mind just hypotheses) that they be true in all respects: nor is it for the astronomer knowingly to suppose false or ingeniously contrived hypotheses, in order that from them he may deduce the celestial movements.[82]

The conclusion is frequently reiterated.

That astronomer well performs his office who predicts with the greatest measure of approximation the motions and situations of the stars: but he does better and is held worthy of the greater praise who in addition to this furnishes us with true opinions concerning the form of the world.[83]

For astronomers ought not simply to enjoy a licence of making any fictions they choose without rational grounds.[84]

The whole array of his errors [sc. Reymer Baer's] is due to his perverse understanding of the term *hypothesis* which to him means the same as *fiction.*[85]

Since in Copernicus there appears that most admirable order that is between a cause and its effects, it must be that that which Copernicus gives is the true cause of the backward turnings of the planets, i.e., that that hypothesis of his is no mere fiction.[86]

As in every discipline, so in astronomy, too, the conclusions that we teach the reader are seriously intended, and our discussion is no mere game.[87]

There is even a sense in which Kepler advocates an astronomy "without hypotheses." Petrus Ramus, the Regius Professor of Astronomy at the University of Paris, had once written:

The fiction of hypotheses is absurd . . . would that Copernicus had rather put his mind to the establishing of an astronomy without hypotheses. . . . And if the fruit of a failing utility may be offered as the reward of such great virtue, I will promise you the Regius Professorship at Paris as the prize for an astronomy constructed without hypotheses; and I will gladly fulfill this promise by resigning my Professorship.[88]

Ramus died in 1572; but Kepler, writing in 1597 and again in 1609, answers him as follows:

Most opportunely, Ramus, hast thou voided this pledge's surety, by leaving both life and office, which didst thou still retain, I would of good right claim it for myself, or for Copernicus. . . . I confess it a most absurd play to explain the processes of nature by false causes, but there is no such play in Copernicus, who indeed himself did believe his hypotheses to be true . . . nor did he only believe, but also proved them true.[89]

Galileo's interpretation of the significance of astronomical hypotheses is also, like that of Kepler, thoroughly realistic in spirit. Such was already his view when in 1606 he composed his *Trattato della sfera* (published in 1656), in which he still employed the Ptolemaic hypotheses. He there wrote:

As to method, the cosmographer is accustomed to proceed in his investigations by four means: the first of which comprises the appearances, otherwise termed phenomena; and these are nothing but the observations of sense, such as we perceive throughout the day. . . . In the second place come the hypotheses; and these are nothing but certain suppositions concerning the structure of the celestial spheres, and such as correspond to the appearances. . . . There follow then, in the third place, geometrical demonstrations; in which, by means of the properties of circles and straight lines, are demonstrated the particular accidents that follow from the hypotheses. And, finally, what has been demonstrated by means of the lines is then by arithmetical calculation reduced and distributed into tables, from which we can without difficulty and at our pleasure find again the disposition of the heavenly bodies at any moment of time. And, since we are dealing with the first principles of this science, we leave aside for the present the calculations and more difficult demonstrations, and deal solely with the hypotheses, attempting to confirm and establish them by means of the appearances.[90]

And later when Galileo accepted the theory of Copernicus it was in the same realistic spirit. He now criticizes those who retain as true and real deferents, equants, and epicycles, assumed by pure astronomers in order to facilitate their calculations but no longer to be retained by such astronomical philosophers who,

. . . in addition to the attempt to "save the appearances" in some fashion, seek to investigate, as the greatest and most admirable of problems, the true constitution of the universe, since that constitution is, and is in one way only, true, real and impossible to be otherwise, and by its grandeur and nobility worthy of being preferred to any other knowable question raised by speculative genius.[91]

In 1615 the celebrated theologian Robert Cardinal Bellarmine wrote to Foscarini a letter that has since become famous. He expressed himself very moderately, pointing out that no legitimate criticism could be made of the theories of Galileo, except in so far as he rashly put forth his new astronomical ideas as though they represented demonstrable physical realities. Galileo and his friends, he wrote, would have been wise if they had limited themselves to speaking *ex suppositione* and not absolutely, as he believed that Copernicus had spoken.

To say that on the supposition that the earth moves and the sun stands still all the appearances are saved better than on the assumption of eccentrics and epicycles, is to say very well—there is no danger in that, and it is sufficient for the mathematician: but to wish to affirm that in reality the sun stands still in the center of the world, and only revolves upon itself without traveling from east to west, and that the earth is located in the third heaven and revolves with great velocity about the sun, is a thing in which there is much danger not only of irritating all the scholastic philosophers and theologians, but also of injuring the Holy Faith by rendering false the Sacred Scriptures. . . . I say that once there is a real demonstration that the sun stands in the center of the world and the earth in the third heaven, and that the sun does not go around the earth, but the earth around the sun, then we must go to work with much thoughtful consideration to explain the passages of Scripture that seem to oppose this view, and better to say that we have not understood these passages, than to say that that which has been demonstrated is false. But I shall not believe that there is any such demonstration until it has been shown to me: nor is it the same thing to demonstrate that supposing the sun to stand in the center and the earth to move in the heaven will save the appearances, and to demonstrate that in truth the sun does stand in the center and the earth moves in the heaven; for I believe that the first demonstration can be given, but concerning the second I have the gravest doubt, and in case of doubt one should not abandon the Sacred Scriptures as they have been expounded by the Holy Fathers.[92]

Replying to this argument in the same year (1615) Galileo warns us against two errors.[93] The first consists in representing the theory of the mobility of the earth as "such an enormous paradox and such a manifest piece of foolishness, that on no account is it to be doubted that neither at present nor at any other time is it susceptible of demonstration, and that it could not even find a place in the mind of any person of good judgment."[94] The second consists in supposing that Copernicus himself spoke merely *ex suppositione*.[95]

Elsewhere Galileo seems to argue that, if there are good empirical reasons for rejecting the Ptolemaic system, then the Copernican system is sufficiently established—as though the latter were the only possible alternative and a sort of *experimentum crucis* were involved.[96] Nevertheless Galileo seems in the end to have recognized, albeit with some reluctance, the justice of Bellarmine's main contention that to show that a hypothesis saves the phenomena, even, it may be, better than any other known to us, is no sufficient demonstration of its truth.

It is true that it is not the same thing to show that the supposition that the earth moves and the sun stands still saves the appearances, and to demonstrate that such hypotheses are really true in nature; but it is quite as much and even more true that by means of the other system commonly received it is impossible to account for these appearances. The latter system is undoubtedly false, just as it is clear that the former, which fits the appearances most excellently, can be true, and that no other greater truth can or ought to be sought in a hypothesis than its correspondence with the particular appearances.[97]

The point was brought up again in a conversation between Galileo and Maffeo Cardinal Barberini (afterward Pope Urban VIII). It is recorded as follows:

The Supreme Pontiff Urban VIII . . . while he was still a Cardinal warned a certain friend of his, one no less conspicuous for his learning than praiseworthy in his religion, to take much thought whether the scheme that he had contrived with regard to the motion of the earth, with a view to saving all the phenomena that appear in the heavens, and all that philosophers commonly accept concerning the motions of the heavens and stars as a result of their diligent observation and reflection, were in accordance with the Scriptures. For granting the whole scheme which that very learned man had invented: he queried whether God was not able, and did not know how, in some other manner so to dispose and to move the spheres or the stars that all the phenomena that appear in the heavens, and everything that is commonly said concerning the motions, order, position, distance, and disposition of the stars, might be saved. "If you deny it," said the Most Holy Father, "you must prove that it implies some contradiction to suppose that these things can happen otherwise than according to the scheme that you have invented. For God in his infinite power can do whatsoever implies no contradiction: and since the wisdom of God is not less than his power, if we grant that God could have done it, then we must assert that he knew how to do it. But if God might have arranged, and knew how to arrange, these matters otherwise than according to that scheme of yours, so that everything we have mentioned is accounted for, we ought not to confine the divine power and wisdom to this one way." Having heard this the learned man made no reply.[98]

Was Galileo convinced? Or was his silence diplomatic? He returns to the point at the close of his *Dialogo sopra i due massimi sistemi del mondo*, where Simplicio is talking about the reason for the ebbing and flowing of the sea.

I confesse that your conceit seemeth to me far more ingenious than any of all those that I ever heard besides, but yet neverthelesse I esteem is not true and concluding. . . . I know that both you being asked, Whether God, by his infinite Power and Wisdome might confer upon the Element of Water the reciprocal motion which we observe in the same in any other way, than by making the containing Vessel to move; I know, I say, that you will answer, that he might, and knew how to have done the same many ways, and those unimaginable to our shallow understanding: upon which I forthwith conclude, that this being granted, it would be an extravagant boldnesse for any one to goe about to limit and confine the Divine Power and Wisdome to some one particular conjecture of his own.

Salviati replies:

This of yours is admirable, and truly Angelical Doctrine, to which very exactly that other accords, in like manner divine, which whilst it giveth us leave to dispute, touching the constitution of the World, addeth withall (perhaps to the end, that the exercise of the minds of men might neither be discouraged, nor made bold) that we cannot find out the works made by his hands. Let therefore the Disquisition permitted and ordain'd us by God assist us in the knowing, and so much more admiring his greatnesse, by how much lesse we find our selves too dull to penetrate the profound Abysses of his infinite Wisdome.[99]

This, it must be admitted, has too ironical a flavor to be taken quite literally. Yet there can be little doubt that Galileo saw the difficulty. For he himself, as we have seen, said it: "It is not the same thing to show that the supposition that the earth moves and the sun stands still saves the appearances, and to demonstrate that such hypotheses are really true in nature"; albeit he is thoroughly convinced that the objection can be met:

. . . but it is quite as much and even more true that by means of the other system commonly received it is impossible to account for these appearances. The latter system is undoubtedly false, just as it is clear that the former, which fits the appearances most excellently, can be true, and that no other greater truth can or ought to be sought in a hypothesis than its correspondence with the particular appearances.

4. CONCLUSION

As we look back over the long and complicated course of Renaissance speculation concerning the method of hypothesis, three matters seem to call for emphasis.

First, in the controversy over the propor interpretation of astronomical hypotheses both parties stood for something of the truth, and both were subject to some measure of illusion. The antirealists were more conscious of the limitations of the scientific method, but tended at times to exaggerate those limitations. The realists were more conscious of the necessary unity of the sciences and of the legitimacy of the claims of the sciences to establish objectively valid and reliable conclusions, but were prone to overestimate the certitude and finality of such conclusions.

In the second place, along with the increasing tendency, upon which we have chiefly dwelt, to adopt a realistic interpretation of the hypotheses of astronomy, there goes also a growing realization, assisted largely by the breakdown of the traditional distinction between celestial and terrestrial matter, and indeed of the formerly unquestioned authority of the Aristotelian physics generally, of the thoroughgoing unity of all the natural sciences. Men are thus being led progressively to the recognition (1) that the method of hypothesis hitherto chiefly developed in connection with astronomy is also the true method of physics and of all natural science whatsoever; (2) that the conclusions reached by astronomers must, if they are to be valid, form together with the conclusions of the other physical sciences a single consistent system having one and the same basis and one and the same kind and degree of certitude.

In the third place, there is latent throughout the discussion a problem that occasionally comes more or less to the fore—the question, namely, as to the nature of the results that this one method of the natural sciences, now seen to be so largely identical with the astronomical method of hypothesis, is capable of attaining. Three questions are really here involved. (1) In the view of things that we attain through the use of this method, do we ever succeed in rendering with perfect exactitude and finality the objective order of nature, or are our results always and inevitably partial and approximate and therefore at every stage subject to progressive revision and correction with a view to the achievement of a closer approximation? (2) Does our method suffice to assure us of the absolute *certitude* of our results, or does it succeed in attaining only some degree, however high, of *probability* which is always subject to alteration in the light of further evidence from whatever source? And (3) if our hypothetical method gives us after all only probability for our results, where then are we to look for that entire certitude and finality without which few progressive scientists in this period are yet ready to rest content? No longer can the appeal be to the superior certitude of physics, if its methods and results are simply on all fours with those of astronomy, to confer upon conclusions of the latter science the objective validity its own proper method is incapable of supplying. We must needs seek a new source of certitude for the results of natural science as a whole.

For the solution of this problem the seventeenth century looked prevailingly in the direction typified by the effort of Descartes. Like so many men of that age he strongly tended to feel that conclusions that fall short of absolute certitude, or are in any way merely probable, provisional, approximate, or "for the most part," have no place in any genuine science. He recognized, too, that merely empirical procedure by the hypothetical method is inadequate to guarantee any such result. Hypotheses may be suggested by experience, tested and revised by reference to experience until they are brought into a reasonable conformity therewith. But this scarcely suffices to establish them in any truly scientific manner. They must receive a completely cogent demonstration before they can properly be admitted as scientifically valid conclusions.

It is consequently to the method of the mathematical disciplines that we must look. There we find rigorous deductions from intuitively evident and absolutely indubitable first principles. The natural sciences can only attain results of similar cogency by adopting the same procedure. Only thus can they become genuinely scientific. However, even in the hands of its founder the Cartesian method was doomed to failure. But it was not until the eighteenth century that it was at last widely and freely confessed that natural science must once and for all renounce the attempt to discover any source of certitude independent of the judicious application of its own proper empirical and hypothetical method, or any kind of certitude other than that reasonably high probability that can alone result from the use of this method.

CHAPTER THREE

Francis Bacon's Philosophy of Scienec

I. THE GREAT INSTAURATION

ALMOST ALL OF Bacon's philosophical writings are closely connected with the mission which he appears to have regarded as imposed upon him by his natural gifts, that of serving [1] humanity by providing it with a scientific method that would "kindle a light in nature" and restore science to its proper place of servant to the will of man, rather than merely delight to his intellect. This attempt to restore science, which Bacon entitles the *Instauratio magna*, was never completed; and there is little reason to believe either that his understanding of what he did write of it was any clearer or sounder than his actual words, or that he had worked out those parts of it which he announced, but did not write, in any greater detail than actually appears in the descriptive announcements themselves. *The Great Instauration* was to have consisted of six parts, but only the first two were at all fully written out; the other four are represented only by short works or fragments, the fitting of which under some definite one of the announced divisions of the *Instauration* is in some cases rather conjectural.[2] The *Instauration* is preceded by a "Plan of the Work," in which its parts are enumerated and the respective character of each described. These parts are as follows:

1. *The divisions of the sciences.* This is a "summary or general description of the knowledge which the human race at present possesses" and also an account of "things omitted which ought to be there," since "there are found in the intellectual as in the terrestrial globe waste regions as well as cultivated ones." The *De augmentis scien-*

tiarum is what Bacon gives as constituting this part of the *Instauration.*

2. *The New Organum, or directions concerning the interpretation of nature.* What the intellect needs, Bacon tells us, is not the syllogistic logic, which only empowers man to overcome an opponent in argument, but rather something that will enable him "to command nature in action." This will consist of "a form of induction which shall analyze experience and take it to pieces, and by a due process of exclusion and rejection lead it to an inevitable conclusion."

3. *The phenomena of the universe, or a natural and experimental history for the foundation of philosophy.* Since no excellence of method can supply the mind with the material of knowledge, we must go for it to the facts themselves. The facts to be included in the sort of natural history that science is in need of, however, must be such "as to give light to the discovery of causes," rather than selected on the basis of their promise of immediate practical utility. They must be facts of deliberate experiment, rather than of the observation of uncontrolled nature. They must be so accurately and objectively described as to inform the mind, rather than to strike it with wonder; and the manner of any new experiment performed must be stated in sufficient detail to enable others to check its results. The *Parasceve,* the *Sylva sylvarum,* and various *Histories* are all that are to be found in Bacon's writings belonging to this division.

4. *The ladder of the intellect.* This is to consist of examples of inquiry and invention according to the Baconian method, which can be used by the intellect as types and models of the method and will render it familiar in concrete form. This division of the *Instauration* seems to be unrepresented in Bacon's writings, except for the short fragment called *Filum labyrinthi.*

5. *The forerunners, or anticipations of the new philosophy.* This part, which is "for temporary use only," is to include such things as Bacon has himself discovered, not according to the new method, but "by the ordinary use of the understanding in inquiring and discovering," and which are therefore conclusions by which he does not bind himself. The contents of this part are to provide, as it were, "wayside inns, in which the mind may rest and refresh itself on its journey to more certain conclusions." According to Spedding, such works as the *De fluxu et refluxu maris* and *De principiis* probably belong to this division.

6. *The new philosophy, or active science.* This is to consist of the results which will eventually be attained by humanity when it has learned to avail itself of the method that Bacon advocates.

The *Novum organum*, however, contains the central ideas of Bacon's system, of which the whole of the *Instauration* is only the development.[3] We shall therefore pass at once to the examination of the views of scientific method contained in that celebrated work.

2. THE METHODOLOGICAL PRECEPTS OF THE FIRST BOOK OF THE NOVUM ORGANUM

The first book of the *Novum organum* is in large part devoted to the presentation of the doctrine of the "Idols," which "is to the Interpretation of Nature what the doctrine of the refutation of sophisms is to common logic."[4] The monitions which the discussion of the Idols contains concerning the many weaknesses, errors, and prejudices by which the human mind is hampered in its investigations of nature need not detain us since they do not constitute an intrinsic part of Bacon's method, but rather a preparation of the mind for the reception and proper use of that method. There are, however, scattered through the first book, many remarks that are of much value in making clear the aims Bacon had set himself and the general nature of the method he desired to advocate, as contrasted with the aims and method of the traditional science with which he wished to break. The first book of the *Novum organum*, as compared with the second book, may very roughly be said to constitute a fair example of the contrast, which so often strikes one in Bacon's writings, between the soundness of his insights when he deals with topics in abstract fashion only and the defectiveness of his translations of such insights into concrete terms.

A. *Pragmatism*

In the first aphorisms of the first book, Bacon declares emphatically that the test of real knowledge, as distinguished from what has hitherto passed for knowledge, is ultimately a pragmatic one. Thus he says: "Human knowledge and human power meet in one; for where the cause is not known, the effect cannot be produced. Nature to be commanded must be obeyed; and that which in contemplation is as the cause is in operation as the rule." [5] Again: "Truth therefore and utility are here *ipsissimae res* and works themselves are of greater value as pledges of truth than as contributing to the comforts of life." [6]

The chief reason for the failure of the sciences to bring forth real knowledge is, he tells us, "that while we falsely admire and extol the powers of the human mind, we neglect to seek for its true helps." [7] He, on the contrary, is fully conscious of the weakness of the senses and of the understanding when left to their spontaneous modes of activity, but he believes that when the faculties of man are provided with an adequate instrument, such as the method that he intends to formulate will constitute, it will become as easy for these weak human powers to obtain real knowledge as it is for the most unskilled hand to trace a true circle when provided with a compass.

B. *Gradual inductions*

No adequate method, Bacon declares, has up to his time been formulated for the investigation of nature. The syllogism, which merely "commands assent . . . to the proposition, but does not take hold of the thing," [8] is useless, and induction by simple enumeration is of no avail either, for it is childish: " . . . its conclusions are precarious, and exposed to peril from a contradictory instance; and it generally decides on too small a number of facts, and on those only which are at hand." [9]

Another form of induction is needed which will not fly from the senses and particulars to the most general axioms and, considering the truth of these as settled and immovable, proceed from them deductively; but will, on the contrary, rise from the senses and particulars by a gradual and unbroken ascent and arrive at the most general axioms last of all.[10] This sort of induction

> . . . must analyze nature by proper rejections and exclusions; and then, after a sufficient number of negatives, come to a conclusion on the affirmative instances; which has not yet been done or even attempted, save only by Plato, who does indeed employ this form of induction to a certain extent for the purpose of discussing definitions and ideas.[11]

C. *Hypothesis*

The results even of this sort of induction are, moreover, to be regarded only as hypotheses and must be tested by comparison with facts. In the aphorism immediately following the one just quoted, Bacon says:

> But in establishing axioms by this kind of induction, we must also examine and try whether the axiom so established be framed to the measure of those particulars only from which it is derived, or whether it be larger and wider. And if it be larger and wider, we must observe whether, by indicating to us new particulars it confirm that wideness and largeness as by a collateral security; that we may not either stick fast in things already known, or loosely grasp at shadows and abstract forms; not at things solid and realized in matter. And when this process shall have come into use, then at last shall we see the dawn of a solid hope.[12]

D. *Experimentation*

The data from which we must start in our inductions are to be not only observations of natural facts as they spontaneously occur, but also and preferably of natural facts as brought out by experiments performed under known conditions: [13] "When by art and the hand of man she [nature] is forced out of her natural state, and squeezed and moulded," constrained and vexed, she betrays her secrets more readily than in her natural freedom.[14] Again:

> For the subtlety of experiment is far greater than that of the sense itself, even when assisted by exquisite instruments; such experiments, I mean, as are skillfully and artificially devised for the express purpose of determining the point in question. To the immediate and proper perception of the sense therefore I do not give much weight; but I contrive that the office of the sense shall be only to judge of the experiment, and that the experiment itself shall judge of the thing.[15]

But experiments may be of two sorts. "The mechanic, not troubling himself with the investigation of truth, confines his attention to those things which bear upon his particular work," and the experiments he performs, being thus only experiments of fruit (*Experimenta Fructifera*), are for the most part of little use to science.[16] What science needs as its foundation, on the contrary, is

> . . . a variety of experiments, which are of no use in themselves, but simply serve to discover causes and axioms; which I call "*Experimenta Lucifera*," experi-

ments of *light* . . . experiments of this kind have one admirable property and condition: they never miss or fail. For since they are applied not for the purpose of producing any particular effect, but only of discovering the natural cause of some effect, they answer the end equally well whichever way they turn out; for they settle the question.[17]

Even experiments of light, however, are, Bacon believes, of little avail if the experimenter

. . . confine his aim and intention to the investigation and working out of some one discovery and no more; such as the nature of the magnet,[18] the ebb and flow of the sea . . . whereas it is most unskillful to investigate the nature of anything in the thing itself; seeing that the same nature which appears in some things to be latent and hidden is in others manifest and palpable.[19]

Such, in its essentials, is the doctrine of method set forth in the first book of the *Novum organum*. That doctrine can hardly fail to be acknowledged as for the most part admirable by any modern reader who is allowed to interpret its terms in the light of what scientific procedure is today. We must, however, now pass to the examination of the rather disappointing interpretation that Bacon himself gives of the doctrine in the second book of the *Novum organum*. A bare statement of the method there concretely described will first be placed before the reader, and afterward a discussion will be given of the meaning of some of the terms used by Bacon and of such other matters as it may be necessary to introduce in order to render the nature of his doctrine completely clear.

3. THE DOCTRINE OF THE SECOND BOOK OF THE NOVUM ORGANUM

Our observations of nature, Bacon tells us, must first be arranged in the form of tables, for our knowledge of particular facts "is so various and diffuse, that it confounds and distracts the understanding, unless it be ranged and presented to view in a suitable order." [20] From these tables, the understanding may then be able to gather by induction what it seeks—namely, the "form" of whatever "nature" [21] it is investigating.

Three different sorts of tables will be necessary:

1. *Table of essence and presence*. This means that, a nature being

given (the form of which is in question), we must first of all have a muster or presentation before the understanding of all known instances that agree in the same nature, though in substance the most unlike.[22] Thus, if heat is the nature under investigation, we shall compile a table of instances agreeing in the presence of heat, comprising the rays of the sun, "fiery meteors," "burning thunderbolts," bodies under the influence of friction, organic matter in fermentation ("green and moist vegetables confined and bruised together"), "quick lime sprinkled with water," and so on.

2. *Table of deviation, or of absence in proximity.* We must, that is to say, "make a presentation to the understanding of instances in which the given nature is wanting; because the Form . . . ought no less to be absent when the given nature is absent, than present when it is present." But, as such a list would be endless, the absence of the given nature should be "inquired of in those subjects only that are most akin to the others in which it is present and forthcoming." [23] Thus, an instance similar to that of the rays of the sun, but in which heat is absent, is that of the rays of the moon and stars; again, sheet lightning is similar to the thunderbolt, but not accompanied by heat, and so forth.

3. *Table of degrees or comparison.* This means that "we must make a presentation to the understanding of instances in which the nature under inquiry is found in different degrees, more or less." [24] As instances forming a proper part of such a table, Bacon lists both observations such as that certain substances—for example, sulphur, naphtha, rock oil—are "more strongly disposed" to heat than others, and observations of a more significant nature, such as that "animals increase in heat by motion and exercise" and that "an anvil grows very hot under the hammer."

The "Presentation of Instances to the Understanding" having been made by the three tables, "induction itself must be set at work; for the problem is, upon a review of the instances, all and each, to find a nature such as is always present or absent with the given nature and always increases and decreases with it." [25] It is not given to man, however, to find such a nature at once by inspection of the tables; he must proceed gradually by "rejection or exclusion of the several natures which are not found in some instance where the given nature is present, or are found in some instance where the given nature is absent, or are found to increase in some instance where the given nature decreases, or to decrease when the given nature increases." [26] Thus, the fact that both dense and tenuous bodies are capable of being hot excludes denseness and tenuousness as possible forms of heat, and so on. Were this process of gradual exclusion carried out thoroughly, nothing would remain but the form of heat at its conclusion. Bacon says, however, that he does not profess in the instance to have carried out thoroughly the method he is describing, the discussion he gives of heat being intended solely by way of illustration; nor can that method be carried out thoroughly at first, owing to the confusedness of our notions of natures.[27] He nevertheless hazards a provisional induction, or, as he calls it, a "first vintage," on the basis of his tables and of the process of exclusion that he has partly performed, to the effect that "heat is a motion, expansive, restrained, and acting in its strife upon the smaller particles of bodies." [28] He then goes on to a discussion of the "other helps of the understanding," of which he gives a catalogue under nine headings, as follows: "Prerogative Instances," "Supports of Induction," "Rectification of Induction," "Varying the Investigation According to the Nature of the Subject," "Prerogative Natures," "Limits of Investigation," "Application to Practice," "Preparations for Investigation," and "Ascending and Descending Scale of Axioms." Only the "Prerogative Instances," however, are discussed by Bacon; the other eight of these helps of the understanding were never set forth and in all probability never in any detail thought out by him.

Prerogative instances is the name that Bacon gives to such instances of a phenomenon as are, for some reason or other, of especial value in scientific investigation. The word "instance," however, as used by him has a very broad signification. Of the prerogative instances, Bacon distinguishes twenty-seven sorts, which he classifies according to the kinds of utility they possess.[29] They are useful either for (A) the obtaining of information or (B) purposes of operation. Those that have informative value assist either (i) the senses or (ii) the understanding. Those that are of value for operation either (i) point out practice, (ii) measure practice, or (iii) facilitate practice. Again, while some of the instances need not be inquired into until some particular nature is under investigation, others (such as: conformable, singular, deviating, bordering, of power, of quantity, intimating, polychrest, and magical) must be collected at once, for they have a general or disciplinary value; they "either help and set right the understanding and senses, or furnish practice with her tools in a general way."

The nature and particular use of each of the various sorts of instances as above classified will now briefly be set forth. The fact may be mentioned in passing that the illustrations subjoined by Bacon in more than one case appear to fit under other of his instances as easily as under that for which he intends them. The analogies from which he takes the names of his various instances are for the most part too fanciful to lead to the desideratum of a logically tight division.[30]

A. *Instances of value for information*

i. *Instances that assist the senses.* To these, Bacon gives the general name of "Instances of the Lamp":

a. *Instances of the door* "strengthen, enlarge and rectify the immediate action of the senses." Microscopes, telescopes, measuring rods, and the like constitute such helps to the senses (II, 38 and 39).

b. *Summoning instances* "are those which reduce the non-sensible to the sensible; that is make manifest things not directly perceptible by means of others which are." Bacon enumerates seven reasons owing to which an object may escape the senses; and the illustrations are therefore most diverse, as may be judged from the fact that among them are to be found the diagnosis of the inner condition of the human body by the state of the pulse, urine, and so forth, as well as the observation of the sun by means of the reflection of it in some medium that will so diminish the brightness of it as to bring it within the measure of what the eye can bear (II, 40).

c. *Traveling instances,* which "indicate the continued process or series of those things and motions which are for the most part unobserved except in their end or period." Thus, if we are investigating the vegetation of plants, we should observe the seed through all the stages of germination and growth (II, 38 and 41).

d. *Substitutive instances,* "which supply information when the senses entirely fail us." These instances consist either in the "observation of some cognate body which is perceptible" and from which we can draw inferences by analogy to the imperceptible object with which we are directly concerned; or in the observation of a number of things which, although they do not completely exhibit what we seek, exhibit gradual approximations to it (II, 42).

e. *Dissecting instances,* which, reminding us of the exquisite subtlety of nature, stir up and awaken our attention and observation; for instance, "that a little saffron tinges a whole hogshead of water" (II, 43).

ii. *Instances that assist the understanding.* These include:

a. *Solitary instances,* which "accelerate and strengthen the process of exclusion." They are instances "which exhibit the nature under investigation in subjects which have nothing in common with other subjects except that nature; or, again, which do not exhibit the nature under investigation in subjects which resemble other subjects in every respect except in not having that nature." Thus, prisms have nothing in common with flowers and other colorful objects except the exhibition of colors, and two flowers of the same species that differ in color agree in everything else (II, 22).

b. *Migratory instances,* "in which the nature in question is in the process of being produced when it did not previously exist, or on the other hand of disappearing when it existed before." These assist the understanding, not only by accelerating and strengthening exclusion, but particularly by narrowing down the group of natures among which the form is to be found. Thus, the nature whiteness is produced in transparent glass by pounding it into powder (II, 23).

c. *Striking instances* are another class of instances that are of value in thrusting the form of a nature upon the attention. "They are those which exhibit the nature in question naked and standing by itself," or in a striking degree. Thus, a striking instance of the fact that expansion belongs to the form of heat is an air thermometer (II, 24).

d. *Instances of companionship* have the same general function of "bringing the affirmative of the form within narrow limits." They are furnished by concrete things of which the nature investigated is an inseparable companion or, contrariwise, an irreconcilable foe. Thus, heat is an inseparable companion of flame (II, 33).

e. *Instances of limit,* again, subserve the same end by pointing out how far nature may act or be acted upon without passing into something else. They are instances in which the nature investigated is maximally or minimally present—for example, in spirit of wine, weight is at its minimum (II, 34).

Instances that assist the understanding by exalting it and leading it to genera and common natures (II, 52) are the following:

f. *Clandestine instances,* "which exhibit the nature under investigation in its lowest degree of power"—for example, in fluids, consistency is at a minimum. It is not easy to say precisely what difference Bacon saw between these and instances of limit (II, 25).

g. *Singular instances,* which are instances of concrete bodies "not agreeing with other bodies of the same kind," such as the magnet among stones. The requirement that the form shall include such deviating cases as well as the ordinary cases may point the way to the form (II, 28).

h. *Deviating instances* are also deviations, but of individuals, while singular instances are deviations of species (II, 29).

i. *Bordering instances* also, which, like that of the flying fish, exhibit species that appear intermediate between two others, might be classed with singular instances. They reveal natural possibilities (II, 30).

j. *Instances of alliance,* which exhibit the unity of natures supposed to be heterogeneous; for instance, when the artificial heat of a conservatory acts on plants as the heat of the sun, the fancied distinction between the species of heat is broken down (II, 25).

k. *Constitutive instances,* which furnish an approach to the form by constituting one of its species—for example, an approach to the form of that which excites memory is furnished by noting the species of causes of remembering, such as sameness of place (II, 26).

l. *Conformable instances,* which are the cases of resemblance of concrete things—for example, the roots of plants are like their branches. They are of little use for the discovery of forms. They, however, rather than diversities, constitute the proper material for natural histories (II, 27).

The understanding, again, may be assisted by having the falsity of some supposition concerning the forms and the causes of things made evident. This is accomplished by:

m. *Instances of the finger post* (*crucial instances*), which are for the most part artificially designed by man. They are contrived to exclude either one of two natures which appear equally capable of belonging to the form that is being sought in a given case. An instance where one only of the two rival supposed characters of the form is found conjoined with the nature investigated excludes the possibility of the other's belonging to the form of that nature (II, 36).

n. *Instances of divorce* are such as prove the separability of one nature from another that was supposed inseparable from it (II, 37).

B. *Instances that are of value primarily for operation*

i. *Instances that point out a source of action.* These are:

a. *Instances of power,* which are constituted by the most perfect achievements of human artifice and are to artifice what singular instances are to nature. That is, they excite and raise the understanding to the discovery of forms; but they do so more effectively than singular instances "because the method of creating and constructing such miracles of art is in most cases plain," and we thus know where we should begin our new investigations (II, 31).

b. *Intimating instances,* which point out to what we should aspire, if means be given—that is, what is useful to man (II, 49).

ii. *Instances that measure practice.* These are mathematical instances, which are of four sorts:

a. *Instances of the rod,* which are those that reveal the distances at which the powers and motions of things take effect (II, 45).

b. *Instances of the course,* which "measure nature by periods of time" (II, 46).

c. *Instances of quantity,* "which measure virtues according to the *quantity* of the bodies in which they subsist, and show how far the *mode* of the virtue depends upon the *quantity* of the body" (II, 47).

d. *Instances of strife,* which indicate which of two virtues is stronger and prevails, and which is weaker and gives way. Bacon here gives a classification of motions under nineteen headings, stating that of these some are stronger than others (II, 48).[31]

iii. *Instances that facilitate practice.* These are:

a. *Polychrest instances,* which "relate to a variety of cases and occur frequently," and thus save labor. These are constituted by the general ways by which man operates on natural bodies, such as compression, the application of heat or cold, and so on (II, 50).

b. *Magical instances,* which are those where "the material or efficient cause is scanty or small as compared with the work and effect produced," so that the latter appear miraculous (II, 51).

4. NATURES, FORMS, AND THE TASK OF SCIENCE

The substance of Bacon's statements concerning the method of induction being now before the reader, something must next be said of what Bacon conceived to be the nature of the problems of science that were to be dealt with by means of that method.

A. *The task of science*

In the first nine aphorisms of Book II, Bacon endeavors to define as precisely as possible "the mark of knowledge"—that is, the sort of information science must seek to obtain. The first aphorism gives a statement of it as follows:

> On a given body to generate and superinduce a new nature or new natures is the work and aim of Human Power. Of a given nature to discover the form, or true specific difference, or nature-engendering nature, or source of emanation (for these are the terms which come nearest to a description of the thing) is the work and aim of Human Knowledge. Subordinate to these primary works are two others that are secondary and of inferior mark; to the former, the transformation of concrete bodies, so far as this is possible; to the latter, the discovery, in every case of generation and motion, of the *latent process* carried on from the manifest efficient and the manifest material to the form which is engendered; and in like manner the discovery of the *latent configuration* of bodies at rest and not in motion.

Science may then be divided into metaphysics, which has for its task the investigation of formal and final causes—the practical application of its results being magic; and physics, which has for its task the investigation of efficient and material causes, and of latent processes and latent configurations—the practical application of the results of physics being mechanics.[32] From various places, we obtain explanations of the meanings of the terms used by Bacon in the first aphorism quoted above.

B. *Natures*

The natures of which science may seek to know the forms are either simple or complex. Examples of what Bacon regards as simple natures are yellowness, transparency, opacity, tenacity, vegetation, malleability, ductility, volatility, fixity, fluidity, solution.[33] Elsewhere [34] he gives a list in which things, to some of which he has already referred as simple natures, are called "configurations of matter," so that he apparently means the two terms to be interchangeable. The list is as follows:

The *Configurations of Matter* are Dense, Rare; Heavy, Light; Hot, Cold; Tangible, Pneumatic; Volatile, Fixed; Determinate, Fluid; Moist, Dry; Fat, Crude; Hard, Soft; Fragile, Tensile; Porous, Close; Spirituous, Jejune; Simple, Compound; Absolute, Imperfectly Mixed; Fibrous and Venous, simple of structure, or equal; Similar, Dissimilar; Specific, non-specific; Organic, Inorganic; Animate, Inanimate.

This list is followed by a list of "simple motions, in which lies the root of all natural actions, subject to the conditions of the configurations of matter." A more elaborate list of these is given in the *Novum organum*, II, 48.

Complex natures, or substances, on the other hand, would be a lion, an oak, gold, water, air, and so on.[35] To inquire into the forms of such natures would be "neither easy nor of any use," for they are infinite in number; and moreover they are complexes of simple natures, which "are not many and yet make up and sustain the essences and forms of all substances." [36] Knowledge of the forms of simple natures,[37] therefore, is all we need and all that is worth attempting. Bacon thus conceives of the world, or at least of the knowable world,[38] as a congeries of simple natures—that is, of qualities, variously combined—a qualitative atomism somewhat akin to that of Anaxagoras [39] and to the later analysis of Berkeley. Of these simple natures, some are more primitive, more real than others, and constitute the forms of the others, which forms it is the task of science to discriminate by means of the inductive method already set forth.

But a great difficulty, to which passing notice has already been given, is introduced by the fifteenth aphorism of the first book, which reads:

> There is no soundness in our notions whether logical or physical. Substance, Quality, Action, Passion, Essence itself, are not sound notions: much less are Heavy, Light, Dense, Rare, Moist, Dry, Generation, Corruption, Attraction, Repulsion, Element, Matter, Form and the like; but all are fantastical and ill-defined.

In the next aphorism, he refers to such notions as "but wanderings, not being abstracted and formed from things by proper methods." In the fourteenth, he states that our only hope of obtaining notions that are not "confused and over-hastily abstracted from the facts . . . lies in a true induction." [40] The difficulty is thus that, while the task of science is to isolate such simple natures as are forms of other simple natures, our notions of simple natures are all unsound, fantastical, ill-defined; and that sound notions of simple natures are obtainable, like true axioms, only as the result of an inductive process. And while Bacon tells us what the inductive process is by which the *axioms* are to be obtained, he on the contrary says nothing as to the inductive process by which sound *notions* are to be gathered,[41] the antecedent possession of such notions, however, being necessary for the induction of axioms. He tells us nothing of it, that is, unless his recommendation to vex and constrain nature by experimental conditions under which she will utter her secrets is regarded as the process for the induction of sound notions. But this is hardly a plausible conjecture.

In the second book of the *Novum organum* (aphorism 19), Bacon returns to this unsoundness of our notions, declaring that, on account of it, the process of exclusion cannot be made accurate; and he then says that he will devise and supply more powerful aids for the use of the understanding. Of these, however, he gives nothing beyond the prerogative instances, which in some cases may be of use in correcting our notions but do not constitute the needed method for the induction of them. The probability appears to be that, by a sound notion, he meant only a notion so abstracted from facts as to be well-defined; that, in so far as he thought of an inductive process yielding such notions, it was that of Plato, which he mentions; [42] and that, as concerns the terms in which the definition would have to be framed, he was inclined to select them from among the so-called "primary qualities," and particularly the primary qualities of the minute particles of bodies.[43] It is true that, according to the first aphorism of the second book, the investigation of latent processes and configurations is referred to physics and distinguished from the investigation of forms, which is referred to metaphysics. But from *Novum organum*, II, 20, where the hypothesis as to the form of heat is stated in terms of the minute particles of bodies, and from other passages, it is clear that the two investigations are not meant by Bacon to be in any way *opposed*. *Both* are concerned with processes and configurations *of minute particles;* and metaphysics, which is concerned with the "fundamental and universal laws" of nature, merely pushes the inquiry further than physics, which stops at her "particular and special habits."[44] Most definite on this point is the passage[45] where the kinds of results which the physical and the metaphysical inquiries, respectively, yield are concretely described in terms of the investigation of whiteness.

C. *Forms*

Some support is lent to the hypothesis already stated, regarding Bacon's understanding of "sound notions," by an inquiry into the meaning that he attaches to the term "form." Following are some of the principal statements (in the *Novum organum*) from which that meaning must be gathered:

> Forms are figments of the human mind, unless you will call those laws of action forms [I, 51].
>
> Forms or true differences of things (which are in fact laws of pure act) [I, 75].
>
> For though in nature nothing really exists beside individual bodies, performing pure individual acts according to a fixed law, yet in philosophy this very law, and the investigation, discovery, and explanation of it, is the foundation as well of knowledge as of operation. And it is this law, with its clauses, that I mean when I speak of *Forms* [II, 2].
>
> Whosoever is acquainted with Forms, embraces the unity of nature in substances the most unlike [II, 3].
>
> The form of a nature is such that given the form the nature infallibly follows. . . . Again, the Form is such, that if it be taken away the nature infallibly vanishes [II, 4].
>
> . . . the Form of a thing is the very thing itself, and the thing differs from the form no otherwise than as the apparent differs from the real, or the external from the internal, or the thing in reference to man from the thing in reference to the universe [II, 13].
>
> When I speak of Forms, I mean nothing more than those laws and determinations

of absolute actuality, which govern and constitute any simple nature, as heat, light, weight, in every kind of matter and subject that is susceptible of them. Thus the Form of heat or the Form of light is the same thing as the law of heat or the law of light. Nor indeed do I ever allow myself to be drawn away from things themselves and the operative part. And therefore when I say (for instance) in the investigation of the form of heat "reject rarity," or "rarity does not belong to the form of heat," it is the same as if I said "It is possible to superinduce heat on a dense body"; or "It is possible to take away or keep out heat from a rare body" [II, 17].

Now from this our First Vintage it follows that the Form or true definition of heat (heat, that is, in relation to the universe, not simply in relation to man) is in few words as follows: Heat is a motion, expansive, restrained, and acting in its strife upon the smaller particles of bodies. . . . Viewed with reference to operation, it is exactly the same thing. For the direction is this: If in any natural body you can excite a dilating and expanding motion, and can so repress this motion and turn it back upon itself, that the dilation shall not proceed equably, but have its way in one part and be counteracted in another, you will undoubtedly generate heat. . . . Sensible heat is the same thing; only it must be considered with reference to the sense [II, 20].

How Bacon in these passages is able to speak of the form of a thing both as a law of action and as a nature which is "the very thing itself" has been shown by Robert Leslie Ellis (in his General Preface to the *Philosophical Works*, section 8). The three essential points of his interpretation are as follows:

1. For Bacon, a "substance" (*not* in the sense of complex nature, but in the sense of substratum of attributes) is the formal cause, the *causa immanens*, of the attributes that are referred to it.

2. Ellis believes that, although Bacon had not brought to full consciousness the distinction between the so-called "primary" and "secondary" qualities of bodies, he had nevertheless made it; and that he regarded the essential attributes of substances (the attributes of substances viewed in relation to the universe) as consisting only of primary qualities.

3. The primary qualities, in terms of which the form of a thing is to be defined, are considered by Bacon the causes [46] of the secondary qualities—that is, of what the thing is in relation to our sensations.[47]

The reference in the last passage quoted to "sensible heat," as distinguished from heat defined in terms of motion, is directly in line with this interpretation of Bacon's "forms," which receives support from many other passages also. The use of the word "direction" in the same passage (and elsewhere in the *Novum organum*) to designate the form as viewed with reference to practice ties together the doctrine of the *Novum organum*, according to which science aims at the discovery of forms, and that of the *Valerius terminus*, according to which science aims at the "freeing of directions" for operation.[48]

Ellis remarks that these "directions" for operation appear to be Bacon's original notion, and that their immediately practical import seems to have carried over in his mind to the "forms" by which he replaces them in the *Novum organum*. It is true that some change of motion or configuration is the only immediate external expression the human will ever has, and that it is thus necessary for man to translate into terms of motion or configuration any change of sensible nature he may wish to effect. But, since the changes that man can bring about directly are only changes in the *gross* motion or configuration of things, while the forms seem to be regarded by Bacon as matters of *minute* motions and configurations, there is reason for Ellis' statement that a knowledge of the latter would not *ipso facto* confer the power of superinducing them. However, this probable inutility of a knowledge of forms seems equally a character of "free directions," as exemplified in the *Valerius terminus*. The difference between the directions of the *Valerius terminus* and the forms of the *Novum organum* seems to be one of name only and not of meaning, as Ellis suggests. The description of directions in the eleventh chapter of the *Valerius terminus* is substantially reproduced in the fourth aphorism of the second book of the *Novum organum*, where Bacon states in so many words that, by a rule of operation that is certain, free, and leading to action, and by a true form, he means the same thing.

Another difficulty arises from the fact that if, as appears to be Bacon's belief, the forms are to be defined in terms of the motions and configurations of the *minute* particles of bodies which "escape the sense," no method of sifting bare perceptual data could exhibit such forms. It is to be noted, however, that in the first aphorism of the first book of the *Novum organum* Bacon speaks of man's *observing* nature in fact *or in thought;* and that *experimentation,* on which he lays so much stress, may well be regarded as part at least of a method of *observation in thought*—that is, a method of observing aspects of nature not accessible to bare perceptual examination (observation "in fact").

The inductive approach to forms is furnished by the principle that, where the form is present, so is the sensible nature; and that, where the form is absent, the sensible nature is likewise absent. As Ellis remarks, "there is nothing in this criterium to decide which of two concomitant natures is the form of the other" (General Preface, section 9). But, although Ellis does not appear to notice the point, a sufficient criterion of distinction [49] is constituted by the fact—which he himself brings out—that Bacon believes that only primary qualities can be used in defining forms, the secondary being apparently for the most part those of which the form is sought. Bacon's demand that the form be a nature that is "convertible with the given nature, and yet is a limitation of a more general nature, as of a true and real genus," [50] confirms this; for the meaning of "more general nature" is easier to construe in terms of primary than of secondary qualities,[51] and the discussion of such a "more general nature" given by Bacon at the end of the eleventh chapter of the *Valerius terminus* is so explicit as to leave very little doubt that he intended the meaning of the term to be construed in terms of the former:

> . . . to make a stone bright or to make it smooth, it is a good direction to say, make it even; but to make a stone even, it is no good direction to say, make it bright or make it smooth; for the rule is that the disposition of any thing referring to the state of it in itself or the parts, is more original than that which is relative or transitive towards another thing. *So evenness is the disposition of the stone in itself, but smooth is to the hand and bright to the eye* [italics ours].

Of course Bacon does not explicitly say that this more general nature is to be a primary quality; but one can hardly expect that he should, since it is not contended that he achieved completely clear consciousness of the distinction between primary and secondary qualities—only that his statements and examples take *implicit* account of it pretty clearly and consistently.

In concluding this examination of Bacon's notion of form, it must be stated that, undoubtedly, Bacon himself had but a hazy consciousness of what he meant by a form. Therefore, no *clear* statement of the meaning of the term can be taken as a *reproduction* of his thought, but on the contrary must very definitely be taken as an attempt to *construct*, on the basis of the material constituted by his statements, a thought latent in them indeed, yet to which he himself never fully attained.

5. GENERAL REMARKS ON BACON'S DOCTRINE OF METHOD

A. *Gradual inductions*

According to Whewell, Bacon's greatest merit is, as already stated, his insistence upon a gradual passage from concrete facts to broader and broader generalizations. But, although there is indeed little ground for doubt of the soundness of Bacon's teaching in this respect, nevertheless, in the light of what scientific method is today and of the influence that Bacon's views have had upon the thought of later methodologists, the merits of his doctrine of method appear to attach rather to other aspects of that doctrine than to the element of it stressed by Whewell, which is really a corollary of any practicable method that makes observable fact both the starting point and the test of scientific knowledge.

B. *Experimentation*

One of these other aspects of Bacon's teaching is to be found in his repeated insistence that recourse be had to experiments carefully devised by man, rather than to mere observation. Bacon has often been charged with having overlooked the fact that a hypothesis, a conception to be applied to experience, is an indispensable prerequisite even for the collection of data. Thus H. W. Blunt [52] says: "And collection implies selective hypotheses. Has Bacon forgotten this?" and Ellis declares the discovery of forms (in approximately the above interpretation of the term) impossible by the method of exclusion. He and Whewell [53] agree in the belief that, at least in the immense majority of cases, such "forms" can be found in nature only after they have been conceptually created—that is, invented—by the investigator, and cannot be gathered from empirical observations as final residues of any such automatic and infallible process of sifting as Bacon believed his *Exclusiva* to constitute. There is little doubt that, if we consider mainly Bacon's own tables and natural histories, this criticism is justified; but if, on the other hand, we give due weight to his demand that recourse be had to experiment rather than to bare observation, the charge is at least partially disposed of by the fact that there can be no such thing as an experiment without some hypothesis, however vague. We thus have here one more case of Bacon's precepts being sounder than his practice, and we must remember that it is the former much more than the latter that have arrested the attention of those whom he has influenced. Such men, indeed, have often read between the lines, and Bacon's greatness may perhaps most truly be said to lie in having written lines between which so much that was true and vital was later so plainly to be read.[54]

C. *Principles of agreement and difference*

Another aspect of Bacon's doctrine, and one the importance of which can hardly be overestimated, is to be found in the prominent recognition he gives to the principles of agreement, difference, and concomitant variations as instruments of empirical analysis (interpretation). Whatever may be thought of the material to which he proposed to apply them, there is no doubt that he was fully conscious of the great power of these instruments; indeed this appeared to be what he regarded as his own great insight. Apart from the *use* that was being made of these principles by men of science in his day, and indeed by common men in common affairs at all times, it remains Bacon's very great merit to have been the first to give of them a clear and impressive *formulation*,[55] to which Herschel's formulation is directly traceable and, through Herschel's, Mill's.

Owing to Bacon's quasi-Berkeleyan apperception of nature as a complex of perceptual qualities, revealed either by actual sense experience or by the representative imagination properly controlled, these principles of analysis, as used by him *for the isolation of the elements of such a complex*, assume the form of a systematization of the *psychological* process of discrimination by constancy among varying concomitants, and of its less familiar converse, discrimination by variation among constant concomitants.[56] It is perfectly true that, as Sigwart has shown,[57] Bacon's use of the principles of agreement and difference can be thrown into syllogistic form. But this can be done also with any other use of them that has ever been made. Would Blunt then be prepared to say on this account that these principles are not principles of induction, as he appears to mean is the case with them as used by Bacon? [58] Blunt's assertion, moreover, that this instrument of Bacon's "is an instrument which no scientific enquirer has ever employed to any purpose" [59] seems quite indefensible, since it is (as already noted) to Bacon's conception of the *material* he had to deal with that the barrenness of his investigations is due, and *not* to the *instrument* he proposed to use to deal with that material. The instrument in itself is an excellent one and—in one or another of the particular forms that it takes, according to the kind of material dealt with—is fruitfully employed by scientific inquirers every day. Sigwart's formal analysis of Bacon's use of the principles, however, does show that Bacon was mistaken concerning the infallibility and the quasi-automatic character he believed the method to insure, since this character would be present only if the disjunctive major of the deduction were strictly established, and it never is. But to say that it never is strictly established is only to say that the results of induction are never more than probable, and this is universally admitted of the conclusions reached by the application of the principles of agreement and of difference. This merely probable character, moreover, is present no less in the results of their experimental than of their statistical application, the latter being the only one that might claim —although only in a very restricted sense indeed—to "place all wits and understandings nearly on a level." [60]

D. *Hypothesis*

A third aspect of Bacon's doctrine, which is to be found not merely between his lines but also to a considerable degree in them, and to which but scant justice has been generally done, is his formulation of the method of hypothesis. Sufficiently numerous and explicit quotations have already been given, in connection with the exposition of the first book of the *Novum organum*, to make it clear that the method of hypothesis is beyond question to be found in Bacon's work. It is quite probable, indeed, that Bacon's original idea was to proceed merely by means of tables and exclusions. But, when he came to try this out, he found it impossible to go very far with it for two reasons: (1) the tables were too incomplete, and (2) the notions in terms of which the observations were to be stated were too ill-defined. Moreover, he knew of no way to induce well-defined notions except the Platonic, and for that also more abundant and diverse facts were required. Therefore does he say that the most pressing task of all is the gathering of "Histories," without which the sharpest understanding and the best method avail nothing. Although Bacon never abandons his belief that the method of tables and exclusions is theoretically sufficient, yet, when it comes to practice on the basis of such data as are actually at hand, he both declares and acts upon the doctrine that the results of that method are to be taken as provisional hypotheses only, which are to be tested by comparison of their consequences with facts. The first vintage is such a hypothesis, and his statement in connection with it, that "truth will sooner come out from error than from confusion," is tantamount to a declaration that it is not truth that the tables and exclusion *in practice* are expected to produce directly so much as some *means of order*, which is just what "working hypotheses" are.[61] And much of the doctrine of prerogative instances constitutes precisely an enumeration of such facts as he believed to be of especial value in forming or testing hypotheses—for example, the famous *instantiae crucis*.[62]

The importance of the method of hypothesis in Bacon's view, and the relation to the method of tables and exclusions that it came to have in his thought as time passed, can best be perceived from the passage at the end of the *New Atlantis* where he describes the employments and offices of the workers in Salomon's house. The passage is also of interest as illustrating Bacon's ideal of scientific collaboration:

> For the several employments and offices of our fellows; we have twelve that sail into foreign countries, under the names of other nations (for our own we conceal); who bring us the books, and abstracts and patterns of experiments of all other parts. These we call Merchants of Light.
>
> We have three that collect the experiments which are in all books. These we call Depredators.
>
> We have three that collect the experiments of all mechanical arts; and also of liberal sciences; and also of practices which are not brought into arts. These we call Mystery-men.
>
> We have three that try new experiments, such as themselves think good. These we call Pioneers or Miners.
>
> We have three that draw the experiments of the former four into titles and tables, to give the better light for the drawing of observations and axioms out of them. These we call Compilers.

We have three that bend themselves, looking into the experiments of their fellows, and cast about how to draw out of them things of use and practice for man's life, and knowledge as well for works as for plain demonstration of causes, means of natural divinations, and the easy and clear discovery of the virtues and parts of bodies. These we call Dowry-men or Benefactors.

Then after divers meetings and consults of our whole number, to consider of the former labours and collections, we have three that take care, out of them, to direct new experiments, of a higher light, more penetrating into nature than the former. These we call Lamps.

We have three others that do execute the experiments so directed, and report them. These we call Inoculators.

Lastly, we have three that raise the former discoveries by experiments into greater observations, axioms and aphorisms.[63]

Now it is easy to show that the procedure Bacon thus describes can be interpreted as substantially in accordance with that actually in use today. The air of strangeness it has is due chiefly to the division of labor and the specialization of functions that it introduces. The work of science, for the most part, has been and still is being carried on individualistically, although it may well be questioned whether research, systematically organized, and planned and directed by the few men pre-eminently fitted by previous training and endowment for that function, would not in some departments at least be more effective. The collection of meteorological data and of material for statistical treatment constitutes at least partial illustration of what Bacon describes, and Blunt [64] calls attention to the success of the Germans in the chemical industries which has resulted from a quasi-nationalization of scientific effort. The specification by Bacon of "meetings and consults" of the whole body of investigators makes it clear that he recognized the *coordination* of labor to be no less important than the division of it, and therefore shuts the door to an easy objection to the plan described.

If we leave out the multiplicity of investigators as constituting not an essential part of the method set forth, but only *one* possible way of putting it into practice, we find that the method can be presented as describing pretty accurately what the individual investigator does do. The information supplied by the merchants of light, the depredators, and the mystery men in Bacon's scheme may fairly be said to correspond to the information that the individual investigator acquires in his general education in school and college, in his travels, in his miscellaneous reading, and elsewhere. To the experiments of the pioneers may be said to correspond the miscellaneous simple experiments which, while still in his period of training, he may perform as various detached ideas (hypotheses) occur to him in connection with the heterogeneous information he is acquiring. When the student begins to specialize, with some intention of doing original research, the work of the compilers begins in him also. It is doubtless largely an automatic, psychological process of collating his information according to likenesses and differences, which, as it goes on, may become gradually somewhat more deliberate and conscious, although it is hardly likely to assume the objective tabular form contemplated by Bacon, or, if at all, not until some definite problem has been selected for investigation. Such selection, and the formulation of hypotheses (first vintage) concerning the solution of the problem, corresponds to the office of the dowry men. To that of the lamps corresponds the process of deducing the consequences of these hypotheses and of determining what experiments will be suitable for testing them. Performing them corresponds to the task of the inoculators, and the formulation of the final outcome of the investigation to that of the interpreters. To the "meetings and consults" corresponds the fact that the whole process is present in the consciousness of one man, who coordinates all its parts.

Little doubt can be entertained that the *New Atlantis* was written several years after the *Novum organum*. Rawley, in his *Life of Bacon*, mentions it among the works that were written by Bacon during the last five years of his life, and Spedding assigns it to the year 1624, while the *Novum organum* was published in 1620. The account of the system of scientific investigation just discussed may therefore fairly be taken as representing Bacon's maturest opinion on the subject of the relation between the method of tables and exclusion and that of hypothesis and verification. On the basis of that account, and of the other passages quoted from the first book of the *Novum organum* on the subject of hypothesis, the reader may now be left to judge whether, as Blunt would have it, all Bacon's references to hypothesis are indeed but of the nature of "warnings or grudging recognitions." [65] Such a characterization hardly seems to do them justice.

CHAPTER FOUR

The Role of Experience in Descartes' Theory of Method

I. THE IDEAL OF MATHEMATICAL CERTITUDE

IN A FAMILIAR passage of the *Discourse on Method* Descartes tells of the profound sense of dissatisfaction which he experienced when, in the year 1616, having come to the end of his formal education, he looked back upon the whole course of instruction that he had undergone and attempted to take stock of the results.[1] In particular, he found himself plunged into utter skepticism concerning the whole of the traditional teaching on physical science, for he reflected that in this field there was scarcely a single question on which men of equal ability did not hold contrary opinions.

> But whenever two men come to opposite decisions about the same matter one of them at least must certainly be in the wrong, and apparently there is not even one of them in the right; for if the reasoning of the second was sound and clear he would be able so to lay it before the other as finally to succeed in convincing his opponent also.[2]

Now, as a matter of fact, Descartes' scholastic preceptors themselves fully recognized the limitations and defects of their own physical science; and Descartes, when he so pointedly contrasts the mere probabilities of the current teaching in natural science with the absolute certitude of mathematical demonstrations, is no more than following their example.[3] But they were accustomed to maintain, following Aristotle, that in matters of physics absolutely demonstrative proofs are impossible, and that we must therefore here content ourselves with no more than probable conclusions. Descartes, however, found himself unable to put any confidence whatever in conclusions that were merely probable. A "science" that did not possess the demonstrative certitude which he admired in mathematics he thought unworthy the name.[4] And when he himself came back at a later date to the systematic study of physical problems, it was with the conviction, begotten of his reflections upon the method of mathematics, that this method, with all its rigor of demonstration and certitude of results, was capable of application also in the sphere of the natural sciences.

Accordingly we find Descartes frequently describing physics as a true science after the mathematical model, and on occasion seeming to hold it capable of purely a priori deduction in its entirety, and with absolute certainty, from self-evident principles innate in the human reason. For example, he writes:

> As for physics, I should believe myself to know nothing of it if I were able to say only how things may be without demonstrating that they cannot be otherwise; for having reduced physics to the laws of mathematics it is possible to do this, and I believe that I can do it in all that little that I believe myself to know.[5]

Elsewhere Descartes tells us how he "observed certain laws" the ideas of which God has "imprinted on our minds," and how, "by considering the sequence of these laws, it seems to me that I have discovered many truths more useful and more important than all I had formerly learned, or even hoped to learn." [6] Of these he had originally intended to expound in *Le monde* all that had to do with "the nature of material objects"; [7] and, writing of the project to Mersenne, he says, "I am resolved to explain all the phenomena of nature, that is to say the whole of physics." [8] And finally, in *Le monde* itself Descartes wrote, "those who shall have sufficiently examined the consequences of these truths and of our rules will be able to know effects by their causes; and, to explain myself in scholastic terminology, will be able to have demonstrations a priori of everything that can be produced in this new world." [9]

2. THE LIMITS OF PURE A PRIORI PHYSICS

So far Descartes seems to be taking Aristotle's *ideal* of a perfect science altogether *au sérieux*, as literally realizable, and in strict mathematical form, even in our knowledge of the physical world.

But, as we shall shortly see, he really knows quite well how impossible it is actually to carry through any such ambitious program as that of a physics that should be completely demonstrable, down to its very last details, in a purely a priori fashion. We accordingly find him forced by the exigencies of the facts so to interpret and to qualify such extreme claims as he sometimes seems to make for his physics, as really to come out, however hesitatingly and reluctantly, with a theory of natural science, and of the limits of scientific certitude, not far removed from that of Aristotle himself.

After all, the passages in which Descartes even appears explicitly to maintain the a priori demonstrative nature of the *whole* of physics are relatively few in number. More often he is content to attribute this character to the *primary* physical principles alone. Thus he writes, "The foundations of my Physics . . . are nearly all so evident that it is only necessary to understand them in order to accept them, and . . . there is not one of them of which I do not believe myself capable of giving demonstration." [10] And elsewhere he tells us that the demonstration here in question, "which I hope to do some day," is the proof of "the principles of Physics by Metaphysics." [11] Or again:

> . . . the order which I have followed is as follows: I have first tried to discover generally the principles or first causes of everything that is or can be in the world without considering anything that might accomplish this end but God Himself who has created the world, or deriving them from any other source excepting from certain germs of truths which are naturally existent in our souls.[12]

In fact, however, he does occasionally go somewhat further than this. For even the explanatory principles which he uses in the *Météores*, which, wishing for the moment to avoid any exposition of the fundamental principles of his physics,[13] Descartes is there content to characterize merely as hypotheses or "suppositions," *can* be derived by a priori deduction from the primary physical principles,[14] and are therefore mediately derivable, like these primary principles themselves, from the first principles of metaphysics.[15]

In the light of what follows it will perhaps appear doubtful whether even here Descartes is thinking of a *purely* a priori deduction, in the strictest sense of the term; but at any rate he certainly nowhere makes any perfectly explicit and unqualified claim that the powers of this sort of deduction extend any *further* than this, however much such a claim may at first sight seem to be implied in such more extreme general statements as those quoted above.

In fact these general statements, even considered in themselves, are susceptible of an interpretation (and in view of the ensemble of his doctrine we must take this to be the only interpretation that Descartes himself would mean seriously to defend) that does not really involve any such extravagant view of the matter. For there are two quite different senses in which it might be held that the whole of physics is demonstrable deductively from purely a priori principles. It might be held, namely, on the one hand, that it is possible, starting exclusively with principles self-evident a priori, and without ever introducing any merely empirical premises, to proceed from these primary truths alone, by purely deductive reasoning, to the discovery and proof of all the truths of physics down to the last detail, in such wise that we should have a demonstration purely a priori of the absolute necessity of every fact and phenomenon of nature. On the other hand, it might be held that it is the general explanatory principles of physics *alone* that are deducible in this *purely* a priori fashion. We could, on this view, know by purely a priori reasoning that the explanation of every fact and phenomenon of nature is to be found in terms of these general explanatory principles, and nowhere else; but we could *not* discover by any process of pure a priori deduction from these principles alone that just this or that *particular* fact or phenomenon and no other, caused in just this or that *particular* fashion, and no other, must necessarily occur.

To put it in other terms, the "laws of nature" may, on this view, contain, in a sense, the *explanation* of every particular natural occurrence. But they only "explain" it by showing how, given the truth of these laws, this particular occurrence is the necessary consequence of some previous particular occurrence or set of occurrences. The particular occurrence is thus deducible from the laws, but only when additional premises are introduced, viz., knowledge concerning *other* particular occurrences. According to this way of viewing the matter, a knowledge of particular natural phenomena is never deducible *wholly* a priori, even if the laws that govern and explain its occurrence are themselves known independently of any merely empirical validation. With regard to wholly particular occurrences the laws serve only to confine and to limit the possibilities. They tell us that, whatever particular events may occur, they must necessarily (given the truth of the laws) fulfill certain fixed functional relationships. But apart from this the laws do not suffice to determine the occurrence of the events. In a word, the history of the universe is not deducible from the laws of nature *alone*. It involves a factor of contingency that eludes all determination by the universal. It is deducible, at best, from the laws of nature *plus* certain empirical information about particular occurrences. Yet it would still remain true that, given a sufficient amount of such purely empirical information, the whole history of the physical universe could *then* be derived deductively from the a priori principles.

Now if we examine Descartes' more extreme statements of his view in the light of this distinction, it is clear that they are quite susceptible of being interpreted as involving no more than would be implied in this latter view. In the first of the statements quoted above, for example, he actually says, after all, only that he believes, with regard to "all that little I believe myself to know" in physics, that he is able not merely "to say how things may be," but also to demonstrate "that they cannot be otherwise." That is, not only *do* particular natural occurrences happen in accordance with the general laws, but they *must* do so, since these laws are not themselves of merely empirical origin but are demonstrated purely a priori from self-evident principles. But it is not *explicitly* stated that it can be demonstrated that they cannot be otherwise, by deducing them from these necessary universal laws *alone*. In the second of our passages Descartes claims merely to be able, in terms of certain a priori laws, "to *explain* all the phenomena of nature, that is to say, the whole of physics" (italics ours). And in the passage from *Le monde* he professes indeed to have demonstrations a priori, but only of "all that *can* be produced in this new world" (italics ours), i.e., the a priori laws, as we said above, serve, with regard to particular occurrences, only to confine and limit the possibilities.

3. THE EMPIRICAL BASES OF PHYSICS

We now proceed to show not only that (as we have already indicated above) these more extreme statements of Descartes are *susceptible* of such an interpretation as that just given, but further, that considered in the light of the rest of his own statements about his theory of science, and in the light of his own practice as a scientific inquirer, they positively *require* such an interpretation. For, despite all appearance to the contrary, the physical sciences, in the Cartesian conception of the matter, rest upon a solidly empirical basis.

It is true that Descartes sometimes has harsh and disparaging things to say about experience. He refers, for example, to "the fluctuating testimony of the senses"; [16] or writes:

We must note then that there are two ways by which we arrive at the knowledge of facts, viz. by experience, and by deduction. We must further observe that while our inferences from experience are frequently fallacious . . . none of the mistakes which men can make . . . are due to faulty inference; they are caused merely by the fact that we found upon a basis of poorly comprehended experience, or that propositions are posited which are hasty and groundless.[17]

But three things are here to be remembered. In the first place, such passages from Descartes may be paralleled with equally severe expressions from Francis Bacon himself. For example, "By far the greatest impediment and aberration of the human understanding proceeds from the dullness, incompetence, and errors of the senses." [18] Or, "There remains but mere experience, which, when it offers itself is called chance; when it is sought after, experiment. But this kind of experience is nothing but a loose fagot, and mere groping in the dark. . . ." [19] Again Bacon speaks of "the uncertain light of the sense that sometimes shines, and sometimes hides its head" or of "collections of experiments and particular facts, in which no guides can be trusted, as wanting direction themselves, and adding to the errors of the rest." [20]

In fact neither Bacon nor Descartes had any confidence in opinions founded upon the uncriticized experience of the senses. For correcting the errors consequent upon a naïve trust in the testimony of such experience both place their reliance upon the adoption of a right *method* of conducting scientific inquiry. And both thus seek an intimate alliance of experience and reason. Thus Bacon says:

. . . the true labour of philosophy . . . neither relies entirely or principally on the powers of the mind, nor yet lays up in the memory the matter afforded by the experiments of natural history and mechanics in its raw state, but changes and works it in the understanding.[21]

We have good reason, therefore, to derive hope from a closer and purer alliance of these faculties (the experimental and the rational) than has yet been attempted.[22]

And thus we hope to establish forever a true and legitimate union between the experimental and the rational faculty, whose fallen and inauspicious divorces and repudiations have disturbed everything in the family of mankind.[23]

And Descartes, in similar vein, remarks not only upon "the small faith that we ought to put in observations not accompanied by true reason,"[24] but also upon the impossibility of doubting the truth of his explanations because "in all this reason accords so perfectly with experience";[25] and even more significantly writes to Plempius, "I use that sort of philosophizing in which there is no principle that is not mathematical and evident, and its conclusions confirmed by true experiments."[26]

In the second place, we must add to these expressions the fact that Descartes is known to have been throughout his life both a keen and persistent observer of natural phenomena and an indefatigable experimenter.[27] And finally we must note how Descartes again and again insists upon the indispensability for physical science of a wide empirical foundation in careful observation and experiment. For example, he tells us[28] how in 1619, when he first conceived the project of reconstructing the edifice of physics anew, in accordance with his recently developed theory of method, he realized that the time was not yet ripe for the accomplishment of this program. The plan could be carried into effect only after he should first have elaborated his metaphysics (for the fundamental laws of physics were to be deduced from the primary principles of metaphysics), and, moreover, only after he should first of all have employed "much time in preparing myself for the work by . . . accumulating a variety of experiences fitted later on to afford materials for my reasonings."[29] In a later passage Descartes notes that he actually carried out this latter part of his preparation: "I made various observations and acquired many experiences."[30] And still later he describes how he proceeded in the actual construction of his new system of physics: "First, I have tried to discover generally the principles or first causes of everything that is or can be in the world, without considering anything that might accomplish this end but God Himself who has created the world, or deriving them from any source excepting from certain germs of truths which are naturally existent in our souls." But "then, when I wished to descend to those which were more particular," he found it impossible to proceed "if it were not that we arrive at the causes by the effects, and avail ourselves of many particular experiments." [31]

Descartes' account of the way in which he applied his method to the examination of the problem of the circulation of the blood, moreover, affords positive confirmation of the view stated above concerning the limitations under which alone he may be understood to have maintained that the natural sciences are constructed by a priori deduction from the first principles of metaphysics. For in dealing with this problem he demonstrates deductively (or, as he himself puts it, by "mathematical demonstration"),[32] in accordance with general mechanical principles, in just what way, *given certain empirical data derived from anatomical dissections*,[33] the circulation of the blood takes place. And he himself refers to the empirical data in question as constituting an integral element in the demonstration:

> . . . this movement which I have just explained follows as necessarily *from the very disposition of the organs*, as can be seen by looking at the heart, and from the heat which can be felt with the fingers, and from the nature of the blood of which we can learn from experience, as does that of a clock from the power, the situation, and the form of its counterpoise and of its wheels [italics ours].[34]

Similarly in connection with the problem of the nature of the magnet Descartes writes as follows:

> He who reflects that there can be nothing to know in the magnet which does not consist of certain simple natures, will have no doubt how to proceed. He will first collect all the observations with which experience can supply him about this stone, and *from these* he will next try to deduce the character of that intermixture of simple natures which is necessary to produce all those effects which he has seen to take place in connection with the magnet. This achieved, he can boldly assert that he has discovered the real nature of the magnet in so far as human intelligence *and the given experimental observations* can supply him with this knowledge [italics ours].[35]

There thus remains, according to Descartes himself, an inexpugnable and indeed indispensable empirical factor in the very heart of his a priori deductive physics. It follows a fortiori that, whatever part mathematical methods (in the narrower and more technical sense) may play in the Cartesian physics, there can be no real question of any "reduction of physics to mathematics" or of any conceivable deduction of physics from purely mathematical premises alone. And Descartes makes it quite plain that he recognizes this fact. For in the second part of the *Discourse,* after briefly characterizing his "universal mathematics" [36]—a mathematical discipline (more fully developed in the *Rules for the Direction of the Mind* [37]) which should be capable of resolving the problems not only of pure mathematics (arithmetic and geometry) but also of those other sciences then denominated "mixed mathematics," [38] especially "Astronomy, Music, Optics, Mechanics" [39]—he tells us how, as soon as he was confronted with the problem of actually applying the procedures of this method to such *physical* sciences, he realized that it would be necessary first to accumulate "a variety of experiences fitted later on to afford matter for my reasonings." [40] In other words he realized that the *matter* of his reasonings could not here be deduced a priori from mathematical principles, but must be furnished by experience.

He reverts to the same point in the third part of the *Discourse* where he says:

> I set aside some hours from time to time which I more especially employed in practicing myself in the solution of Mathematical problems according to the Method, or in the solution of other problems which though pertaining to other Sciences [i.e., to "mixed mathematics"], I was able to make *almost* similar to those of mathematics, by detaching them from all principles of other sciences which I found to be not sufficiently secure [i.e., from the doctrines of scholastic physics] [italics ours].[41]

To make them *quite* mathematical was impossible, for

> Sciences such as Optics, Meteorology, etc., appeal to notions which are not reducible to the abstract numerical relationships upon which the universal mathematics works. In this sense the complete mathematization of physics is impossible, and it is so in a way by definition; for as soon as mathematics cease to be pure, they require a datum to which they are applied and which they interpret, but which they accept without themselves being able to justify.[42]

Elsewhere also Descartes insists that physical science, although in a certain sense mathematical and geometrical, is not so in any unqualified sense. Its demonstrations are by no means simply on all fours with those of Euclid. The geometry of the mathematicians is an *abstract* geometry; physical science can on the contrary employ only a *concrete* geometry. Thus he writes to Mersenne:

I have decided to abandon only abstract geometry, that is to say the inquiry into questions which only serve to exercise the mind; and that in order to have so much the more leisure to cultivate another sort of geometry, which takes as its problem the explanation of the phenomena of nature.[43]

Again:

You ask me whether I hold that what I have written of refraction is a demonstration; and I believe that it is, at least as far as it is possible to give one in this matter . . . and as much as any other question of Mechanics or Optics, or of Astronomy, or any other matter which is not purely geometrical or arithmetical has ever been demonstrated. But to require of me geometrical demonstrations in a matter which depends upon Physics is to wish me to do the impossible. And if we are to call demonstrations the proofs of the geometers alone, we must then say that Archimedes never demonstrated anything in Mechanics, nor Witelo in Optics, nor Ptolemy in Astronomy, etc., which, however, we do not say.[44]

In the following passage Descartes seems at first sight to speak in the contrary sense: "I receive no principles in physics which are not also received in mathematics."[45] But mathematics must here be understood in the wide sense which includes "mixed" mathematics, and therefore *mechanics* as well as arithmetic and geometry. Liard also quotes Descartes as saying, "My entire physics is nothing but geometry."[46] But geometry must here be taken in the sense of *concrete* geometry.

In the light of the foregoing it is not so surprising as it might otherwise at first sight appear to find Descartes denouncing in no uncertain terms, and with what might well be the voice of Francis Bacon himself, "those Philosophers who, neglecting experiences [*experimentis neglectis*] think that truth will spring from their brains, like Pallas from the brain of Zeus."[47]

But it is time to examine more closely the role Descartes assigns to experience in connection with the problems of physics. What is the function of those large accumulations of empirical observations which Descartes, like Bacon,[48] is so anxious to have at his disposal, and without which he regards it as useless to embark upon any physical inquiry?[49]

1. The a priori development of the fundamental laws of nature serves, as we have seen, to set definite limits to what is physically possible. Nothing can happen that is not in accordance with these laws, and nothing that requires for its explanation any other laws. But

within the limits thus prescribed there still remain open for the physical universe practically infinite possibilities. If we were to attempt to set forth in detail *all* that can be deduced from (i.e., all that can be shown to be in accordance with, and explicable in terms of) these laws, we should have upon our hands a task not only endless, but useless, since most of these possibilities are never, so far as we know, actually realized. What we desire is to explain the phenomena of *our* universe—to know how *these* can be accounted for in terms of the fundamental principles that govern all natural happenings. Only empirical observation can tell us what these phenomena are, and only it, accordingly, is capable of setting the problem we wish to resolve. Descartes explains the point clearly:

The principles which I have discovered are so vast and so fertile that much more follows from them than we see happen in this visible world, and even much more than our minds can ever exhaust in thought. But we place before our eyes such a brief history of the principal phenomena of nature (whose causes are here to be investigated); not indeed in order to use these as grounds for proving anything: for we desire to deduce the explanation of effects from their causes, not that of causes from their effects; but merely in order that from among the innumerable effects which we judge capable of being produced by these same causes, we may determine our minds to the consideration of some rather than others.[50]

2. In the early stages of the development of a science, Descartes holds, it is better to make use, for this purpose, "simply of those [sc. experiments] which present themselves spontaneously to our senses, and of which we could not be ignorant, provided we reflected ever so little, rather than to seek out those which are more rare and recondite."[51] This is the single point upon which Descartes criticizes the method of Bacon. The latter is constantly insisting upon the superior value, as a foundation for science, of detailed and artificial experiments, as contrasted with everyday "experiences." Speaking of his project of a "natural history" he writes:

With regard to its collection; we propose to show nature not only in a free state, as in the history of meteors, minerals, plants, and animals; but more particularly as she is bound, tortured, pressed, formed, and turned out of her course by art and industry . . . for the nature of things is better discovered by the torturings of art, than when they are left to themselves. . . . The kind of experiments to be procured for our history are much more subtle and simple than the common; abundance of them must be recovered from darkness, and are such as no one would have inquired after, that was not led by constant and certain tract to the discovery of causes.[52]

In Descartes' view of the matter, information based upon elaborate experiments of this sort is useless and confusing until a later stage of development has been attained. The defect of Bacon's project of "natural history" consists in the fact that Bacon failed to discriminate between the sort of empirical information that is of service at the beginning of inquiry into the problems of the physical universe, and that which is appropriate only to the more advanced stages of the investigation. Descartes expressed his attitude in the matter in a letter to Mersenne, as follows:

I had forgotten to read a paper that I just found in your letter, in which you inform me . . . that you wish to know a means of making useful experiments. As to that I have nothing to say, after what Verulam has written concerning it, except that, without being too curious in searching out all the little details concerning a matter, it would be necessary chiefly to make general collections of all the most common things, such as are most certain, and can be known without expense. As that in all shells the spirals turn in the same direction, and to know whether it is the same on the other side of the equator. That the bodies of all animals are divided into three parts, head, chest, and abdomen, and the like; for these are the things which serve us without fail in the search for the truth. As for the more particular experiments, it is impossible not to make many that are superfluous, and even false, if one does not know the truth of things before making them.[53]

The main point is contained in the last sentence. It is one of the principles of the Cartesian method that we should preserve in every inquiry the order of the simple to the complex. We should solve the simpler and easier problems before attacking the more difficult and more complex. Here is a case in point. It will be difficult or impossible for us to find the right explanation of more recondite phenomena unless we have first begun by explaining such facts of nature as are easily accessible. Moreover, once we have discovered the right principles to explain the latter—which is relatively easy to do—we shall find that the more detailed phenomena are readily explicable in terms of the same principles. As Descartes puts the matter:

As to those sciences . . . some are deduced from common objects of which everyone is cognizant, and others from rare and well thought out experience. And I confess likewise that it would be impossible for us to treat in detail each one of these last. . . . But I shall believe myself to have sufficiently fulfilled my promise if, in explaining to you the truths which may be deduced from common things known to each one of us, I make you capable of discovering all the others when it pleases you to take the trouble to seek them.[54]

There is no use piling up miscellaneous experimental data until we are in possession of the fundamental principles that serve to explain the general fabric of nature. To make elaborate experiments is expensive, difficult, and troublesome, besides consuming much time. It is better to wait until they are really needed, and we are equipped with sufficient knowledge of nature in general to attack their solution with some hope of success.[55]

In Descartes' own procedure, therefore, as he describes it to us in the sixth part of the *Discourse*, he first gave an a priori deduction of the primary principles of all physics, and next explained as consequences of these principles all such phenomena of nature as everyday experience makes us acquainted with.

After that I considered which were the primary and most ordinary effects which might be deduced from these causes; and it seems to me that, in this way, I discovered the heavens, the stars, an earth, and even on the earth, water, air, fire, the minerals, and some other such things which are the most common and simple of any that exist and consequently the easiest to know.[56]

3. But once the explanation of such relatively obvious phenomena is fairly well advanced, and we wish to proceed to the explanation of matters of detail, we must then have recourse to many detailed experiments especially designed to reveal the more recondite facts we have still to explain.

Then when I wished to descend to those which are more particular, so many objects of various kinds presented themselves to me, that I did not think it was possible for the human mind to distinguish the forms or species of bodies which are on the earth from an infinitude of others which might have been so if it had been the will of God to place them there, or consequently to apply them to our use, if it were not that we arrive at the causes by the effects, and avail ourselves of many particular experiments.[57]

4. It is not enough, however, to have, on the one hand, the general principles of explanation, and, on the other, the facts to be explained. For we frequently find that the particular phenomena in question might be explained, in perfect accordance with the general principles, in either of two quite different ways. So far as the principles them-

selves tell us, the phenomena may be produced in either way, for they can in either case equally be "deduced" from the principles. But in which way are these phenomena *actually* produced? Only experience, according to Descartes, can tell us. Here, then, is another function of experience in the Cartesian theory of science—to decide, as between two different ways of "deducing" the phenomenon to be explained, both equally possible a priori, which alone accords with the actual procedure of nature. What is needed is, in fact, an *experimentum crucis*, or *instantia crucis*, to use the Baconian expression,[58] i.e., we must seek by careful experimentation *some other* phenomenon that can be accounted for only provided one of the alternatives, and not the other, is actually realized. Thus Descartes writes:

> In subsequently passing over in my mind all the objects which have ever been presented to my senses, I can truly venture to say that I have not there observed anything which I could not easily explain by the principles which I had discovered. But I must also confess that the power of nature [59] is so ample, and so vast, and these principles are so simple and so general, that I observed hardly any particular effect, as to which I could not at once recognize that it might be deduced from the principles in many different ways; and my greatest difficulty is usually to discover in which of these ways the effect does depend upon them. As to that I do not know of any other plan but again to try to find experiments of such a nature that their result is not the same if it is to be explained by one of these methods, as it would be if explained in the other.[60]

The facts to which Descartes appeals as against Harvey's version of the theory of the circulation of the blood in the fifth part of the *Discourse* constitute, in his mind, a series of just such crucial experiments.[61] And he sums up the matter elsewhere as follows:

> . . . experiments themselves often give us occasion to deceive ourselves, when we do not sufficiently examine all the causes which they may have. . . . But in order to be able to tell which of these two causes is the true one, it is necessary to consider other experiments which cannot be brought into accordance with both.[62]

5. Finally, the testimony of experience must again be invoked, according to Descartes' view of the matter, at the end of any scientific inquiry, in order to verify and to confirm the results attained. Thus he speaks of "all the experiments necessary to me in order to support and justify my reasoning." [63] Again he writes to Morin, "and I had to make use of these balls . . . in order to submit my reasons to the

test of the senses, as I always try to do." [64] And finally, "I use that kind of philosophizing in which there is no principle that is not mathematical and evident, and its conclusions confirmed by true experiments." [65] In fact it is an added confirmation even of the primary principles of physics themselves, capable of a priori deduction though they be, that they are borne out by experience. Descartes writes in this sense to Huygens:

> What I find most strange is the conclusion of the opinion that you have sent me, viz., that what will prevent my principles from being accepted in the School is that they are not sufficiently confirmed by experience. . . . For I marvel that notwithstanding I have demonstrated in detail almost as many experiments as there are lines in my writings, and having given a general explanation in my *Principles* of all the phenomena of nature, have explained by the same means all the experiments that can be made in connection with inanimate bodies, and on the contrary not one of them has ever been well explained by the principles of the vulgar philosophy, those who follow the latter do not cease to reproach me with lack of experiments.[66]

In the *Principia* itself Descartes refers to the "laws of mechanics, confirmed as they are by certain and daily experience." [67] Again, he admits to a correspondent that an experimental proof that the transmission of light is not instantaneous would ruin his entire philosophy.[68] And Liard points out how Descartes regularly accompanies the a priori demonstration of the fundamental laws of mechanics with illustrations of their truth drawn from experience.[69]

4. DESCARTES ON HYPOTHESES

In so far as Descartes admits the utility and the necessity of crucial experiments in physical inquiries, it is already evident that he is at least tacitly admitting the existence of an element of hypothesis in the situation. The "many different ways" in which "any particular effect" may be "deduced" from the principles are really so many hypotheses, all in accordance with general mechanical laws, but between which only further experiment can decide which is the one that gives the true cause. But the obvious problem now arises, how under these circumstances can we ever *know* that we have discovered the true cause? For how can we ever be sure that we have excluded all even of the mechanically possible alternatives? If this problem cannot be resolved, the "absolute certainty" which Descartes desires for the whole of his physics will after all extend only so far as the general principles themselves, and not to the details of the explanations offered of particular phenomena. That Descartes is aware of the difficulty is already evident from a remark quoted above: "experiments themselves often give us occasion to deceive ourselves, when we do not sufficiently examine all the causes they may have." And it is noteworthy that he does not here expect, and nowhere else ventures explicitly to claim, that this difficulty can be in any way met by means of a priori deduction. Here and elsewhere it is to *experience* once more, and especially to experiment, that he looks for the solution.

Sometimes, indeed, for certain limited purposes, and on certain occasions, Descartes is content to offer particular mechanical explanations as *mere* hypotheses such as *would* suffice to explain the given phenomena if they were true, but which, for all we know to the contrary, may be pure fictions. On such occasions Descartes claims for the hypotheses he offers not *truth*, but merely *possibility*. In the *Regulae*, for example, he recommends that we consider all physical realities as possessing primary qualities only. But he represents this simply as a methodological convenience. We need by no means necessarily deny that they actually possess other qualities also; but it will be an aid to clear and consecutive reasoning about them if we imaginatively represent to ourselves the differences, say, between the various colors of things, as though these were reducible simply to geometrical differences in the things themselves. This hypothetical representation may not answer to the real nature of the physical world; but it will be most convenient to our thinking about it.[70] Here the hypothesis is explicitly put forward as a fiction, an abstract way of viewing things which, taken abstractly, may well be false, but which will enable us to think more clearly, and to deduce results that accord with the phenomena of experience. So, too, the hypothesis concerning the creation of the world briefly summarized in the *Discourse*[71] is there, in order to avoid theological complications, advanced explicitly as a convenient fiction. In the French version of the *Principia* it is even characterized as "absolutely false";[72] but it is the difficulty of reconciling his view of the matter with the revealed account in Genesis that here determines Descartes' attitude.

There are passages, moreover, where Descartes seems to go even further and to contend that we have no right to require or to expect that any explanatory hypothesis in physics be shown to be more than a *possible* explanation of the phenomena in question. He appears to hold that such hypotheses should be required to satisfy two conditions only: they must assume as principles nothing that is in manifest contradiction with the facts of experience; and in the deductions that lead from these principles to their consequences there must be no fallacies.

For we are satisfied, in such matters, that authors, having presupposed certain things that are not manifestly contrary to experience, for the rest shall have spoken consistently and without committing any paralogisms, even though their suppositions were not precisely true. For I might demonstrate that even the definition of the center of gravity given by Archimedes is false, and that there is no such center; and the other things which he supposed also are not precisely true either. As for Ptolemy and Witelo, they make suppositions much less certain, and nevertheless we should not, for all that, reject the demonstrations which they have deduced therefrom.[73]

From this point of view there are only two legitimate grounds of objection to a hypothesis in physics. For, says Descartes:

. . . know that there are only two ways of refuting what I have written, of which the one is to prove by some experiments or reasons that the things I have assumed are false; and the other is to prove that what I have thence deduced cannot be thence deduced. . . . But as for those who content themselves with saying that they do not believe what I have written because I deduce it from certain suppositions which I have not proved, they do not know what they are asking, nor what they ought to ask.[74]

Again, toward the close of the *Principia* he writes:

But here it may be said that although I have shown how all natural things can be formed we have no right to conclude on this account that they were produced by these causes. For . . . doubtless there is an infinity of ways in which all things that we see could be formed by the great Artificer without it being possible for the mind of man to be aware of which of these means he has chosen to employ. This I most freely admit; and I believe that I have done all that is required of me if the causes I have assigned are such that they correspond to all the phenomena manifested by nature without inquiring whether it is by this means or by others that they are produced.[75]

And he concludes by quoting Aristotle's familiar dictum in the *Meteorology*. Descartes here actually appears to concede that it is not *possible* for the human mind to know that the causes which he has supposed are the true ones. But in view of what he goes on to say in the immediately succeeding chapters he must not here be interpreted too literally. This extreme admission must be taken to be for the purposes of the argument with the supposed objector. Descartes' point really is that even though this were conceded, and the causes which he has assigned to the phenomena were admitted to be mere hypotheses, he has nevertheless done all that is required *for practical purposes* if he has given a set of workable hypotheses—"it will be sufficient for the usage of life to know such causes." [76] And this seems to be the general explanation of such passages. It is "sufficient" in such matters to have any possible hypothesis in terms of which the actual phenomena can be predicted. More is not required for practical purposes.

At any rate Descartes is not really prepared to rest permanently in any such view of the matter. Whatever his occasional concessions to less realistic views of hypotheses, he never finally acquiesces in such interpretations. He may dally with them for the moment, when it suits his purpose to do so, but his thirst for certitude and his conviction of the possibility of attaining it are too strong to permit him to give them any permanent credence. He is really thoroughly convinced that his hypotheses are far more than mere fictions or abstract possibilities. In his view the real physical world actually contains none but primary qualities; the astronomical and physical hypotheses by which he explains the phenomena are literally and physically true.

In fact he insists that even on an empirical basis his hypotheses can be securely established as constituting the true explanation of the phenomena of nature. Just as we cannot take seriously and literally his occasional declarations concerning the complete a priori deducibility and absolute certainty of the whole of physics, so we cannot, in view of the *ensemble* of his doctrine, accept these occasional expressions of extreme empiricism, which if taken strictly amount to a denial of any certitude, or even of any grounded probability for any of the explanatory hypotheses of physics.

If the explanatory principles laid down at the beginning of the *Dioptrique* and the *Météores* are described expressly as "suppositions," it is not because Descartes thinks that he has not even in those very treatises sufficiently demonstrated them. He calls them "suppositions" for the moment, but only because he does not desire just then to enter into the proof of them by deduction from the primary principles of physics, and ultimately from metaphysics. "And I have not named them hypotheses with any other object than that it may be known that while I consider myself able to deduce them from the primary truths which I explained above, yet I particularly desired not to do so." [77] But they have none the less received a perfectly adequate demonstration of another sort. The nature of this demonstration Descartes describes as follows:

As to the hypothesis that I made at the beginning of the *Meteors,* I could not demonstrate it a priori without giving my whole physics; but the experiments that I have deduced from it necessarily, and which cannot be deduced in the same fashion from any other principles, seem to me sufficiently to demonstrate it a posteriori. . . . As, in fact, it is not always necessary to have reasons a priori in order to persuade one's hearers of a truth; and Thales, or whoever it was, who first said that the moon receives its light from the sun, doubtless gave of this no other proof than that by making this supposition we explain very easily all its different phases: which was sufficient to bring it about that henceforth that opinion was received by the world without contradiction. And the connection of my thoughts is such that I dare hope that my principles will be found as well proved by the consequences that I derive from them, when they shall have been noticed sufficiently to render them familiar and to cause them to be considered all together, as the fact that the moon borrows its light is proved by its waxing and waning.[78]

Or, as he puts it in the *Discourse:*

If some of the matters of which I have spoken in the beginning of the *Dioptric* and *Meteors* should at first sight give offense because I call them hypotheses and do not appear to care about their proof, let them have the patience to read these in entirety, and I hope that they will find themselves satisfied. For it appears to me that the reasonings are so mutually interwoven, that as the later ones are demonstrated by the earlier, which are their causes, the earlier are reciprocally demonstrated by the later which are their effects.[79]

At first sight this sort of demonstration may seem to involve a vicious circle. But such is not really the case; for, "since experience renders the greater part of these effects very certain, the causes from which I deduce them do not so much serve to prove their existence as to explain them; on the other hand, the causes are proved by the effects."[80] Or, as Descartes explains to an objector, there is no circle involved

> . . . in explaining effects by a cause and then proving the cause by them; for there is a great difference between *proving* and *explaining*. To which I add that one may use the word *demonstrate* to signify both the one and the other, at least if one takes it according to common usage, and not in the special signification which Philosophers give it.[81]

But how, after all, is this "proof" of the "cause" to be accomplished? The truth of a hypothesis cannot be established merely by showing that the consequences legitimately deduced therefrom are actually verified in the facts of experience. The nerve of the proof must lie elsewhere. It is indicated in a passage where Descartes says that "the experiments that I have deduced from it necessarily, and which cannot be deduced in the same fashion from any other principles, seem to me sufficiently to demonstrate it a posteriori."[82] He puts the same points elsewhere as follows:

> I examined all the principal differences which may be found between the figures, sizes and motions of various bodies which only their smallness renders insensible, and what sensible effects may be produced by the diverse ways in which they are combined. Afterward when I met with like effects in the bodies which our senses perceive, I thought they might have been so produced. And then, when it seemed to me impossible to find in the whole extent of nature any other cause capable of producing them, I believed that they had infallibly been so produced.[83]

There is obvious exaggeration here. How could Descartes really so assure himself that "in the whole extent of nature" there is "no other cause capable of producing" the effects in question, as to warrant him in concluding "that they had infallibly been so produced?" Nevertheless he has a real point to make in this connection. It is not merely that from the given hypotheses can be deduced just the particular effects under investigation. The essential fact is that one and the same set of hypotheses serves to explain a wide range of diverse phenomena, extending beyond the particular effects in question. When this is found to be the case, we can scarcely doubt that we have reached the true cause.

> Although it is true that there are many effects to which it is easy to fit different causes, one for each, it is not always so easy to fit one and the same cause to many

different effects, unless it is the true cause from which they proceed; there are even effects which are so many and diverse that it is sufficient proof of their true cause to give even one from which they can be clearly deduced; and I claim that all those of which I have spoken are of this number.[84]

Moreover, once the initial hypotheses have been validated by the conformity of some of their consequences with a sufficiently wide variety of experimental data, they may be used in turn to prove the reality of any of their consequences that may be called into question. For "know that each of these effects may also be proved by this cause, in case it be brought into doubt, and the cause have been proved by other effects." [85] Nor is any circle involved "in proving a cause by several effects which are known otherwise, and then reciprocally proving certain other effects by this cause." [86]

In fact this a posteriori demonstration of true causes is applicable to, and sufficient of itself to put beyond all reasonable doubt, the truth of thc primary and most fundamental principles of physics themselves. If all the phenomena of nature can be explained in terms of one and the same single set of principles, this fact tends of itself to establish the truth of those principles. For

. . . it can hardly be that causes from which all phenomena are clearly deduced are not true.[87]

They who observe how many things regarding the magnet, fire, and the fabric of the whole world, are here deduced from a very small number of principles, although they considered that I had taken up these principles at random and without good grounds, they will yet acknowledge that it could hardly happen that so much should be coherent if they were false.[88]

Hence the great superiority, in Descartes' view, of his own over the Scholastic physics. The Scholastics are accustomed to imagine a different cause to explain each different sort of effect. Descartes makes use of a single supposition only—that bodies are composed of minute particles too small to be perceived. Now then:

If the suppositions of others be compared with mine, that is to say all their *real qualities*, their *substantial forms*, their *elements* and like things, of which the number is almost infinite, with that single one, that all bodies are composed of certain parts . . . and finally if what I have deduced from my suppositions, concerning vision, salt, winds, clouds, snow, thunder, the rainbow, and like things, be compared with what the others have deduced from theirs concerning the same matters I hope that will suffice to persuade those who are not too much preoccupied that the effects which I explain have no other causes than those whence I have deduced them. . . .[89]

Moreover, the truth of these first principles, since it rests upon a wider empirical basis, is thus established with a greater certainty than the particular cause of any particular effect can ever be. "I think the same particular effect can be explained in several different ways; but I think the possibility of things in general can be explained in one way only, which is the true way." [90] That is to say, given a single system of principles which profess to limit and circumscribe the possibilities of nature (as do the principles of the Cartesian physics), the fact that no natural phenomenon is anywhere empirically discoverable which is not fully explicable in accordance with these principles demonstrates their truth in a particularly cogent fashion.[91] It is on this point that Descartes' criticism of the procedure of Galileo turns. He praises him for applying mathematical methods to the examination of physical problems, but nevertheless remains dissatisfied with his results because he failed to show systematically how all natural phenomena can be explained in terms of a single set of first principles. "Without having considered the first causes of nature he has merely sought for the causes of certain effects and has thus built without foundation." [92]

Although these are the most fundamental, they are not, however, the *only* empirical considerations that enable us to distinguish true causes. In so far as our explanatory hypotheses involve the supposition of entities not immediately accessible to sense perception, they will scarcely be true, or even intelligible at all, if they are not conceived on the analogy of what we *do* perceive. To proceed otherwise is in fact to commit the fallacy of explaining *ignotum per ignotius.*

Nor do I think that anyone who uses his reason will deny that we do much better to judge of what takes place in small bodies which their minuteness alone prevents us from perceiving, by what we see occurring in those that we do perceive (and thus explain all that is in nature, as I have tried to do in this treatise), than, in order to explain certain given things, to invent all sorts of novelties, that have no relation to those that we perceive (such as are first matter, substantial forms, and all the great array of qualities which many are in the habit of assuming, any of which it is more difficult to understand than all the things which we profess to explain by their means).[93]

And Descartes further makes the claim that his "new" system of physics is, precisely on the ground of its preservation of this analogy, pre-eminently that of common sense. In fact he lays it down "that there are no principles in this treatise which are not accepted by all men; and that this philosophy is not new, but is the most ancient and most common of all." [94] "They will be found to be so simple and so conformable to common sense, as to appear less extraordinary and less paradoxical than any others which may be held on similar subjects." [95] For "no one ever doubted that bodies were moved and have diverse magnitudes and figures, according to the diversity of which their motions also vary, and that from mutual collision those that are larger are divided into many smaller, and thus change their figure. We have experience of this. . . ." [96] Consequently, "I hope that those who avail themselves only of their natural reason in its purity may be better judges of my opinions than those who believe only in the writings of the ancients." [97]

5. MORAL VS. ABSOLUTE CERTAINTY

But, when all is said and done, Descartes recognizes that a posteriori proofs, however high a probability they may establish that our hypotheses embody "true causes"—and they may go so far as to make it so improbable that any other causes are the true ones as to make this "seem incredible"—give us, after all, only a "moral certainty that all the things of this world are such as has here been shown they may be." [98] Can he rest satisfied with this result? We know how he longs for an absolute certitude in physics comparable to that of the purely mathematical disciplines. We know how desperately he desires to conclude that the hypotheses that he has been led to assume as to the causes of natural phenomena, whether in astronomy or in physics, represent not only *de facto* physical actuality, but also metaphysical and mathematical necessity; not merely probable truth or "moral certainty," but absolute certainty, conferred upon them by rigorous demonstration from self-evident first principles of reason.

Can he really claim so much? We know already that he believes he can, so far at least as concerns the primary principles of physics—the most general laws of nature. A posteriori proof is in itself capable of establishing these with "moral certainty," but Descartes believes that these at least can also be demonstrated a priori with absolute certainty. But what of the *particular* mechanical hypotheses that he has elaborated as explanations of particular natural phenomena? Are these also capable of being raised to this height of infallible truth? In the concluding chapter of the *Principia* we find him wrestling with this problem. And here it is highly instructive to compare the varying expressions of the Latin and the French versions.[99]

The title of the chapter begins boldly (Fr.), "we possess even more than a moral certainty of it" (sc. that "all the things of this world are such as has been here shown they may be"). But Descartes is conscious of the difficulty concerning particular mechanical hypotheses, and is obliged immediately to qualify his original statement (L. only) —"there are some [i.e., presumably, the fundamental principles] even among natural things which we judge to be absolutely and more than morally certain." To this class belong such truths as are attested by clear and distinct perception, and thus ultimately guaranteed by the metaphysical fact of God's existence and veracity. "Of this character are the demonstrations of mathematics, the knowledge that material things exist, and all evident reasonings concerning them" (L.). More particularly, "it [sc. this more than moral certitude] extends to all the things that can be demonstrated concerning these bodies, by the principles of mathematics or by others as evident and certain" (L.). "To which class it seems to me that the things I have written in this treatise should be admitted" (Fr.). But too much is here implied, and a qualification is immediately introduced—"*at least the principal and more general*" (Fr.). "And I hope they will be" (Fr.), or "perhaps they will be" (L.), "by those who consider that they are deduced in a continuous series from the primary and most simple principles of human knowledge" (L.). But Descartes knows quite well that particular mechanical hypotheses cannot be deduced wholly a priori from metaphysical principles, for (1) they follow from a priori principles only when these are taken together with a knowledge of certain data that can be given only empirically, and (2) it is always impossible *absolutely* to exclude the possibility of other mechanical hypotheses equally "deducible" from the a priori principles. And thus he is obliged to qualify once more, and to conclude as follows: "especially if it be sufficiently understood" (L.) "that the heavens are fluid" (Fr.)—note the intrusion of empirical data—"for these points [sc. concerning the fluidity of the heavens] being admitted, all the rest, at least the more general that I have set down concerning the world and the earth, will appear to be scarcely capable of being understood otherwise than in the way I have explained them" (L.). The French here contains even more reserves and qualifications:

> . . . this single point [sc. the fluidity of the heavens] being admitted as sufficiently demonstrated by all the effects of light [note the empirical basis of this "demonstration"], and as a consequence of all the other things that I have explained, I think that it ought also to be recognized that I have proved by mathematical demonstration [but note that Descartes has already admitted that the premises of this "mathematical demonstration" are in part of empirical origin] everything that I have written.

But once more this is going too far. Descartes has not always in every particular case been able to exclude the possibility of mechanical hypotheses alternative to his own. Hence a further restriction—"at least the more general points concerning the fabric of the heaven and the earth and in the way in which I have stated them: for I have been careful to propose as doubtful all those that I thought were so."

This is the final word in the matter. Descartes' purely a priori physics fails, even on his own showing, quite to come off.

6. THE PLACE OF "INDUCTION" IN DESCARTES' THEORY OF METHOD

It remains for us briefly to consider one further function assigned to experience in the Cartesian theory of method. For Descartes, like Aristotle, maintains that the ultimate principles of physics themselves, and indeed of all knowledge whatsoever, have, in one sense at least, their origin in experience of particulars. However much he may speak in this connection of "innate ideas," and however much the guarantee of the truth of the primary principles is to be found in the self-evidence of a rational intuition, rather than in any multiplication of empirical instances, the fact nevertheless remains, that for Descartes, as for Aristotle, the process by which these "innate ideas" are actually brought into explicit consciousness is one which as a matter of fact begins with experience of particulars. Thus Descartes characterizes it as an "error" to suppose that "the knowledge of particular propositions must always be deduced from universals": one who really understands how we ought to seek truth will see that "in order to find it we must always begin with particular ideas in order to arrive at those that are general, although one can also reciprocally, once having found those that are general deduce from them others that are particular." [100] For example, the truth "that everything that thinks is, or exists" has been "learned from the experience of the individual—that unless he thinks he cannot exist." Of any details of the process by which we thus pass from an individual experience to the general truth there exemplified Descartes has practically nothing to say. He recognizes, indeed, the precariousness of the inductive "leap." We run a risk of error "as often as we judge that we can deduce anything universal and necessary from a particular or contingent fact." [101] As for the rest, he can only say that "our mind is so constituted by nature that general propositions are formed out of knowledge of particulars." [102] And this simple proposition may thus be said to contain practically his whole theory of inductive generalization.

Descartes in fact believed, like Aristotle, that the human mind possesses a further inexplicable power, through the exercise of a rational intuition, of distinguishing, within the contingent complications of particular phenomena as revealed to us in experience, those relations that are truly universal and necessary precisely because they hold between the ultimate and simple universal natures into which the complicated particulars can be analytically resolved. "This mental vision extends both to all those simple natures, and to the knowledge of the necessary connections between them." [103] The mind, according to Descartes, is able to do this sometimes directly and sometimes only by the aid of certain processes preparatory to and subordinate to intuition proper. One of these processes is that to which he gives the name of analogy or "imitation." It enables us to reach a conclusion regarding the necessary relations embodied in a particular concrete experience, the complication of which is such that the "natural light" is unable immediately to unravel it, from a consideration of some simpler and more familiar experience, real or depicted in the imagination, where the same relations appear in a form more easily grasped by the mind.

Note that the only reason why preparation is required for comparison . . . is the fact that the common nature we spoke of does not exist equally in both, but is complicated with certain other relations or ratios. The chief part of human industry consists merely in so transmuting these ratios as to show clearly a uniformity between the matter sought for and something else already known.[104]

Berthet [105] cites from Poisson a passage that seems to throw some light upon the procedure that Descartes here has in mind. It is a question of the colors observed in the rain. For such relatively complicated objects as drops of rain we first substitute analogous but geometrically simpler objects, namely, balls of glass. We then distinguish between the shape, the movement, etc., of the balls of glass, attributing to the shape the effects of the shape, to the movements the effects of the movements, to rest and various dispositions of the balls the effects that depend upon these. In each case these relations of cause and effect are necessary and universal relations between simple natures and are such that reason discerns them, abstracting from the complication that comes from their mutual interference in the concrete phenomena. Having discovered these relations in the simpler case of the glass balls, the reason is then able to trace them also in the more complicated case of the raindrops.[106]

Another auxiliary process is that which Descartes knows by the name of "enumeration." At first sight this process suggests to a modern reader the idea of induction by complete or incomplete enumeration. In fact Descartes himself distinguishes between a complete and an incomplete enumeration, and gives examples that seem to be of an inductive process. Moreover, he calls it "enumeration or induction." [107] A further examination of his treatment of the matter shows, however, that such an interpretation of Descartes' meaning would scarcely be warranted, and that as a matter of fact "enumeration" is for him hardly an inductive process at all, but rather, like his "analogy," a process merely auxiliary to intuition or deduction. It consists essentially in a rapid running over in thought of the steps of a long or complicated deduction, which serves to reassure us of its validity by a sort of comprehensive glance at the course of the demonstration as a whole. "It is therefore necessary that my thought run over them repeatedly, until I pass so quickly from the first to the last as to seem, almost without the aid of memory, to embrace the whole in a single intuition." [108] The deduction "is perceived by intuition when it is simple and clear, but not when it is complex and involved. Then we give it the name of enumeration and induction, because it cannot be comprehended in one whole at a glance of the mind, but its certainty depends in some measure on the memory." [109]

For in all cases where we have deduced propositions immediately from one another, if the inference has been evident, they have already been reduced to a true intuition. But if we infer a proposition from various and disconnected premises, it frequently happens that our intellectual capacity is not such that it can embrace them all in a single intuition; in this case we must be satisfied with the certainty of this operation [sc. of "enumeration" or "induction"].[110]

Such an enumeration must in all cases be "sufficient," [111] but it need not necessarily be in all cases "complete" or "distinct." Sometimes completeness is necessary.

For if I wish to prove by enumeration how many genera there are of corporeal things, or of those that fall anywise under the senses, I shall not assert that they are so many and no more, unless I know for certain that I have comprehended all of them in the enumeration and distinguished each from each.[112]

In other cases such perfect completeness may however be dispensed with.

But if in the same way I wish to show that the rational soul is not corporeal, it will not be necessary that the enumeration be complete, but it will be sufficient if I collect all bodies at once into a certain number of classes in such wise that I shall be able to demonstrate that the rational soul cannot belong to any of them. If, finally, I wish to show by enumeration that the area of a circle is greater than the area of any other figure of equal perimeter, it is not necessary to consider all figures, but it is enough to demonstrate this conclusion for certain particular ones only, and to conclude the same by induction for all the rest.[113]

The last example sounds like an induction in our sense of the term, and perhaps something of the sort is what Descartes, for the moment, really had in mind. But, if we remember that the illustration is drawn from mathematics, we shall be led to wonder whether this is indeed the case. Is not enumeration even here rather an auxiliary to the intuition of universals or simple natures, and of their necessary connections? This interpretation at least seems to be implied, for instance, in the following passage: "But if we arrange all the particulars in the best order, they will be reduced for the most part to determinate classes of which it will be sufficient to take one only for exact inspection, or some item of each, or some rather than others." [114]

That Descartes' "enumeration" or "induction" is not our induction is also evident from his use of terms. Thus "induction" is used synonymously with "deduction," and to "induce" (*inducere*) for to "deduce" or to "infer." [115] He speaks of enumeration as one kind of "illation" and contrasts it with "simple deduction" [116] as that kind of deduction "to which we give the name of enumeration or induction." [117] And he tells us indifferently that the two sole methods of arriving at certain knowledge are "intuition" and "deduction," [118] or that they are "intuition" and "induction," [119] and in this passage "deduction" and "induction" seem plainly to be regarded as synonymous.

CHAPTER FIVE

Thomas Hobbes and the Rationalistic Ideal

THE THEORY OF Thomas Hobbes (1588–1679) concerning the nature and the methods of science is built upon no one clear and consistent basis, but embraces a number of mutually incompatible elements which he was unable to reconcile with one another, but no one of which he was prepared wholly to reject.

His early association with Francis Bacon (during the years 1621–26) seems to have had practically no influence upon Hobbes's mature philosophy. In his biographical sketch of Hobbes, R. Blackbourne writes:

> At that time he became a friend of Francis Bacon, Baron Verulam, Viscount St. Albans, Lord Chancellor of England, and promoter of *Atlantis*. Bacon was greatly charmed by Hobbes's manner; Bacon helped Hobbes to translate some of his writings into Latin, and used to say that no one understood the magnitude of his [Bacon's] thoughts as well as Thomas Hobbes.[1]

The occasional resemblances between their views that can be pointed out seem in all important respects to be rather coincidental than derivative; and even Bacon's influence on a very early phase of Hobbes's philosophy of nature, pointed out by Max Köhler,[2] does not extend, as one would expect, to Bacon's theory of induction. Although his philosophy exhibits in some of its phases a strongly empirical, even a phenomenalistic strain,[3] he seems never to have achieved any appreciation of inductive theory or the experimental method.[4]

Hobbes's really serious interest in matters of science seems only to have been aroused when at the age of forty he first made the acquaintance of Euclid's *Elements*. He was particularly impressed with the method of geometry,[5] although its content, too, came to play a crucial role in his systematic natural philosophy. He became convinced that the geometrical method of demonstration by rigorous deduction from certainly known premises was the only procedure capable of yielding results of that absolutely cogent certainty which in his view alone warranted for one's conclusions the title of science or philosophy. The "rationalistic" method of geometry thus came to seem the only true basis for a genuinely scientific knowledge of the world.

Geometry, besides providing the method, is also a crucial part of the content of truly scientific knowledge. From a few absolutely certain first principles Hobbes looks for a deductive derivation by cogent demonstrations, *more geometrico*, of the principles of geometry, mathematics, mechanics, physics, ethics, and politics.[6] Except for the basic principles and definitions, then, geometry is foundational. In fact, it even sometimes seemed to him that all the other sciences, mathematical, physical, and moral, should be looked upon simply as further extensions of geometry; [7] and in the *Leviathan* he once goes so far as to remark that geometry "is the only science that it hath pleased God hitherto to bestow on mankind." [8]

The method of science, Hobbes begins, is partly analytical, partly synthetical: the process that goes from sense to first principles, or definitions (*philosophia prima*), is analytical; the process of derivation from first principles (all else) is synthetical.[9] By the analytical method, Hobbes tells us, we will work back from the particulars of sense to the "universal notions" of things; to the causes of such accidents as are common to all bodies, that is, to all matter.[10] By the process of analysis, or abstraction, we find that there are only three of these universal notions or causes: motion, which is the universal cause; body, which undergoes motion; and space, wherein motion occurs. "All is body, motion and space; and those things which seem not to be expressible in terms of them are yet, in fact, to be explained wholly by them." [11]

The knowledge we have of universals and their causes, Hobbes continues,[12] is, in the first place, by their definitions. For example, if one has a true conception of *motion* then he cannot help knowing that *motion is the privation of one place, and the acquisition of another;* or if one conceives *place* aright, then he cannot be ignorant of its definition, namely, *that place is that space which is possessed or filled adequately by some body*. The knowledge we have of universals and their causes, next, comes through their "generations" or "descriptions," as Hobbes says; e.g., *that a line is made by the motion of a point, superficies by the motion of a line, one motion by another, etc.*[13] After this point is reached, scientific inquiry becomes synthetic, or, as Hobbes says, at this particular spot, compositive. One can do no better, perhaps, than read Hobbes's own account of this involved and chainlike procedure:

It remains, that we enquire what motion begets such and such effects; as, what motion makes a straight line, and what a circular; what motion thrusts, what draws, and by what way; what makes a thing which is seen or heard, to be seen or heard sometimes in one manner, sometimes in another. . . . First we are to observe what effect a body moved produceth, when we consider nothing in it besides its motion; and we see presently that this makes a line, or length; next, what the motion of a long body produces, which we find to be superficies; and so forwards, till we see what the effects of simple motion are; and then, in like manner, we are to observe what proceeds from the addition, multiplication, subtraction, and division, of these motions, and what effects, what figures, and what properties, they produce; from which kind of contemplation sprung that part of philosophy which is called *geometry*.

From this consideration of what is produced by simple motion, we are to pass to the consideration of what effects one body moved worketh upon another; and because there may be motion in all the several parts of a body, yet so as that the whole body remain still in the same place, we must enquire first, what motion causeth such and such motion in the whole, that is, when one body invades another body which is either at rest or in motion, what way, and with what swiftness, the invaded body shall move; and, again, what motion this second body will generate in a third, and so forwards. From which contemplation shall be drawn that part of philosophy which treats of motion [mechanics].

In the third place we must proceed to the enquiry of such effects as are made by the motion of the parts of any body, as, how it comes to pass, that things when they are the same, yet seem not to be the same, but changed. And here the things we search after are sensible qualities, such as *light*, *colour*, *transparency*, *opacity*, *sound*, *odour*, *savour*, *heat*, *cold*, and the like; which because they cannot be known till we know the causes of sense itself, therefore the consideration of the causes of *seeing*, *hearing*, *smelling*, *tasting*, and *touching*, belongs to this third place; and all those qualities and changes, above mentioned, are to be referred to the fourth place; which two considerations comprehend that part of philosophy which is called *physics*. . . .

After *physics* we must come to *moral philosophy*; in which we are to consider the motions of the mind, namely, *appetite*, *aversion*, *love*, *benevolence*, *hope*, *fear*, *anger*, *emulation*, *envy*, *&c.*; what causes they have, and of what they be causes. And the reason why these are to be considered after *physics* is, that they have their causes in sense and imagination, which are the subject of *physical* contemplation.[14]

Some pages later he reviews this whole structure of science, as he sees it, in this desirably succinct way:

He that teaches or demonstrates any thing, proceed[s] in the same method by which he found it out; namely, that in the first place those things be demonstrated, which immediately succeed to universal definitions (in which is contained that part of philosophy which is called *philosophia prima*). Next, those things which may be demonstrated by simple motion (in which geometry consists). After geometry, such things as may be taught or shewed by manifest action, that is, by thrusting from, or pulling towards. And after these, the motion or mutation of the invisible parts of things, and the doctrine of sense and imaginations, and of the internal passions, especially those of men, in which are comprehended the grounds of civil duties, or civil philosophy; which takes up the last place.[15]

An appraisal of Hobbes's philosophy of science must, of course, start with his first principles, or definitions—or, more particularly, with his analytical method of arriving at them, the obscurity of which procedure, as we shall see, arouses a good deal of puzzlement. But, if it is not clear how he gets to his first principles, it is at least clear how he thinks he does not get to them. Although the analytical procedure goes "backward" logically from the particulars of sense to first principles, or definitions, nevertheless Hobbes is quite insistent that this procedure is not what we would call inductive inference; from the first he insisted that mere experience of fact can never be the source of any universally certain truth:

This taking of signs by *experience*, is that wherein men do ordinarily think, the difference stands between man and man in *wisdom*, by which they commonly understand a man's whole ability or *power cognitive*; but this is an *error:* for the signs are but *conjectural*; and according as they have seldom or often failed, so their *assurance* is more or less; but *never full* and *evident:* for though a man have always seen the day and night to follow one another hitherto; yet can he not thence conclude they shall do so, or that they have done so eternally: *experience concludeth nothing universally*.[16]

So, too, in the *Leviathan* we learn that "*knowledge of fact* . . . is nothing else but sense and memory"[17]—a matter of "experience" (which is defined as "much memory, or memory of many things").[18] Upon "the experience of time past" there may be founded a certain "presumption of the future." [19] This arises as follows: "Sometimes a man desires to know the event of an action; and then he thinketh of some like action past, and the events thereof one after another; supposing like events will follow like actions." [20] But this sort of knowledge is extremely precarious. It is true that "by how much one man has more experience of things past, than another, by so much also he is more prudent, and his expectations the seldomer fail him." He is able to make a better use of "signs":

> A *sign* is the evident antecedent of the consequent; and contrarily, the consequent of the antecedent, when the like consequences have been observed, before: and the oftener they have been observed, the less uncertain is the sign. And therefore he that has most experience in any kind of business, has most signs, whereby to guess at the future time; and consequently is the most prudent.[21]

This "applying the sequels of actions past, to the actions that are present . . . with most certainty is done by him that has most experience." But, Hobbes adds, "not with certainty enough. And though it be called prudence, when the event answereth our expectation; yet in its own nature, it is but presumption"; and "such conjecture, through the difficulty of observing all circumstances, [may] be very fallacious." [22] So, too,

> . . . there is a presumption of things past taken from other things, not future, but past also. For he that hath seen by what courses and degrees a flourishing state hath first come into civil war, and then to ruin; upon the sight of the ruins of any other state, will guess, the like war, and the like courses have been there also. But this conjecture, has the same uncertainty almost with the conjecture of the future; both being grounded only upon experience.[23]

Such probable knowledge, based upon generalization from observed facts, constitutes, of course—we should say today—the very essence of the inductive procedure. But Hobbes was so wedded to the rationalistic ideal of demonstrative certainty and infallibility that he was unwilling to take seriously anything that fell short of this perfection. We find, therefore, in Hobbes's writings no account whatever of the rules or standards in accordance with which such inductive generalizations should be made and no discussion of the various circumstances which may in a given case render such a generalization more or less probable. According to G. C. Robertson, "The word 'Induction' which occurs in only three or four passages throughout all his works (and these again in minor ones) is never used by him with the faintest reminiscence of the import assigned to it by Bacon." [24] Hobbes, in fact, disparagingly remarks, "But induction is not a statement of a claim except where every particular is enumerated." [25] And concerning Hobbes's views on experimentation Leslie Stephen observes, "He often said that if people who tried such a farrago of experiments were to be called philosophers, the title might be bestowed upon apothecaries and gardeners and the like." [26] Hobbes writes, "For if experimentations of natural phenomena are to be called philosophy, then pharmacists are the greatest physicians of all." [27]

If Hobbes does not come by his first principles, or definitions, by way of induction, the question remains, then, how does he come by them? What *is* the process by which they are discovered and formulated? To this question Hobbes answers, as we have seen, that it is by a rational *analysis* of phenomena, a sort of reasoning that proceeds analytically from the observed facts of experience to the discovery of the universal causal principles by which they are to be explained.[28] But what does this procedure precisely amount to? Hobbes says that we set out

> . . . from such knowledge of things as is had by way of sense. This is a knowledge of things as they appear in fact—which is to say, in all that complex variety of sensible aspects that makes them singular. For knowledge of cause, on the other hand, the aspects or appearances of things must be taken separately rather than as they are in fact joined together. Now, though less known *to us* than the singular things of sense, universals, as the common aspects of things may be called, are (in the Scholastic phrase, after Aristotle) better known *to nature* or in themselves. The first step, then, in the actual work of philosophy is to make an analysis of sensible objects into their "parts" as thus understood.[29]

Hobbes, however, makes an extraordinary omission, for he neglects to specify what is, after all, fundamental—namely, the directions for the methodological conduct of such analysis. As G. C. Robertson points out:

> With what things it [an analysis] should begin, if with any rather than with others, and by what steps the more should be reached from the less universal till the most universal notions are clearly recognised to be such, is in no way indicated. Here, if nowhere else, some reference might be expected to Bacon's theory of Induction,

had that made any impression upon Hobbes' mind; but there is not the faintest reminiscence of it. He is only sure that it is the sensible world in which he, with his contemporaries, is interested, that it is from a consideration of the actual things of sense that any attempt at understanding must set out. The understanding itself, however, is a purely rational determination, and proceeds in him without any of that anxious looking backwards and forwards to sense-experience which is the note of the Baconian method.[30]

In this phase of his theory, according to which, beginning with the observed facts of sense, we rise somehow by analysis to the apprehension of certain universals that are thereupon perceived by an immediate rational intuition to constitute the true causal explanation of the phenomena, Hobbes is simply an Aristotelian *malgré lui.*

Equally characteristic of this aspect of Hobbes's thought, and equally Aristotelian in essence, is his insistence that the definitions with which science starts [31] must, so far as possible, state the causes or modes of generation of the entities defined. In the case of the ultimate "simple ideas" reached by the process of analysis the causal definition will not be possible; here one must be contented to give a mere "explication" of their meaning.[32]

The next question after, How does Hobbes get his first principles? is, How does he justify them? What are the grounds of the necessity and certainty he ascribes to them? Again, this problem gives Hobbes considerable trouble. His first notion seems to have been to seek an a priori justification of the first principles in a direction suggested by his nominalistic theory of knowledge or of reasoning as computation with signs. In this account of the matter he represents the first principles of science as being simply definitions, and these definitions are nothing more than arbitrary conventions as to how we intend to use terms. We know the first principles to be valid because we *make them to be true* by an arbitrary convention. "A *name* or appellation therefore is the *voice* of a man *arbitrary*, imposed for a *mark* to bring into his mind some conception concerning the thing on which it is imposed." [33] "We cannot from experience conclude . . . any proposition *universal* whatsoever, except it be from remembrance of the use of names imposed arbitrarily by men." [34] Science, therefore, is not knowledge of the facts of experience but knowledge of the uses of *names* and the "consequences of names." "Reason . . . is nothing but *reckoning*, that is adding and subtracting, of the consequences of general names agreed upon for the *marking* and *signifying* of our thoughts."[35] As a consequence it of course follows that "error" is simply absurdity and self-contradiction—a failure to abide by the sense in which we have agreed to use our words.

Seeing then that truth consisteth in the right ordering of names in our affirmations, a man that seeketh precise truth had need to remember what every name he uses stands for, and to place it accordingly, or else he will find himself entangled in words.[36]

When we reason in words of general signification, and fall upon a general inference which is false, though it be commonly called *error*, it is indeed an *absurdity*, or senseless speech. For error is but a deception, in presuming that somewhat is past, or to come; of which, though it were not past, or not to come, yet there was no impossibility discoverable. But when we make a general assertion, unless it be a true one, the possibility of it is inconceivable.[37]

Upon this phase of Hobbes's thought G. C. Robertson justly comments as follows:

Truth is, in fact, a certain fixed or consistent way of using the names imposed by ourselves or others; and as names, we saw, are arbitrarily imposed, Hobbes is forward, when he is in the vein, to draw the conclusion that truth is in all strictness the creation, and even the arbitrary creation, of speaking men. It follows that Science, which consists of names, in this sense, truly joined into propositions, and of propositions, in the same sense, truly joined into syllogisms, does not, like Experience, give a knowledge of facts or reality, but only of the use of names. . . . There is no way of overcoming the difficulty here in Hobbes' theory of general knowledge. . . . Truth, with Hobbes, is so much an affair of naming, and naming is so much an arbitrary process, that science, for any reason he gives to the contrary, becomes altogether divorced from fact. If a true proposition is made true in the act of giving names, and remains true as the same names continue to be joined together, what need that it have any objective application at all?[38]

Hobbes, it is true, never squared his "conventional" interpretation of language with the view that (scientific) language makes an objective reference; but unfortunately it is clear that he held the latter view as well as, and at the same time as, the former. He writes, "It is true (for example) that *man is a living creature*, but it is for this reason, that it pleased man to impose both these names on the same thing."[39] The truth of the proposition apparently is completely a matter of how we use language. But, of course, when he says *we impose names on the same things* he transgresses his view that "a name is simply an arbitrary mark for recalling a *thought*"; and by this external reference he breaks out of the "prison of names." Hobbes again: "Wherefore, in every proposition three things are to be considered, *viz.*, the two names, which are the *subject*, and the *predicate*, and their *copulation;* both which names raise in our mind the thought of one and the same thing; but the copulation makes us think of the cause for which those names were imposed on that thing." [40] On this passage D. G. James comments:

But to say this is indeed to give away the game; if there *is* such a cause, the doctrine of caprice and arbitrariness has been yielded up; or rather would have been yielded up, had Hobbes been bent less on power than on truth. Or again he says: "And therefore those propositions only are *necessary*, which are of sempiternal truth, that is, true at all times. From hence also it is manifest, that truth adheres not to things, but to speech only, for some truths are eternal; for it will be eternally true, *if man, then living creature;* but that any *man*, or *living-creature*, should exist eternally, is not necessary" (E. I. p. 38). But if necessary propositions arise from arbitrary naming, what purpose is there in speaking of eternal truth? And if truth adheres to speech only, how can it be eternal? Hobbes's bland persistence in self-deception never deserts him. . . . The more we read Hobbes the less we think of his doctrines in an intellectual point of view; but also, the more fascinated we become by *him*, in a human point of view. He enthrals us the more as a figure in human history in proportion as we observe the contradictions in his thought which he passionately overlooks.[41]

Hobbes himself at times, however, could not help obscurely feeling that he was dealing with more than the relations between "names" arbitrarily defined. Science seemed to be dealing with the actual facts of experience; and in fact Hobbes had from the first a highly dogmatic conviction of the objective applicability and truth of his first principle (that the causes of all natural phenomena are to be found in motions). In *De corpore*, for example, he presents this principle in terms of rational intuition rather than linguistic conventionalism, as a self-evident truth concerning objective reality of experience rather than as an arbitrary convention with regard to the use of names:

The causes of natural phenomena (in the instances where there are certain causes) are made clear of themselves or by observation of nature (as is said); so that no method is needed; for the cause of all is one universal cause, namely motion . . . (for as a thing gives way either from a stationary position or from motion, it cannot be understood except by the motion). . . .[42]

Hobbes unfortunately, however, is not very clear on this matter of rational intuition or self-evidence either, for in chapter ix [43] he attempts to give a *demonstration* of this principle! Moreover, while he writes in *De corpore* that "the nature of definition consists in this, that it furnish an idea of the true thing; for principles are known of themselves or else they are not principles," [44] yet the definitions of mathematics are still represented as arbitrary in *Principia*,[45] though in the same passage their truth is said to be "evident to the natural light."

There is one further matter which must be attended to on the level of "first principles, or definitions," namely, this locution itself which sounds so queer and self-inconsistent to a modern ear. Hobbes nevertheless clearly insists that the indemonstrable first principles of science are all definitions—"Moreover those same principles are merely definitions" [46]—and for these reasons: postulates and axioms cannot be admitted as first principles because the former have to do not with knowing but with doing and the latter are really demonstrable from definitions.[47]

Hobbes's analysis of geometry, as we have seen, follows his discussion of first principles. In his analysis of geometry, the study of "simple motions," Hobbes entertains a queer conception of space, which, while it appears no more acceptable, appears less queer at any rate after one digs out the conflict in Hobbes's thought which the "queer" conception was designed—although not too clearly, perhaps —to resolve.

In his chapter "Of Place and Time" [48] Hobbes asks us to "feign" that the world is annihilated and then to inquire what would remain for anyone to consider as a subject of philosophy. Now, Hobbes says, if we remember or have a "phantasm" of anything that was in the world before the supposed annihilation, and remember not that the thing had such and such characteristics but only that it existed outside the mind, then we have a conception of *space*. However this conception of space, we are startled to discover, makes space *imaginary*, a phantasm of the mind. Hobbes defines *space* in this way: "SPACE *is the phantasm* [memory or idea] *of a thing existing without the mind simply;* that is to say, that phantasm, in which we consider no other accident, but only that it appears without us." [49] It seems more than passingly strange, of course, for Hobbes to make space imaginary but maintain the objective reality of body and motion! But the motivation for introducing the notion of imaginary space begins to emerge in the next chapter. Suppose, Hobbes writes, some one of the annihilated things to be placed again in the world, created anew.

The *extension* of a body, is the same thing with the *magnitude* of it, or that which some call *real space*. But this *magnitude* does not depend upon our cogitation, as imaginary space doth; for this is an effect of our imagination, but *magnitude* is the cause of it; this is an accident of the mind, that of a body existing out of the mind.[50]

It is necessary . . . that this new-created or replaced thing do not only fill some part of space . . . or be coincident and coextended with it, but also that it have no dependance upon our thought. And this is that which, for the extension of it, we commonly call *body*; and because it depends not upon our thought, we say is *a thing subsisting of itself*; as also *existing*, because without us; and, lastly, it is called the *subject*, because it is so placed in and *subjected* to imaginary space, that it may be understood by reason, as well as perceived by sense. The definition, therefore, of *body* may be this, a *body is that, which having no dependance upon our thought, is coincident or coextended with some part of space*.[51]

On these statements, D. G. James justly comments as follows:

Hobbes's motive is clear. Space is an accident of the mind; the innumerable figures of geometry we *make* in this accident, and their generation is thus known to us; therefore we have truly philosophical knowledge of them. But happily, what is not an accident of the mind, namely, body in its extension (*quod appellari solet, propter extensionem quidem, corpus*) is subject to this accident; and therefore, considered as extension or quantity, body too is open to our philosophical investigation. And thus it comes about that the mind in search of science *simpliciter*, by deduction from the definitions of space and motion, is able wholly to understand, as it is able to understand a geometrical theorem, the structure of the universe; it can arrive at a "science of causes of the διότι." [52]

The difficulty which Hobbes felt, and which his space gymnastics were supposed to overcome, came about in this way. Hobbes, of course, reflecting the views of his time, felt that geometry had empirical content (referred to extended figures, etc.); and he also, of course, felt that its principles and results were necessarily and certainly true. His problem was how to square these claims. Here, as James indicates, and as we have seen Hobbes do in the case of his own first principles, he relied on linguistic conventionalism to get the certainty and necessity, and on the alleged coincidence of space and extension to bring in the required external reference. However, again, he did not always try to salvage the certainty and necessity of mathematics in this fashion. In Hobbes's letter to Cavendish, quoted by F. Tönnies,[53] he characterizes the first principles of mathematics as definitions made true by mere convention; but by 1655 he had come to think of these principles as self-evident truths concerning objective realities, albeit in one and the same passage he talks confusedly of the principles as both arbitrary and "evident to the natural light." [54]

After geometry in Hobbes's system, as we have seen, come mechanics, physics, ethics, and politics. When in the course of developing deductively this system of philosophy he came to treat in detail of the phenomena of physics, psychology, and politics, and to seek for the *particular* motions from which as causes these phenomena can be derived, Hobbes found himself in the greatest difficulty. It was no longer possible to proceed purely deductively. It then became evident to him that it is impossible to derive the phenomena merely from the general principles of geometry and mechanics, and he began to realize that the most that can be done in this realm is to discover the possible or at best the probable causes of the phenomena. In other words one must here be content to deal with *hypotheses*. Consequently, we find Hobbes accepting the idea that there can after all be a *science* that deals with objective facts, and moreover that this science cannot demonstrate its conclusions with perfect certainty. That is to say, Hobbes here begins to appreciate the importance and the peculiarity of the inductive side of science. The contrast in this respect between the mathematical and the physical sciences was already in his mind by 1644, where we find him noting, in an unpublished manuscript quoted by Tönnies,[55] that, whereas the first principles of geometry are definitions of terms and can therefore be known to be true with certainty, the explanatory principles employed by the physical sciences are merely hypotheses or suppositions with regard to the truth of which we cannot be certain. For it may happen that the phenomenal effect can be validly explained on the basis of a supposed motion, and yet this hypothesis may nevertheless not be true. Consequently, the most that can reasonably be expected of a physical scientist is that the motions which he supposes in his hypotheses as the cause of the observed phenomena shall be intrinsically conceivable, that the observed phenomena can be shown to follow necessarily if such causes were really in existence, and finally that nothing contrary to the phenomena can be deduced from them.

Already in 1640 Hobbes had given a definition of hypothesis or "supposition," had indicated the method of its testing, and had noticed that it can never attain to more than probability.[56] He now begins to understand in some measure the place of such hypotheses in science. Hence the altered definition of philosophy in the last part of the *Leviathan*—"By PHILOSOPHY is understood *the knowledge acquired by reasoning, from the manner of the generation of any thing, to the properties: or from the properties, to some possible way of generation of the same*" [57]—so out of accord with the account given in the earlier chapters of the same work,[58] where mere knowledge of fact based upon experience is said to constitute no part of *science*, the procedure of which is purely deductive,[59] and the conclusions of which are infallible.[60] The hypothetical nature of the physical sciences, and the contrast between them and mathematics, which proceeds by rigorous demonstration from absolutely certain premises, is again pointed out in several passages of *De corpore*.[61] Hobbes sometimes explains this difference as follows: a priori demonstrations are possible in geometry because we are here dealing with entities (geometrical figures) which we construct for ourselves as we please. In *physics*, however, the causes of the objects with which we deal are not within our power, and we are therefore confined to a posteriori demonstrations of what these causes *may* be.[62] Hence, "in natural causes all you are to expect, is but probability"; [63] for "the doctrine of natural causes hath not infallible and evident principles" and is characterized by "the absence of rigorous demonstration." [64] Yet, as we might expect by now, Hobbes is not after all quite prepared to acquiesce wholly in this view of the matter. He still feels at times that at least some part of physics can be deduced a priori from the principles of geometry and pure mechanics.[65] Hobbes constantly stumbled against the problem of the true relation of physics to geometry but did not succeed in solving it. The relation remained obscure to him, Tönnies writes,

. . . because he could not decide to content himself, as against the plain absurdities of the old physics and of the whole world-view bound up with it, with merely presumptively universal propositions, but believed—with so many of his contemporaries and successors—that to the so-called necessary truths of the scholastic philosophy he must oppose truly necessary ones. Accordingly he would not concede that Galileo's law of inertia was only an ingeniously invented abstraction, and after all of hypothetical character; and would not content himself with the fact that this law, inasmuch as all the calculations based on it were confirmed by experience, thereby gained in constant progression an increasing probability, and therefore might be *treated* as a certainty; nor, consequently, would he admit that the proposition that all change is local movement, which alone makes possible the application of the laws of mechanics to the universe, must always remain, in so far as it is meant to apply to actualities, a postulate or a regulative principle, which may indeed be of overwhelming probability, so long as no change of any other kind except by means of such movement is made evident to the senses, and so long as no other kind of transfer of motion except by means of contact is shown to be possible, i.e., actual. With such reservations Hobbes would not have believed that he could successfully combat the scholastic philosophy.[66]

With regard to the methods to be employed in the invention and verification of hypotheses Hobbes has very little to say. He contents himself with laying down a few simple precepts. In the first place, the hypothesis employed to explain the phenomena must be intrinsically possible and conceivable; [67] and, in the second place, it must be possible to conclude from the truth of the hypothesis to the necessity of the observed phenomena.[68] The nature of a *cause* he defines as follows: "*A cause is the sum or aggregate of all such accidents, both in the agents and the patient, as concur to the producing of the effect propounded; all which existing together, it cannot be understood but that the effect existeth with them; or that it can possibly exist if any one of them be absent.*" [69] He insists that every phenomenon must have a cause.[70] In fact it is impossible even to imagine such a thing as an event without a cause.[71] The entire cause of any effect is, of course, always sufficient to produce it.[72] Any part of the cause short of the whole is necessary but not sufficient to produce the effect, and is therefore termed a "*causa sine qua non*" or "*necessarium per hypothesin.*" [73] The entire cause, being sufficient to produce the effect, produces it therefore also *necessarily*.[74] In one passage Hobbes maintains that the cause of a given event can consist in nothing less than the whole of the preceding state of the universe; [75] but elsewhere he admits that something less than this may legitimately be regarded as the cause.[76]

Finally, in the *De corpore* Hobbes gives a brief account of the method according to which, in his view of the matter, the cause of any given effect is to be sought.[77] This method again is altogether rationalistic in character. It consists simply in the rational analysis of the given situation, and a separation in thought of those factors which the intellect perceives to be essential to the production of the effect from those which it perceives to be unessential. That all the elements essential to the constitution of the complete cause have been included is insured by the intellectual recognition which at last supervenes of the necessity of the given event, once the operation of these causal factors is presupposed.

CHAPTER SIX

Isaac Newton and the Hypothetico-Deductive Method

1. THE GENERAL NATURE OF NEWTON'S ACHIEVEMENT

NEWTON was not primarily a student of the theory of scientific method, but an active experimental investigator. We are consequently not surprised to discover that there are comparatively few passages in his works in which he discusses, in any connected or systematic fashion, his views of the subject. He was, however, by no means unconscious of the nature of the methods he was employing, and scattered through his writings there are numerous remarks concerning the matter, from which it is not too difficult to gather his doctrine. When his teaching is thus reconstructed, moreover, it can scarcely fail, despite certain lingering confusions, to strike us as representing a very notable advance. Not only is there abundant evidence that Newton has profited to the full by all that his predecessors have had to say on questions of method, but to a remarkable degree he shows himself capable of separating the gold from the dross, utilizing all that is valuable in their views, and at the same time freeing himself from all that is extravagant, one-sided, or perverse. Moreover, in the clearness and fullness of his grasp of the subject, he goes considerably beyond anything they have to offer. Indeed to such an extent was he in advance of his age that his followers themselves by no means always grasped the true significance of a point of view that even to us of today still seems quite essentially modern.

Newton's achievement is twofold. In the first place he has completely emancipated himself from the current notion that our understanding of nature can ever reach embodiment in an absolutely certain and definitive science. And in the second place he gives us for the first time a thoroughly clear and well-balanced account, to say nothing of a brilliant exemplification, of the analytic-synthetic method of mathematical physics.

2. NEWTON AND THE THEORY OF SCIENTIFIC CERTITUDE

In respect of his frank abandonment of the possibility for natural science of any absolute certitude and finality Newton must, however, share the honors with his contemporary Christian Huygens, who had independently arrived at the same position. Huygens expresses his conclusions in the matter in the Preface to his *Treatise of Light* (published in 1690, but representing work completed some years before that date), in a passage which also in other respects embodies views akin to those of Newton. One finds in this subject, he writes, speaking of the physics of light,

> . . . a kind of demonstration which does not carry with it so high a degree of certainty as that employed in geometry: and which differs distinctly from the method employed by geometers in that they prove their propositions by well-established and incontrovertible principles, while here principles are tested by the inferences which are derivable from them. The nature of the subject permits of no other treatment. It is possible, however, in this way to establish a probability which is little short of certainty. This is the case when the consequences of the assumed principles are in perfect accord with the observed phenomena, and especially when these verifications are numerous; but above all when one employs the hypothesis to predict new phenomena and finds his expectations realized.[1]

With the spirit of this passage Newton is really in thorough agreement at every point. To one approaching his writings for the first time, however, this agreement will not always be readily apparent; and in fact much confusion has often been felt with regard to Newton's true position in the matter. A careful study can scarely fail to disclose the real state of his mind; but there are at the same time features of his work that may easily give rise to misapprehensions.

In the first place, the mathematical and deductive form with which Newton increasingly [2] invested his works undoubtedly belies the true spirit of his method, and is only too likely to mislead the reader into supposing that his theory of science is substantially one with that of Descartes. Nothing could really be further from the truth. In fact there are not wanting signs that Newton, apart from the influence of external circumstances, would never spontaneously have chosen to throw his exposition into the strict geometrical form of a system which, starting with a set of definitions and axioms, proceeds to a series of lemmas, propositions, and corollaries. In this connection Bloch [3] points to the frequency and length of the *scholia* in the *Principia*. These sections, which interrupt the logical sequence of the demonstration for a variety of purposes, testify to the embarrassment Newton feels at being confined within that rigid schematism; and it is significant that in one of them he tells us how it was by means of certain experiments that he has just described, and in advance of any formulation of the theory of the matter, that he really discovered, concerning the resistance of fluids, facts that at an earlier stage he has already expounded as necessary consequences of certain theoretical principles.[4]

In the *Opticks*, too, Newton finds the mathematical form in which the work is cast so unsuitable that he is obliged to intercalate into his series of deductions sets of experiments that supply the necessary premises of the demonstrations that follow.[5] Moreover, in a notable passage of the third book of the *Principia* he writes:

> Upon this subject I had, indeed, composed the third Book in a popular method, that it might be read by many; but afterward, considering that such as had not sufficiently entered into the principles could not easily discern the strength of the consequences, nor lay aside the prejudices to which they had been many years accustomed, therefore, to prevent the disputes which might be raised upon such accounts, I chose to reduce the substance of this Book into the form of Propositions (in the mathematical way), which should be read by those only who had first made themselves masters of the principles established in the preceding Books.[6]

Newton here tells us, in effect, that demonstration *more geometrico* is so little essentially necessary to the subject that another form would actually be clearer to most readers, and that he had himself at first found it natural to adopt another style of exposition. Why then, we may ask, if this is the true state of the case, did he persist in throwing his work into this more or less uncongenial and misleading form? He himself supplies the answer in the latter portion of the passage just quoted. He had found, namely, by bitter experience, that to present his conclusions merely as a result of experimental investigation did not suffice to produce conviction of their truth. Only by casting his argument in a severely deductive form could he hope to carry with him his more reluctant readers. Hence the procedure that he actually adopted.[7]

In other words, as Bloch sums up the matter,[8] the geometrical mode of exposition is not for Newton a definitive form insuring a grasp of absolute fact. It is rather a means of persuading those whom dogmatic prejudice renders incredulous, by the use of a language which they find clearer than that of the facts themselves.

Similar polemical motives led Newton frequently to insist in emphatic terms upon the objective character and the certainty of his results as compared with the a priori theorizing of his opponents. It is in this connection that he so forcibly protests that his principles are no mere hypotheses, but are rather certified facts. Thus in the General Scholium at the end of Book III of the *Principia* he tells us that universal gravitation is no "hypothesis": "gravity really does exist"; "I frame no hypotheses"; "hypotheses . . . have no place in experimental philosophy." [9] It is accordingly quite in the spirit of Newton that Roger Cotes in his Preface to the second edition of the *Principia* (1713) insists that the "real existence" of gravity "is clearly demonstrated by observations," that "it is plain from the phenomena that such a virtue does really exist," [10] and characterizes those who, "too much possessed with certain prejudices, are unwilling to assent to this new principle" as "ready to prefer uncertain notions to certain." [11]

Such language, if unduly pressed or taken in too literal a sense, may very easily lead us into a serious misapprehension of Newton's true meaning—just how easily we may see from the case of his own contemporary disciples.

The enthusiasm excited by the theory of gravitation, that of the tides, that of sound, that of colors, was an essentially dogmatic enthusiasm. As often happens in the history of the sciences, the pupils and admirers forced the words of the master, and replaced by brutal assertions truths that he had stated with reservations . . . the contemporaries and the disciples of Newton, what formed at the end of the XVIIth century the Newtonian School, accepted the method of the *Principia* as an instrument possessing absolute value. With a boldness and a tenacity that contributed not a little to the success of the Newtonian doctrine, they affirmed that

Newton alone was in possession of the true rules of experimentation, that Newton alone had known how to draw from them irrefragable conclusions, that he alone had set science on the road to success.[12]

They believed that the *Principia* in its methods and in its results had introduced into physics for the first time an absolute certainty equal to that of algebra or of arithmetic.

Newton himself neither supported nor disavowed these extreme interpreters of his work. Amid the polemics of the day they doubtless powerfully aided his cause. But we must not for all that suppose that he ever allowed them appreciably to affect his own thought. His own much more carefully qualified statement of the matter is to be found, for instance, in his Fourth Rule of Philosophizing:

In experimental philosophy we are to look upon propositions collected by general induction from phenomena as accurately or very nearly true, notwithstanding any contrary hypotheses that may be imagined, till such time as other phenomena occur, by which they may either be made more accurate, or liable to exceptions. This rule we must follow, that the argument of induction may not be evaded by hypotheses.[13]

Here it is very evident that Newton regards his conclusions not in the light of certain and finally demonstrated truths, but rather in the light of provisional approximations, possessed, no doubt, of a high degree of probability, but nevertheless open always to revision or correction as a result of further experience. And such is his real opinion throughout. Thus, in the *Opticks* he writes:

As in mathematicks, so in natural philosophy, the investigation of difficult things by the method of analysis, ought ever to precede the method of composition. This analysis consists in making experiments and observations, and in drawing general conclusions from them by induction, and admitting of no objections against the conclusions, but such as are taken from experiments, or other certain truths. For hypotheses are not to be regarded in experimental philosophy. And although the arguing from experiments and observations by induction be no demonstration of general conclusions; yet it is the best way of arguing which the nature of things admits of, and may be looked upon as so much the stronger, by how much the induction is more general. And if no exception occur from phaenomena, the conclusion may be pronounced generally. But if at any time afterwards any exception shall occur from experiments, it may then begin to be pronounced with such exceptions as occur.[14]

To the same point is the following passage:

In the last place, I should take notice of a casual expression, which intimates a greater certainty in these things, than I ever promised, *viz. the certainty of Mathematical Demonstrations.* I said, indeed, that the science of colors was mathematical, and as certain as any other part of Optics; but who knows not that Optics, and many other mathematical sciences, depend as well on physical sciences, as on mathematical demonstrations? And the absolute certainty of a science cannot exceed the certainty of its principles. Now the evidence, by which I asserted the propositions of colors, is in the next words expressed to be from experiments, and so but physical: whence the Propositions themselves can be esteemed no more than physical principles of a science. And if those principles be such, that on them a mathematician may determine all the phaenomena of colors, that can be caused by refractions, and that by disputing or demonstrating after what manner, and how much, those refractions do separate or mingle the rays, in which several colors are originally inherent; I suppose the science of colors will be granted mathematical, and as certain as any part of Optics.[15]

3. NEWTON ON HYPOTHESES

But what then of Newton's frequently expressed contempt for "hypotheses"? It must already be clear that his attitude is by no means merely equivalent to that common seventeenth-century feeling that what is merely probable or provisional or approximate (no matter in how high a degree) has no place in a genuine science—which must, on the contrary, possess absolute certainty and finality. But it is by no means equally clear, at first glance, just what he does mean. His condemnatory statements are certainly very emphatic and seem to be expressed with a deep conviction of their truth and importance. Yet, on the other hand, we find that "hypotheses" are by no means entirely absent from his own work. We find him, for example, making use on occasion of the supposition that rays of light are really streams of solid particles.[16] Yet at the same time he assures us that this is no fundamental presupposition of his system. It is merely a probable opinion, to the literal truth of which he by no means commits himself —an opinion, moreover, that forms no essential part of his teaching.[17] In another passage he offers with similar reservations An Hypothesis Explaining the Properties of Light:

Were I to assume an hypothesis it should be this, if propounded more generally so as not to determine what light is, further than that it is something or other capable of exciting vibrations in the aether; for thus it will become so general and comprehensive of other hypotheses as to leave little room for new ones to be invented . . . and though I shall not assume either this or any other hypothesis, not thinking it necessary to concern myself whether the properties of light discovered by me be explained by this, or Mr. Hook's, or any other hypothesis capable of explaining them; yet while I am describing this, I shall sometimes, to avoid circumlocution and to represent it more conveniently, speak of it as if I assumed it and propounded it to be believed.[18]

Moreover, whatever his uncertainty with regard to the details of the hypothesis, "it never occurred to him to doubt the existence of a medium which at least performed the function of transmitting light. Amid all their disagreements Newton agreed with Hook to this extent, that there existed an ether, and that it was a medium susceptible of vibrations."[19] In fact Newton propounds the hypothesis of an ether in another connection also. This time he is suggesting, again with some diffidence, a probable explanation of gravitation.[20] And in this case also, despite uncertainty with regard to the details, he was unable, in view of his conviction of the impossibility of action at a distance, ever to relinquish the hypothesis, in some form or another, of an ethereal medium.[21] More than this, he "never gave up hope that experimental evidence might eventually be secured which would establish or definitely overthrow some of these specific conjectures" concerning the nature of the ether.[22] "It was in this spirit and to this purpose that he proposed many of the thirty-one queries attached to the *Opticks*. This judgment of Newton's ethereal hypothesis is interestingly confirmed by the last paragraph of the *Principia*."[23] This paragraph runs as follows:

And now we might add something concerning a certain most subtle Spirit which pervades and lies hid in all gross bodies; by the force and action of which Spirit the particles of bodies mutually attract one another at near distances, and cohere, if contiguous; and electric bodies operate to greater distances, as well repelling as attracting the neighbouring corpuscles; and light is emitted, reflected, refracted, inflected, and heats bodies; and all sensation is excited, and the members of animal bodies move at the command of the will, namely, by the vibrations of this Spirit. . . . But there are things that cannot be explained in a few words, nor are we furnished with that sufficiency of experiments which is required to an accurate determination and demonstration of the laws by which this electric and elastic Spirit operates.[24]

We may add to this the fact that Newton, on occasion, actually employs the term hypothesis to designate propositions that undoubtedly do form an integral part of his theory; [25] to say nothing of his reference to the "Copernican Hypothesis," [26] which he certainly does not reject or regard as superfluous.

Again, so far is Newton's *Opticks* from ignoring hypotheses that the whole argument of the early part of that work consists in showing that current hypotheses concerning light do not explain the facts disclosed by experiment. As a result of successive criticism and rejection of the alternative hypotheses he is led to a new conception of the matter (which we should certainly call a new hypothesis), which is thereupon tested and found to be confirmed by experiment.[27] Furthermore, the appendix to the *Opticks* is wholly composed of a set of Queries that are in effect nothing more or less than tentative hypotheses which he had not yet had time to test experimentally, but which he there proposed as worthy of such investigation. And the fact that in successive editions of the *Opticks* this section of the work was constantly expanded [28] seems to indicate that as the years passed he came to attach to it an ever increasing importance.

The truth of the matter is that Newton's polemic against hypotheses is directed not against any and every use of hypotheses, but rather against certain current forms of their abuse. Among these are the following:

1. In Newton's day many were accustomed to argue for the truth of a theory as follows. There are only a certain number of hypotheses that can possibly explain a given set of phenomena. Now in one way or another it may be possible to show that every one of these hypotheses is false—with one exception. This one exception *must* thereupon be accepted as constituting the true explanation.

Newton protests strongly against such an employment of hypotheses. He points out that this method of proof is extremely fallacious unless we are absolutely sure that we have really enumerated all the possibilities—and who can be altogether sure on such a point? No, the true method of investigating nature is the experimental. We must establish theories not merely by refuting all the suggested alternatives, but rather by positive and direct experimental confirmation. This is the teaching of the following passage:

In the mean while give me leave, Sir, to insinuate, that I cannot think it effectual for determining truth, to examine the several ways by which the phaenomena may be explained, unless where there can be a perfect enumeration of all those ways. You know, the proper method for enquiring after the properties of things, is to deduce them from experiments. And I told you that the theory which I propounded was evinced to me, not by inferring, *it is thus, because it is not otherwise;* that is, not by deducing it only from a confutation of contrary suppositions, but by deriving it from experiments concluding positively and directly. The way therefore to examine it, is, by considering whether the experiments, which I propound, do prove those parts of the theory to which they are applied; or by prosecuting other experiments which the theory may suggest for its examination.[29]

2. Closely connected with this first misuse of hypothesis is the following. If the only way of proving the truth of a theory *were* really the indirect one of *disproving* all the conceivable alternatives, it would follow that no theory could be regarded as adequately established, no matter how good the direct experimental evidence for it, so long as any alternative whatsoever, however devoid of positive grounds, remained possible. Now Newton points out that, no matter how well established by experiment any given theory may be, it will always be relatively easy for an ingenious imagination to construct other hypothetical ways of accounting for the facts—and these hypotheses, however little plausible or probable, cannot be absolutely excluded as at least possible alternatives. But he insists that such considerations should never be allowed to affect our judgment in the least degree. If it were legitimate to reject well-founded theories on *such* grounds, it would forthwith become absolutely impossible for science to reach any results at all.[30] His Fourth Rule of Philosophizing (quoted above) is plainly directed against such a use of hypotheses.[31]

3. A third abuse of hypothesis Newton finds in the procedure of those who, like the Cartesians, having reached their physics by a priori construction, refuse thereupon to recognize the validity of any conclusion contrary to their system, even though it is derived by a legitimate induction from careful experimental investigation. This, in Newton's view, is utterly to invert the true order of things. The

synthetic or deductive method can itself have no legitimate starting point that is not supplied by analysis and induction. He insists upon the priority, for all natural science, of the latter process, and protests strongly against the rejection of its legitimate results merely because they conflict with dogmatic hypotheses.[32]

The conclusion of Newton's critique of hypotheses is, therefore, in sum, that they are incapable of supplying adequate grounds either for the acceptance or for the rejection of any scientific theory, and that whosoever employs them for this purpose is guilty of an abuse. Scientific theories can be legitimately established by experimental evidence only, and they can be overthrown only (1) by showing the insufficiency of the evidence adduced in their favor or (2) by producing adverse experimental evidence.[33]

We now pass to the consideration of Newton's view as to the *legitimate* use of hypothesis. We may perhaps best sum it up by saying that for him the true function of hypothesis is everywhere suggestive rather than dogmatic.[34] Such, for example, is the utility of those hypotheses which find a place in the Queries of the *Opticks*, to which we have already referred. They serve to direct inquiry, to indicate new lines of investigation, and even to suggest new experiments by means of which science may be advanced.[35]

But there is also another use of hypothesis which if Newton does not actively favor he does at least not wholly disallow. Experimental determination of the facts must certainly precede any theories as to their explanation.[36] But once the facts have been duly ascertained Newton by no means entirely forbids the consideration of hypotheses. First of all, we may of course proceed, on the basis of the results experimentally established, to a criticism of current hypotheses and a rejection of all that cannot be reconciled with the facts.[37] But furthermore we may also, if we so desire, permit ourselves to construct hypotheses as to the further explanation of these facts.[38] We have already seen to what extent, and with what reservations, Newton himself occasionally indulged in speculations of this kind. But for the most part, and increasingly in his later years, he seems in his own practice to prefer, as far as possible, to refrain from anything of the sort. The considerations that led him to adopt this policy seem to have been as follows. (1) Even if one is able for oneself to resist the inevitable tendency of such hypotheses to become transformed insensibly into dogmas, one's readers are only too likely to misinterpret them. They will be easily led to suppose that these hypotheses, instead of being merely tentative suggestions having a certain probability, constitute fundamental presuppositions of one's system. (2) Moreover, the inclusion of such questionable material leads in practice to endless and fruitless controversies that had by all means better be avoided. (3) Such dubious speculations, however legitimate from other points of view, form no part of the real business of the positive scientist. It would be therefore better for a scientific treatise to dispense altogether with the consideration of matters more appropriate to the metaphysician.

The latter point is particularly prominent in Newton's mind, and he recurs to it again and again. If he commonly speaks of "attractions and impulsions" as of "forces," "powers," and "causes" of phenomena, he nevertheless frequently warns the reader that these are only conveniences of expression. His real intention is simply to discover by analysis the mathematical relationships that are exemplified in the phenomena, not to ascertain the causes upon which these depend. The nature of his convictions in this matter may be gathered from the following passages:

> I design to give only a mathematical notion of these forces without considering their physical causes and seats.[39]

> I likewise call attractions and impulsions, in the same sense, accelerative, and motive; and use the words attraction, impulse or propensity of any sort towards a centre, promiscuously, and indifferently, one for another, considering those forces not physically, but mathematically: wherefore, the reader is not to imagine, that, by those words, I anywhere take upon me to define the kind, or the manner of any action, the causes or the physical reasons thereof, or that I attribute forces, in a true and physical sense, to certain centres (which are only mathematical points); when at any time I happen to speak of centres as attracting, or as endued with attractive powers.[40]

> I shall, therefore, at present go on to treat of the motion of bodies mutually attracting each other; considering the centripetal forces as attraction; though perhaps in a physical strictness they may more truly be called impulses. But these propositions are to be considered as purely mathematical; and therefore, laying aside all physical considerations, I make use of a familiar way of speaking, to make myself the more easily understood by a mathematical reader.[41]

I here use the word attraction, in general, for any endeavour, of what kind soever, made by bodies to approach each other; whether that endeavour arise from the action of the bodies themselves, as tending mutually to or agitating each other by spirits emitted; or whether it arise from the action of the aether or the air, or of any medium whatsoever, whether corporeal or incorporeal, any how impelling bodies placed therein towards each other. In the same general sense I use the word impulse, not defining in this treatise the species or physical qualities of forces, but investigation of the quantities and mathematical proportions of them, as I observed before in the definitions. In mathematics we are to investigate the quantities of forces, with their proportions consequent upon any conditions supposed; then when we enter upon physics, we compare these proportions with the phaenomena of Nature, that we may know what conditions of those forces answer to the several kinds of attractive bodies. And this preparation being made, we argue more safely concerning the physical species, causes, and proportions of the forces.[42]

It is true that we may consider one body as attracting, another as attracted; but this distinction is more mathematical than natural. The attraction is really common of either to other, and therefore of the same kind in both. . . . And though the mutual actions of two planets may be distinguished and considered as two, by which each attracts the other, yet, as those actions are intermediate, they do not make two but one operation between two terms.[43]

The mathematical laws of phenomena the scientist can hope to discover; [44] but the ultimate natures of things, or the ultimate causes upon which these depend, may be made the subject of endless hypotheses among which it is impossible to decide by any available evidence. Newton declares repeatedly that it is not his intention to determine anything whatever with regard to these matters.[45] Light, for example, is a "something" that is propagated in accordance with certain mathematical laws. That is all to which he will commit himself.[46] Others may busy themselves with such speculations, if they so desire. It is easy to invent hypothetical explanations,[47] and that, too, according to a variety of different principles.[48] Some may even find such hypotheses valuable as furnishing imaginary representations by the aid of which their minds more readily grasp the relations of things. Newton, in fact, does not scruple to throw out frequent suggestions to this end, with appropriate warnings that they are not to be misunderstood.[49] But for the most part Newton himself has very little interest in this sort of thing. He is occupied with more important matters. Why, after all, should we trouble to seek for more ultimate "causes" when the laws that we have discovered suffice to explain all the phenomena? Newton writes, in his First Rule of Philosophizing:

We are to admit no more causes of natural things than such as are both true and sufficient to explain the appearances. To this purpose the philosophers say that Nature does nothing in vain, and more is in vain when less will serve; for Nature is pleased with simplicity, and affects not the pomp of superfluous causes.[50]

If Newton exhibits an especial hostility toward hypotheses involving "occult causes" it is perhaps in part because these have not even the merit of presenting a precise image to the mind. Occult qualities are mere empty names to disguise our ignorance. As Roger Cotes puts the matter:

. . . some have attributed to the several species of things specific and occult qualities, on which in a manner unknown, they make the operations of the several bodies to depend . . . being entirely employed in giving names to things, and not searching into things themselves, we may say, that they have invented a philosophical way of speaking, but not that they have made known to us true philosophy.[51]

The chief passage in which Newton deals with this topic is as follows:

It seems to me farther, that these Particles have not only a *Vis inertiae*, accompanied with such passive Laws of Motion as naturally result from that Force, but also that they are moved by certain active Principles, such as is that of Gravity, and that which causes Fermentation, and the Cohesion of Bodies. These Principles I consider not as Occult Qualities, supposed to result from the specifick Forms of Things, but as general Laws of Nature, by which the Things themselves are form'd: their Truth appearing to us by Phaenomena, though their Causes be not yet discover'd. For these are manifest Qualities, and their Causes only are occult. And the *Aristotelians* gave the Name of occult Qualities not to manifest Qualities, but to such Qualities only as they supposed to lie hid in Bodies, and to be the unknown Causes of manifest effects: Such as would be the Causes of Gravity, and of magnetick and electrick Attractions, and of Fermentations, if we should suppose that these Forces or Actions arose from Qualities unknown to us, and uncapable of being discovered and made manifest. Such occult Qualities put a stop to the Improvement of natural Philosophy, and therefore of late Years have been rejected. To tell us that every Species of Things is endow'd with an occult specifick Quality by which it acts and produces manifest Effects, is to tell us nothing: But to derive two or three general Principles of Motion from Phaenomena, and afterwards to tell us how the Properties and Actions of all corporeal

Things follow from those manifest Principles, would be a very great step in Philosophy, though the Causes of those Principles were not yet discover'd: And therefore I scruple not to propose the Principles of Motions above mention'd, they being of very general Extent, and leave their Causes to be found out.[52]

Thus we see that even gravity, electrical attraction, and the like, inasmuch as we know nothing of their ultimate nature or causes, would be nothing but occult qualities, if it were not for the fact that by these terms we understand simply certain mathematically determinate relationships between phenomena, such, moreover, as can be exactly formulated and subjected to experimental testing.

Some . . . are continually cavilling with us, that gravity is an occult property; and occult causes are to be quite banished from philosophy. But to this the answer is easy: that those are, indeed, occult causes whose existence is occult; and imagined but not proved; but not those whose real existence is clearly demonstrated by observations. Therefore gravity can by no means be called an occult cause of the celestial motions, because it is plain from the phenomena that such a virtue does really exist.[53]

Newton's objection to the "occult qualities" of the Scholastics on the ground that they "tell us nothing" reminds us of the similar objections of the Cartesians. Malebranche, for instance, objects to these qualities that they are "vague, general, and indeterminate ideas which represent nothing in particular to the mind." [54] Aristotle, he tells us,

. . . describes and explains everything by those fine words *genus, species, act, power, nature, form, faculties, qualities, cause per se* and *cause per accidens.* His partisans have difficulty in comprehending that these words mean nothing, and that we are no wiser than we were before when we hear them say that fire melts metals because it has the faculty of dissolving, and that a man does not digest because he has a weak stomach or because his digestive faculty [*faculté concoctrice*] is not functioning well.[55]

These and like ways of speaking are not false, but they really signify nothing. These vague and indeterminate ideas do not lead us into error, but they are entirely useless for the discovery of truth.[56]

In a word these are not "clear and distinct ideas," like extension, figure, and motion, in terms of which alone the Cartesians explain all the phenomena of nature.

Newton's objection, however, really goes further than this, for in his view the mechanistic explanations of the Cartesians are in no better case than are the qualities of the Scholastics. They, too, are "occult" rather than manifest, hypothetical entities lying behind phenomena rather than functional relationships discoverable within phenomena.[57]

In this connection we may pause to inquire what Newton may have meant by those "true causes" (*verae causae*) which he evidently desires to substitute for the occult, and to which he refers, for instance, in the First Rule of Philosophizing (quoted above). The expression has given rise to a good deal of discussion. Herschel, for instance, interpreted the words as signifying "causes recognized as having a real existence in Nature, and not being mere hypotheses or figments of the mind."[58] But Whewell points out that if the rule means, as it has generally been taken to mean, that "in attempting to account for any class of phenomena, we must assume such causes only, as *from other considerations*, we know to exist," then the rule is "an injurious limitation of the field of induction,"[59] inasmuch as it will prevent us from ever becoming acquainted with any new cause.

There seems to be no doubt that when he spoke of "true causes" Newton had in mind principally his own explanation of the motions of the planets by gravitation, as contrasted with the Cartesian explanation by means of vortexes. Yet what the a priori difference is between the two it is not altogether easy to see. We do know, indeed, that some of the bodies we have occasion to observe—those, namely, near the surface of the earth—are attracted to it. But we likewise know that some of the bodies we have occasion to observe are whirled around in vortexes in such media as air or water. Both explanations, therefore, try alike to reach a knowledge of the unknown by assimilating it to some case of the known. Thus vortexes also would belong to the class of "true causes" if by these Newton had meant those that have at some time been observed to produce effects of the same sort as the phenomena for which explanation is sought.[60]

It may appear, however, that what Newton meant by a "true cause" was not one *already known* to exist, but rather one the existence of which was *susceptible of independent corroboration*. The second sentence of the First Rule might lead one to this interpretation. Had Newton said merely that we are to admit no more causes than are sufficient to explain the phenomena, the rule would clearly be at least decidedly inadequate. For several different combinations of causes, each apparently containing nothing superfluous, might be equally adequate to explain some phenomenon. And yet, as between such rival explanations, we do eventually choose one as representing the true cause of the phenomenon in question. And we do so on the basis of evidence for the existence of the cause it supposes, *other than* that constituted merely by its adequacy to explain the phenomenon in question, e.g., on the basis of its explaining some other sort of phenomenon also, or some other aspect of the same phenomenon not originally contemplated. And the principle governing our choice would here plainly be that of simplicity. Whewell accepts Newton's rule in what appears to be very much this sense when he says:

When the explanation of two kinds of phenomena, distinct, and not apparently connected, leads to the same cause, such a coincidence does give a reality to the cause, which it has not while it merely accounts for those appearances which suggested the supposition.[61]

When such a convergence of two trains of induction points to the same spot, we can no longer suspect that we are wrong. Such an accumulation of proof really persuades us that we have to do with a *vera causa*.[62]

Newton's Rule, then, to avoid mistakes, might be thus expressed: "That we may, provisorily, assume such hypothetical cause as will account for any given class of natural phenomena; but that when two different classes of facts lead us to the same hypothesis, we may hold it to be a *true cause*." [63]

It may well be doubted, however, whether such an interpretation does not exhibit more ingenuity in telling us what Newton *should* have meant by his statements, than success in informing us what he actually *did* have in mind when he wrote them. Another possible interpretation of the matter may therefore here be suggested.

In the passage from the *Opticks* dealing with occult qualities, Newton tells us that he employs in their stead "general laws of nature," i.e., formulas merely descriptive of certain relations discernible among phenomena—"their Truth appearing to us by Phaenomena." These, by contrast with occult qualities, are "manifest Qualities, and their Causes only are occult." Now these laws are also characterized as themselves "active Principles . . . by which the Things themselves are form'd" and by which "these Particles . . . are moved." Have we not here the clue to Newton's meaning? If so, the "true causes" of phenomena are determining principles or laws (like the law of gravity) which we can discover by observation and experiment to be actually at work in the world, as opposed to merely hypothetical qualities, or entities (like the Cartesian vortexes). It is at least noteworthy that this version of Newton's meaning seems to be that adopted by Cotes in his Preface to the Second Edition. The followers of the Newtonian philosophy, he there tells us, desire

. . . to follow causes proved by phenomena, rather than causes only imagined, and not yet proved. The business of true philosophy is to derive the natures of things from causes truly existent. . . . Therefore if it be made clear that the attraction of all bodies is a property actually existing *in rerum natura*, and if it be also shewn how the motions of the celestial bodies may be solved by that property, it would be very impertinent for any one to object that these motions ought to be accounted for by vortices.[64]

Those rather have recourse to occult causes who set imaginary vortices, of a matter entirely fictitious, and imperceptible to our senses, to direct these motions.[65]

We are now in a position to understand more fully the true significance of the famous passage on hypothesis in the General Scholium at the end of Book III of the *Principia:*

Hitherto we have explained the phaenomena of the heavens and of our sea by the power of gravity, but have not yet assigned the cause of this power. But hitherto I have not been able to discover the cause of those properties of gravity from phaenomena, and I frame no hypotheses; for whatever is not deduced from the phaenomena is to be called an hypothesis; and hypotheses, whether metaphysical or physical, whether of occult causes or mechanical, have no place in experimental philosophy. In this philosophy particular propositions are inferred from the phaenomena, and afterwards rendered general by induction. Thus it was that the impenetrability, the mobility and the impulsive force of bodies, and the laws of motion and of gravitation were discovered. And to us it is enough that gravity really does exist, and acts according to the laws which we have explained, and abundantly serves to account for all the motions of the celestial bodies, and of our sea.[66]

4. NEWTON AND DESCARTES

It is noteworthy that in this passage Newton exhibits no greater respect for *mechanical* hypotheses than for any others. In the Preface to the 1686 edition of the *Principia* he wrote:

I wish we could derive the rest of the phaenomena of nature by the same kind of reasoning from mechanical principles; for I am induced by many reasons to suspect that they may all depend upon certain forces by which the particles of bodies, by some causes hitherto unknown, are either mutually impelled towards each other, and cohere in regular figures, or are repelled and recede from each other; which forces being unknown, philosophers have hitherto attempted the search of nature in vain; but I hope the principles here laid down will afford some light either to that or some truer method of philosophy.[67]

These and similar expressions[68] have sometimes been interpreted in such a way as to commit Newton to a strict mechanism after the fashion of Descartes.[69] Such an interpretation, however, would be quite erroneous. In the first place, this passage expressed, as Newton himself says, merely a suspicion or a hope, not a dogmatic conviction. And, in the second place, we know from other passages how little he was prepared to acquiesce in the Cartesian reduction of all the phenomena of nature to differing configurations of material particles possessing purely geometrical properties.

He gives us no mechanistic analysis, for example, of the qualities of hardness and impenetrability. These characteristics of bodies, revealed in sensation,[70] he regards as simple and ultimate properties of matter.[71] Again he regards the mechanistic "explanation" of cohesion as absolutely worthless.[72] Further, after remarking that the ancient atomists had at least the virtue of confining their mechanical explanations within the realm of real corporeal things, Newton contrasts their practice in this regard with that of recent physicists who in their passion for mechanism have vainly invented hypothetical substances—ethers, "subtle matter," etc.—in order to force all phenomena into their dogmatic scheme of things.

And for rejecting such a Medium, we have the Authority of those the oldest and most celebrated Philosophers of *Greece* and *Phoenicia*, who made a *Vacuum* and Atoms, and the Gravity of Atoms, the first Principles of their philosophy; tacitly attributing Gravity to some other Cause than dense Matter. Later

Philosophers banish the Consideration of such a Cause out of Natural Philosophy, feigning Hypotheses for explaining all things mechanically, and referring other Causes to Metaphysics: whereas the main Business of Natural Philosophy is to argue from Phaenomena without feigning Hypotheses, and to deduce Causes from Effects until we come to the very first Cause, which is certainly not mechanical.[73]

And, as we have seen, although Newton on occasion finds it natural to think of rays of light as streams of solid corpuscles, in the end he absolutely refuses to commit himself even to the belief that light is a material substance at all, to say nothing of giving any more definite mechanistic account of it.[74]

The great contrast which Newton's whole view of science presents when compared with the Cartesian ideal can hardly fail to be already apparent. We may, however, further enforce this contrast by enumerating a few of the important points of divergence.

In the first place, Newton's view as to the origin and nature of the fundamental principles that stand at the beginning of his exposition, in the deductive form in which it is presented, is utterly different from that of Descartes concerning the fundamental principles of *his* physics. For the latter, at least, the primary principles are justified by the light of the intellect—as clear and distinct ideas. For Newton they are one and all of empirical origin, and result from inductive generalization based on experimental data. We have already seen how Newton insists that the use of the analytic method must always precede that of the synthetic, and how he describes the analytic method as consisting in experimentation, observation of phenomena, and generalizing induction.[75] In the Preface to the 1686 edition of the *Principia* he writes, to the same effect, that "all the difficulty of philosophy seems to consist in this—from the phaenomena of motions to investigate the forces of nature, and then from these forces to demonstrate the other phaenomena." [76] In Newton's view, in fact, even the principles of geometry have an empirical basis—"geometry is founded in mechanical practice, and is nothing but that part of universal mechanics which accurately proposes and demonstrates the art of measuring." [77]

Again, fully to satisfy the Cartesian ideal, a science should reach all its various theorems by deduction from the absolutely certain first principles—the fact that they can be so deduced supplying the fundamental ground for believing them to be true. Even Descartes, in point of fact, is finally forced to concede that no physical science can possibly conform to this standard,[78] and for Newton the case stands throughout far otherwise. In the first place, many theorems owe their immediate origin and proof to experience, and many are only afterward deductively connected with the first principles.[79] In the second place, the fact that the theorems can be deductively derived from the first principles by no means confers upon them any finality. For the principles themselves, like all conclusions derived from experience, stand always open to the correction of further experimental investigation; [80] and this must in turn be true also of the theorems derived from them—they too must be submitted to experimental control. In so far as they continue to be confirmed by further experiment, the grounds for our confidence in their truth are progressively enlarged; and they acquire a status higher than any that mere deduction from first principles could ever alone confer upon them.[81] More than this, such further experimental testing also strengthens the evidence for the truth of the first principles themselves.[82]

In other words, we have no metaphysical guarantee whatever against there appearing exceptions to even our most confidently adopted principles; empiricism is the ultimate test. That this applies to the basic principles of the simplicity and uniformity of nature itself appears from an interesting passage in the *Opticks:*

> That it should be so is very reasonable [i.e., that the theorem of the uniform proportion of the sines applies to all the rays of light], nature being ever conformable to herself; but an experimental proof is desired.[83]

> No deduction from an accepted principle, no matter how general or clearly derived from past phenomena, can therefore pass for absolutely or physically certain, without careful and continued experimental verification.[84]

In fact, if we were to start from our principles, no matter how intrinsically sound, and reason purely mathematically, without further reference to experimental data, we might be landed in palpable absurdities.

> There are problems in which an algebraic treatment would lead to the admission of negative densities, imaginary forces, etc., where nature can admit as real only positive magnitudes. Not every mathematical quantity is capable of a physical interpretation. Hence the necessity of keeping close to the experimental facts.[85]

Science, whether in its first principles or in its derivative theorems, can thus never, in Newton's view, pretend to any absolute finality. It retains to the very end, in some degree, a provisional character.[86]

Not only, however, are the results of science always, for Newton, provisional; they are also always, despite their mathematical form, and the careful use of quantitative methods, only approximately accurate. We all recognize the folly, in the case of mathematical calculations, of carrying them to a pitch of accuracy in excess of what their practical application demands. Newton, too, is quite aware that absolute exactitude is undesirable where an approximation is sufficient for the purpose in hand.[87] The same principle applies to the formulation of mathematical laws in physics. There is no need to state them in a form more accurate than can be tested by the experimental methods at our disposal.[88] Furthermore, it is a part of Newton's view that a mathematical law of nature, considered in its relation to the actually highly complex occurrences that alone are to be found among natural phenomena, really describes not so much how any actually observed phenomena really *do* occur, but rather how they *would* occur in certain cases of an ideal simplicity. It describes a sort of limiting case to which nature is indeed found progressively to approximate, as we consider cases of less and less complexity; but to which, at best, it *only* approximates. In any concrete situation there are always actually at work other factors that must be neglected in stating any one of the simple mathematical laws there involved.[89] These complicating circumstances can afterward perhaps be explained in terms of further simple laws.[90] Thus in the *Principia*,[91] when he studies the law of the propagation of hydrodynamic disturbances, he speaks of an instantaneous propagation. But it is clear that this instantaneous propagation is only a limit that we substitute in thought for real propagations that are only more and more rapid.[92]

5. NEWTON AND FRANCIS BACON

Sir David Brewster in his *Memoirs of the Life of Newton* writes as follows:

Even Newton, who was born and educated after the publication of the *Novum Organum*, never mentions the name of Bacon or his system. . . . When we are told, therefore, that Newton owed all his discoveries to the method of Bacon, nothing more can be meant than that he proceeded in that path of observation and experiment which had been so warmly recommended in the *Novum Organum;* but it ought to have been added, that the same method was practised by his predecessors; that Newton possessed no secret that was not used by Galileo and Copernicus; and that he would have enriched science with the same splendid discoveries if the name and the writings of Bacon had never been heard of.[93]

And Lecky goes so far as to say: "The whole method and mental character of Newton was opposed to that of Bacon, and, as his biographer . . . very forcibly contends, there is not the slightest reason to suppose that Newton owed anything to his predecessor." [94]

It is difficult not to feel a certain exaggeration in these statements. True it is that Newton nowhere makes any direct reference to Bacon; [95] but it is nevertheless hard to suppose that no influence from Bacon ever reached him. In fact it would be very strange indeed if he was not to some extent acquainted with the *Novum organum* itself. Hook and Boyle, with whom Newton was in frequent correspondence, are known as admirers of Bacon. Bacon's works, in the England of that day, were regarded as the fundamental treatise on the logic of science; and it would seem that Newton, during his long years of residence at Trinity College (1661–96), can hardly have failed to learn something of their teaching.[96]

The probability that Newton was to some extent influenced, whether directly or indirectly, by the *Novum organum* is increased by the fact of a curious correspondence between some of the suggestions of that work and certain of Newton's characteristic scientific results. Bacon, in fact, had already hinted at the existence of something in the nature of gravitational attraction [97] and had suspected that its force varies with the distance.[98] He also distinctly foreshadows Newton's theory of the tides,[99] and of colors.[100] Furthermore, it is very striking to find Newton employing tables of instances in a way strongly reminiscent of the procedure recommended by Bacon.[101]

We find also that Newton gives to *experimentum crucis* the same somewhat extended interpretation as had Bacon. It is usually described simply as a means of deciding, as between alternative given hypotheses, which alone answers to the facts of nature. But for Bacon this is not the whole of the matter. If we have already completely enumerated all the possible hypotheses, the experimental refutation of all but one amounts to a demonstration of this one. But if our enumeration, as is often the case, is incomplete, the negative experiments serve to suggest the direction in which it must be completed, and the truly crucial experiment comes only at the end of this process. In Newton's optical researches the crucial experiment occupies a similar position. He first gives an experimental refutation of a number of suggested hypotheses, and as a result of this process is thereupon led to a new suggestion which would not otherwise have occurred to his mind. It is the experiment that verifies this novel hypothesis that Newton, like Bacon, calls *experimentum crucis*.[102]

By way of further illustration of the parallelism between Newton's thought and Bacon's, one may compare with Bacon's account of "prerogative instances" Newton's remark: "for it is not number of experiments, but weight to be regarded; and where one will do, what need many?" [103] Or with Bacon's emphasis upon "gradual inductions" Newton's statement: "By this way of Analysis we may proceed from Compounds to Ingredients, and from Motions to the Forces producing them, and in general, from Effects to their Causes, and from particular Causes to more general ones, till the Argument end in the most general." [104]

In any case, however, whatever the uncertainty of the existence or of the degree of any direct influence of Bacon upon Newton, the foregoing account of the latter's method must already have made it amply clear that the whole spirit of Newton, far from being fundamentally opposed to that of Bacon, is much more closely in accord therewith than with that of Descartes.[105]

There are, nevertheless, important differences between the two men. First of all, for Bacon observation and experiment, together with the consequent inductions, form practically the whole of scientific method; whereas in Newton's view, as we have seen, these procedures, however important and indispensable, form but one phase or stage of the full analytic-synthetic method. In the second place, even within this phase of the method there is a very significant difference between the induction of Newton and that of Bacon. The Newtonian induction, namely, proceeds throughout in quantitative terms; it employs exact measures, and its propositions are cast in numerical and mathematical form. Whatever the virtue of Bacon's precepts concerning the employment of definite measurements and numerical computations,[106] it must be evident that Newton went far beyond his predecessor in his understanding and practice of such procedures. This is nowhere more plainly to be seen than in the contrast between Newton's and Bacon's employment of tables of instances. The phenomena that the latter lists in these tables are never defined in any exact quantitative manner or submitted to any numerical measurement. Consequently the tables, although they may form the basis upon which we can establish the existence of certain vague correspondences between phenomena, never permit us to state these correspondences as mathematical equations. They may show us that certain natural phenomena are functions of certain others, but they can never reveal to us the precise form of those functions. Newton's tables, on the other hand, are always tables of *numerical quantities* determined by exact measurement; and they lead to the discovery of mathematically determinate functional formulas.[107]

Finally, there is another great difference between the two men in the fact that Bacon is still under the illusion that the right method will give us scientific results of absolute and final certainty, an illusion from which Newton is completely free.[108]

6. NEWTON'S RULES OF REASONING IN PHILOSOPHY

We have already quoted the first and the last of the famous four Rules of Reasoning in Philosophy with which Newton introduces the third book of the *Principia;* and we have also seen something of their bearing. The second and third of these rules,[109] together with Newton's comment upon them, run as follows:

Rule II. Therefore to the same natural effects, we must, as far as possible assign the same causes. As to respiration in a man and in a beast; the descent of stones in *Europe* and in *America;* the light of our culinary fire and of the sun; the reflection of light in the earth and in the planets.[110] Rule III. The qualities of bodies which admit neither intension nor remission of degree, and which are found to belong to all bodies within the reach of our experiments, are to be esteemed the universal qualities of all bodies whatsoever. For since the qualities of bodies are only known to us by experiments, we are to hold for universal all such as universally agree with experiments; and such as are not liable to diminution can never be quite taken away.[111] We are certainly not to relinquish the evidence of experiments for the sake of dreams and fictions of our own devising;

nor are we to recede from the analogy of Nature, which uses to be simple, and always consonant with itself.[112]

The First Rule is simply the familiar rule of simplicity, together with the precept about *verae causae* discussed above. The Second Rule is one of various ways of stating the principle of the uniformity of nature; in this case, with special reference to the use of that principle as the foundation of conclusions by analogy. The Third Rule seems to be no more than a reformulation of the first two, with reference to another set of conclusions.[113] It seems obvious that in bringing before our minds these notions of the simplicity and uniformity of nature, Newton has in mind not so much any general theory of the matter, as the utility of these principles for the defense of his own system against various forms of objection.[114] He nowhere gives us any more definite statement concerning either the precise meaning of these rules, or the grounds upon which we are justified in assuming their truth.[115] Neither does he give us any further analysis of analogical arguments, nor any more detailed account of the nature of inductive generalizations, the principles and methods involved, or the canons of their validity.

CHAPTER SEVEN

David Hume on Causation

WHATEVER THE opinion one may hold concerning the intrinsic value of the views on causation propounded in Hume's *Treatise of Human Nature* (1739), that work must at all events be acknowledged as one of very great importance not only in the history of general philosophy, but also in the history of inductive theory. The influence on some of the most noteworthy writers on induction of the philosophical position defined by Hume, and of the doctrine of causation he deduces from it, has been profound. A consideration of his views is therefore in order here even though he does not write on induction specifically, nor make use of the word.

Hume's doctrine of causation may be summarized as follows. All the human mind ever knows is its own perceptions. These are "impressions," i.e., sensations, feelings, and emotions; and the faint images of these we call "ideas"—every *simple* idea being derived from some *simple* impression, which it copies. Now, analysis of our idea of causality finds therein the idea of *contiguity*, and the idea of *succession*, of cause and effect. These two ideas are readily traceable to the external impressions in which they originate. But our idea of *necessary connection*, which is also an intrinsic constituent of our idea of causality, is not similarly traceable to any external impression. In no case of causation is any *connection* observable between cause and effect, but only their contiguity and succession. *Objectively* considered, therefore, causality means only regularity of contiguous sequence. What connection there is, is purely subjective. It is a connection between our thought of one event and our thought of another event: having repeatedly observed that one event followed a certain other, we have thereby acquired a habit of mind, viz., the "propensity, which custom produces, to pass from an object to the idea of its usual attendant."[1] Necessity is a word that refers only to the experience of this propensity in ourselves under such circumstances; it "exists in the mind, not in objects."[2]

With reference to the doctrine just outlined, it is well to note first that Hume's phenomenalism deprives him, as he himself distinctly observes, of the right to the distinction between "perceptions," or "the mind," and "objects."[3] After demonstrating the invalidity of the distinction, however, he declares that he will go on using it none the less.[4] The obvious reason for this is not only, as he observes, that we all instinctively use it, but that he cannot possibly do without it, for without it his conclusions concerning causality *could not even be stated.*

Again, even allowing him the use of the distinction, he himself furnishes us with a principle of objective reference, from the application of which the objective validity of the relation of necessity would follow: "Wherever ideas are adequate representations of objects, the relations, contradictions and agreements of the ideas are all applicable to the objects."[5] It is by means of this principle that he concludes from the only finite divisibility of our *ideas* of space and time, to the only finite divisibility of *objective* space and time. If the principle is good in this case to pass from ideas to objects, why not likewise in the case of causation? Nowhere does Hume contend that the ideas between which the relation of "Necessity" exists are not as adequate representations of objects as our ideas of space and time.

A third point, also worthy of mention, is this: when seeking for the internal impression from which the idea of necessity is derived, Hume tells us that it is "that propensity, which custom produces, to pass from an object to the idea of its usual attendant."[6] We may reply, however, that in a phenomenalistic world such as is Hume's, there cannot be—even in the "subjective" half of that world—such a thing as a "propensity," in the accepted, essentially dynamic meaning of the term. In such a world there can be only perceptions, i.e., experiences, feelings—things in short that *are,* but do not *do.* One of them may indeed be named the feeling "of propensity," but, no matter what it is called, it is just as discrete, just as little capable of any connective activity (whereby to explain the entrance of the idea of the "usual attendant"), just as much itself in need of connection with others, as the feeling of pain, the feeling of warmth, or the feeling of dizziness.

But no matter how numerous the inconsistencies that are to be found in Hume's pages, his position concerning the nature of causation, in what those who are not wedded to phenomenalism have a right to call the *objective* sense of the term, is quite clear-cut: causation is nothing but regular conjunction of phenomena. "The constant conjunction of objects," he tells us, "determines their causation . . . that constant conjunction, on which the relation of cause and effect totally depends."[7] Now, there are several passages in which Hume asserts that it is sometimes possible to discover a causal connection by means of *a single* experiment. How a single experiment can inform one whether two phenomena *have constantly* occurred in succession is by no means apparent. One would on the contrary expect that, once the causal relation is defined as merely such a *de facto* constant conjunction of two phenomena, the only way to ascertain its presence or absence would be to observe the first phenomenon *a considerable number of times*, and note whether or not the second followed constantly. In other words, with causality so defined, one would suppose that only the *extensive* method of investigation, namely the statistical, which counts instances instead of analyzing them, would be applicable; and that the *intensive* method, on the other hand, namely the experimental, which analyzes, instead of counting, its facts, could have no relevancy at all. But let us see what Hume has to say. The passages referred to are, first:

We may attain the knowledge of a particular cause merely by one experiment, provided it be made with judgment, and after a careful removal of all foreign and superfluous circumstances. Now as after one experiment of this kind, the mind, upon the appearance either of the cause or the effect, can draw an inference concerning the existence of its correlative; and as a habit can never be acquired merely by one instance; it may be thought, that belief cannot in this case be esteemed the effect of custom. But this difficulty will vanish, if we consider, that tho' we are here supposed to have had only one experiment of a particular effect, yet we have many millions to convince us of this principle; *that like objects, placed in like circumstances, will always produce like effects*; and as this principle has established itself by a sufficient custom, it bestows an evidence and firmness on any opinion, to which it can be applied.[8]

More explicit still is the following rather long passage from the section entitled "Rules by which to judge of Causes and Effects":

1. The cause and effect must be contiguous in space and time.

2. The cause must be prior to the effect.

3. There must be a constant union betwixt the cause and effect. 'Tis chiefly this quality that constitutes the relation.

4. The same cause always produces the same effect, and the same effect never arises but from the same cause. This principle we derive from experience, and is the source of most of our philosophical reasonings. For when by any clear experiment we have discovered the causes or effects of any phenomenon, we immediately extend our observation to every phenomenon of the same kind, without waiting for that constant repetition, from which the first idea of this relation is derived.

5. . . . where several different objects produce the same effect, it must be by means of some quality which we discover to be common amongst them. For as like effects imply like causes, we must always ascribe the causation to the circumstance wherein we discover the resemblance.

6. . . . the difference in the effects of two resembling objects must proceed from that particular in which they differ. For as like causes always produce like effects, when in any instance we find our expectation to be disappointed, we must conclude that this irregularity proceeds from some difference in the causes.

7. When any object encreases or diminishes with the encrease or diminution of its cause, 'tis to be regarded as a compounded effect, derived from the union of the several different effects, which arise from the several different parts of the cause. The absence or presence of one part of the cause is here supposed to be always attended with the absence or presence of a proportionable part of the effect. This constant conjunction sufficiently proves that the one part is the cause of the other. . . .

8. . . . an object, which exists for any time in its full perfection without any effect, is not the sole cause of that effect, but requires to be assisted by some other principle, which may forward its influence and operation. For as like effects necessarily follow from like causes, and in a contiguous time and place, their separation for a moment shows, that these causes are not complete ones.[9]

The first of the two extracts just quoted is concerned, primarily at least, with the subjective aspect of causation; it tries to show how the *belief* in the existence of a second phenomenon, on the appearance of a first—and that "propensity" to pass from the one to the idea of the other, upon which the belief rests—can become established after a single experiment. Whether the belief so established is no mere belief, but knowledge, i.e., whether the belief represents an objective fact, the passage quoted does not directly consider, but the use of the word "knowledge" at the beginning of it makes it appear that such was Hume's meaning.

The other passage, however, has unambiguous reference to causation in the objective sense. The fifth, sixth, and seventh of the rules quoted, although they receive practically no discussion, are sufficient to show Hume's understanding of the principles of agreement, difference, and concomitant variation. But, according to the explanations contained in the rules, the possibility of discovering by means of these three principles objective causal relations (in the Humean sense of "constant conjunctions of objects") rests on the uniformity of nature. Therefore, to determine whether the principle of the uniformity of nature does provide Hume with the basis of inference that he needs here, we must consider that principle from the standpoint of his philosophy.

At the beginning of Rule Four, he gives a statement of it that is unfortunately worded: "The same cause always produces the same effect, and the same effect never arises but from the same cause." When it is remembered that the words "cause" and "effect" which occur in it have themselves been defined by Hume no otherwise than in terms of the sameness of antecedents and sequents, such sameness being exactly what *constitutes* the "causal" relation, the principle as so worded is seen to be nothing but an empty tautology. Its application would *presuppose* the knowledge that two phenomena stand to each other as cause and effect; and if this were already known the principle would not need to be applied at all.

There are other statements of the principle in Hume's text, however, that are a little more satisfactory in point of wording. One of them has already been quoted: "Like objects, placed in like circumstances, will always produce like effects." But even this, to become free from inconsistencies with Hume's general position, must be modified. Eliminating from it the word "effects," and also the dynamic verb "produce," [10] and replacing, as a strict empiricist must, the future tense by the past, one would have: *Like objects, placed in like circumstances, have always been attended by like sequents.* This only would be a strictly Humean statement of the principle of the uniformity of nature, and in *this* form it could indeed be derived from experience, as he claims it to be. But in this form it is obviously quite useless as a means of inferring a causal relation between two phenomena on the basis of a single experiment. And experience never can furnish a basis for a statement of the principle containing more than the one just given—containing, namely, *that future tense*, which would be needed for the inference Hume wants to make. And this, indeed, was demonstrated at length and with unsurpassable clearness by Hume himself in a previous section of the *Treatise* where he wrote as follows:

> The nature of experience is this. We remember to have had frequent instances of the existence of one species of objects; and also remember that the individuals of another species of objects have always attended them, and have existed in a regular order of contiguity and succession with regard to them.[11] . . . Since it appears that the transition from an impression present to the memory or senses to the idea of an object, which we call cause or effect, is founded on past *experience* and on our remembrance of their *constant conjunction*, the next question is, whether experience produces the idea by means of the understanding or of the imagination; whether we are determined by reason [i.e., by a logical, objectively valid process] to make the transition, or by a certain association and relation of perceptions [i.e., by a psychological, merely subjective process]. If reason determined us, it would proceed upon that principle, *that instances of which we have had no experience, must resemble those of which we have had experience, and that the course of nature continues always uniformly the same.* [But] . . . our foregoing method of reasoning will easily convince us, that there can be no *demonstrative* arguments to prove *that those instances of which we have had no experience resemble those of which we have had experience.*[12]

Nor can this be regarded as established by an argument from probability, for "probability is founded on the presumption of a resemblance betwixt those objects of which we have had experience, and those of which we have had none; and therefore 'tis impossible this presumption can arise from probability. The same principle cannot be both the cause and effect of another." [13]

Thus, the proposition concerning the uniformity of nature that can be obtained from experience is radically different from the proposition that Hume needs for the purpose of his rules and states at the beginning of the fourth rule.[14] The latter, as just shown in his own words, is not and cannot be obtained from experience. His assertion in the fourth rule that it is so obtained is erroneous and in flagrant contradiction with his earlier carefully argued demonstration.

Hume's clear statement of the principles of agreement and difference as means of judging of causal connections is dictated by his common sense, which there gets the better of his metaphysics and leads him to the needed, but unwarranted, formulation of the uniformity of nature just discussed. That he felt more or less clearly the inconsistency between his doctrine of causation and the use of the principles of agreement and difference is perhaps shown by the fact that, while just before coming to his rules he refers to "that constant conjunction, on which the relation of cause and effect totally depends,"[15] in the third rule, on the other hand, he no longer says "totally," but only "chiefly."

Of course, when Hume asserts that no necessary connection but only regularity of conjunction is objectively observable as between cause and effect, he has in mind the distinction—with which he opens section 4 of the *Enquiry concerning Human Understanding* but which he had already made in the *Treatise*[16]—between *relations of ideas* and *matters of fact*. Mathematics and logic are concerned with relations of ideas, and it is among these alone that necessity and impossibility are demonstrable independently of experience. Among matters of fact, on the contrary, only experience tells us what can and what cannot cause what. Here, "necessity" and "impossibility" do not, as in the case of relations of ideas, mean implication and contradiction. Instead, according to Hume, "necessity" means the felt "propensity, which custom produces, to pass from an object to the idea of its usual attendant."[17] And a similarly psychological account of "impossibility," but in terms of felt inhibition instead of felt propensity, could be offered, although Hume does not actually offer it.

The association of the object with its "usual attendant," however, may not have been wholly constant in our experience but only have, in one degree or another, approximated constancy. The propensity, which such experience has produced in us, to pass to the idea of the attendant upon experiencing the object again—i.e., our belief that future experience will resemble past experience—is then weaker in corresponding degree. And the subjective impression, which the feeling of this weaker propensity or belief constitutes, is the impression from which the idea of *probability* as distinguished from that of necessity is derived, where matters of fact are concerned.

It may be noted here, however, that, as against Hume's assertion that necessity is never objectively observable between cause and effect, one may well contend that necessity—though not *logical* necessity, which is what Hume had in mind in his assertion—*is* observable there. That is, besides logical necessity, as between "ideas," there is also *etiological* necessity, as between "facts"—etiological necessity being defined in terms of the particular kind of situation that the experimental method of difference stipulates, and, ideally, being observable in a single experiment without theoretical need of repetition. More specifically, an occurrence *C* can be said to have caused, i.e., etiologically necessitated, a contiguously sequent occurrence *E*, if *C* and *E* were the *only* two changes that occurred in the then existing situation *S*. Etiological necessitation, as so defined, is thus not a dyadic but an irreducibly triadic relation. And the fact is that when we observe or think we observe not just sequence of whatever type, but sequence *of this specific type*, we do term it consequence, and do say that the first event *made* the second occur, i.e., *necessitated* its occurrence. Just this, it may be urged, is what we do mean by necessity as between events, distinguished from necessity as between concepts; so that the assumption that "necessity" has no genuine meaning other than that of logical necessity is wholly gratuitous.

Moreover, as in the case of necessity and of the account of objective, etiological necessity just suggested, so, in the case of *probability*, an account of it in objective terms suggests itself readily as an open alternative to Hume's subjective account—for example, an account such as that which the frequency theory of probability offers; or an account in terms of how close is the resemblance of a present instance to a similar one we observed in the past.

Summing up now our discussion of Hume's contribution, we find that in regard to the history of the term "induction" he contributes nothing, since he does not use it. With regard to inductive methods, he gives in his rules (though as we have seen inconsistently with his own philosophy) the clearest statement up to his time, and for many years after him, of the methods of agreement, difference, and concomitant variations. Again, when he tells us that causal relations are revealed by the observation of constant conjunctions, he brings into prominence the method that underlies statistical investigations. Last, he recognizes explicitly that the validity of generalizations depends on the uniformity of nature; he states clearly the manner in which that uniformity must be formulated in order to warrant generalizations; and he demonstrates that the principle in that form is not obtainable from experience. His later implicit disregard of his own conclusions in this respect in no wise detracts from the value of the earlier discussion in bringing to the focus of attention the important issues that are bound up for inductive theory with the uniformity of nature.

CHAPTER EIGHT

John F. W. Herschel's Methods of Experimental Inquiry

In sir John Frederick William Herschel's *Preliminary Discourse on the Study of Natural Philosophy*, published in 1830, we have, the late Professor Minto points out, "the first attempt by an eminent man of science to make the methods of science explicit." [1] The *Discourse*, however, attempts something more than this. It formulates Herschel's philosophy of science. Although its discussion of scientific method is doubtless its most significant part, it considers also the nature of the objects of the natural sciences, the relation of such objects to the cognitive powers of the human mind, the tasks of these sciences, and the values for human society of the study of science.

The work is divided into three parts. The third, which deals with "the subdivision of physics into distinct branches, and their mutual relations," adds little or nothing to the theory of science contained in the first and second parts and may therefore be left out of consideration here. The first part is entitled "Of the general nature and advantages of the study of the physical sciences." What it has to say concerning the objects studied by these sciences and the questions they ask about those objects will be considered in connection with the contents of Part II. The point of chief interest otherwise in Part I is found in what Herschel has to say about the effects of advances in science on the improvement of the condition of mankind. The "more abundant supply of our physical wants, and the increase of our comforts" are not the only or even the greatest benefits resulting from these advances:

The incalculable advantages which experience, systematically consulted and dispassionately reasoned on, has conferred in matters purely physical, tend of

> necessity to impress something of the well weighed and progressive character of science on the more complicated conduct of our social and moral relations. It is thus that legislation and politics become gradually regarded as experimental sciences; and history . . . as the archive of experiments, successful and unsuccessful, gradually accumulating towards the solution of the grand problem—how the advantages of government are to be secured with the least possible inconvenience to the governed.

The power that we now know a scientific approach to any problem confers may ultimately enable us "to bear down those obstacles which individual shortsightedness, selfishness, and passion, oppose to all improvements, and by which the highest hopes are continually blighted, and the fairest prospects marred." [2]

Part II of the *Discourse*, in seven chapters, sets forth and illustrates "the principles on which physical science relies for its successful prosecution, and the rules by which a systematic examination of nature should be conducted." It is chiefly to this that we must turn for Herschel's philosophy of science, which we propose to outline and examine in what follows.

Herschel's direct predecessor in the attempt to formulate a philosophy of natural science is Francis Bacon, whose best-known work, the *Novum organum scientiarum*, appeared some two hundred years earlier (1620). It should be mentioned at the outset that Herschel was thoroughly familiar with this celebrated work and thought highly of the doctrine contained therein. Indeed, facing the title page of the *Discourse*, we find transcribed the famous first aphorism of the *Novum organum*, "Man, the servant and interpreter of nature, does only so much and understands only so much about the order of nature as he has observed with the mind or the sight; he knows no more, nor can he"; and on the next page there is reproduced an engraving of a bust of Bacon.[3] Laudatory references to Bacon and sympathetic interpretations of the teachings of the *Novum organum* are found in many parts of the *Discourse*, which in the history of the philosophy of science may well be regarded as the direct descendant of the *Novum organum*. Almost every one of the general precepts of method which Bacon formulated but could not adequately illustrate reappears in the *Discourse* and, interpreted in the light of the concrete scientific achievements of the intervening two centuries, is given its true perspective and clearly related to actual and fruitful scientific practice.

It may not be amiss here to call attention to one or two passages in the third chapter of the second part of the *Discourse,* which are of importance in throwing light on the place of Francis Bacon in the history of the philosophy of science. Since Herschel is himself one of the most notable writers on scientific method, his mere opinions as to Bacon's importance in the history of the subject constitute so many facts in evidence concerning the influence Bacon has actually had on one at least of his successors in this field. Referring to the process he like other writers calls *induction,* which he describes in section 95, Herschel says:

It is to our immortal countryman Bacon that we owe the broad announcement of this grand and fertile principle; and the development of the idea that the whole of natural philosophy consists entirely of a series of inductive generalizations, commencing with the most circumstantially stated particulars, and carried up to universal laws, or axioms, which comprehend in their statements every subordinate degree of generality, and of a corresponding series of inverted reasoning from generals to particulars, by which these axioms are traced back into their remotest consequences, and all particular propositions deduced from them. . . .[4]

Further on in the same chapter Herschel writes:

By the discoveries of Copernicus, Kepler and Galileo, the errors of the Aristotelian philosophy were effectually overturned on a plain appeal to the facts of nature; but it remained to show on broad and general principles, how and why Aristotle was in the wrong; to set in evidence the peculiar weakness of his method of philosophizing, and to substitute in its place a stronger and better. This important task was executed by Francis Bacon, Lord Verulam, who will, therefore, justly be looked upon in all future ages as the great reformer of philosophy, though his own actual contributions to the stock of physical truths were small, and his ideas of particular points strongly tinctured with mistakes and errors, which were the fault rather of the general want of physical information of the age than of any narrowness of view on his own part; and of this he was fully aware . . . it is not the introduction of inductive reasoning, as a new and hitherto untried process, which characterizes the Baconian philosophy, but his keen perception, and his broad and spirit stirring, almost enthusiastic, announcement of its paramount importance, as the alpha and omega of science. . . .[5]

Estimates of the importance of Bacon's writings on scientific method have been widely divergent—some judging them negligible and others pronouncing them epoch-making. For lack of space in these pages to adduce the relevant evidence, the writer can only record his opinion that Herschel's estimate of both the merits and the defects of Bacon's contribution is much more just than either of those extremes.

Another influence that can be traced in the *Discourse* is that of the doctrine of method briefly formulated by Newton in a few passages of the *Principia* and of the *Opticks* and so brilliantly applied in his own scientific work.[6] This influence is apparent, for example, in the stress Herschel lays on the method of hypothesis—deduction—verification. It is true that, notwithstanding the common opinion to the contrary, Bacon, before Newton, had described and advocated this method; but, of course, Bacon did not have the clear understanding possessed by Newton of the manner in which this method is to be employed in practice. It is the example of Newton's *use* of it, much rather than the brief description of it in the *Opticks*, or the statements of Bacon, which brought about the adequate recognition of the great importance of that method.[7] Newton's influence is also apparent in the many references in the *Discourse* to the requirement that the causes to which one appeals be *verae causae*. Herschel's interpretation of the meaning of this term, however, is somewhat broader than that which Newton appears to have intended to place upon it.

The examination of Herschel's views we shall now undertake will consist of an exposition of and commentary on the contents of Part II of the *Discourse*. A more systematic treatment than would otherwise be possible will result if we allow ourselves to deviate at times from the order followed by Herschel. We shall therefore arrange our exposition under the following five headings: (1) Observation, Experiment, and Classification; (2) The Uniformity of Nature; (3) The Analysis of Phenomena; (4) The First Stage of Induction; (5) The Second Stage of Induction.

1. OBSERVATION, EXPERIMENT, AND CLASSIFICATION

Part II of the *Discourse* opens with some remarks repeating in effect those made by David Hume in section 4 of his *Enquiry concerning Human Understanding*, where he points out that the objects of inquiry are in all cases either "relations of ideas" or "matters of fact." Inquiry into the first, Hume says, gives us the mathematical sciences, whose conclusions have demonstrated certainty. Inquiry into the second begets the sciences of nature. Their inferences are not demonstrative for they are all based on the relation of cause and effect, and we obtain knowledge of this relation not through reason but only through experience. Herschel similarly contrasts the abstract with the natural sciences. The truths of the abstract sciences, e.g., of mathematics, can be arrived at independently of perceptual observation because such truths are necessary in the sense that denial of them involves one in self-contradiction. But in natural science, where we are concerned with causes and effects, and with laws which "for aught we can perceive might have been other than they are," [8] the great and ultimate source of our knowledge must be human experience.[9] This can be acquired in two ways: either by the mere passive *observation* of facts as they occur—but, when for any reason science is limited to this, progress is usually very slow—or actively by *experimentation*, that is, "by putting in action causes and agents over which we have control, and purposely varying their combinations, and noticing what effects take place." [10] Whenever this has been possible the progress of science has been rapid.

"Experience once recognized as the fountain of all knowledge of nature," we must clear our minds of all prejudices concerning what might or what ought to be the order of nature in any proposed case and determine to stand and fall by the result of direct appeal to facts and of strict logical deduction from them. Such prejudices are of two sorts: prejudices of opinion and prejudices of sense. The first are false opinions which we have come to receive without adequate evidence, and which, from being constantly admitted without dispute, have obtained the strong hold of habit on our minds. The second are not so much errors of sense as judgments in which we mean to report no more than we actually sense, but in which we in fact assert much more —and this more, false—without realizing that we are doing so until appropriate checks are used. These remarks of Herschel's are obviously reminiscent of Bacon's discussion of the "Idols."

The fourth chapter of the second part of the *Discourse* is concerned with "the observation of facts and the collection of instances." Herschel there declares after Bacon that

> . . . whenever . . . we would either analyze a phenomenon into simpler ones, or ascertain what is the course or law of nature under any proposed general contingency, the first step is to accumulate a sufficient quantity of well ascertained

> facts, or recorded instances, bearing on the point in question . . . the more different these collected facts are in all other circumstances but that which forms the subject of enquiry, the better; because they are then in some sort brought into contrast with one another in their points of disagreement, and thus tend to render those in which they agree more prominent and striking.[11]

In the remainder of the chapter, Herschel describes and illustrates the characters such recorded instances must have, to be of scientific value. The facts, and also the circumstances that were present when the facts occurred, must be accurately observed; and the record then made of them must be literal and faithful in the sense of not allowing what are really inferences, however natural or plausible, to creep into the record. Again, precise numerical statements should be given in every case where they are possible, and they should be obtained from actual counting or measurement, since the quantitative reports of the unaided senses are untrustworthy. It is true that even with fixed standards (e.g., of length, weight, etc.) some error unavoidably enters into our measurements, but we can always assign a limit that such error cannot possibly exceed. Moreover, we can increase the accuracy of our measurements almost without limit by repeating them many times and taking their mean.[12] Finally, when our laws have been formulated on the basis of them, we can find out whether our formulations are vitiated by errors of observation or, on the contrary, are trustworthy, by deducing predictions from these laws and observing whether or not the facts verify them.

The fifth chapter is concerned with classification and nomenclature. The number of variety of objects and relations that observation reveals in nature is so great as to confuse and bewilder us unless we limit our observation at any one time to a few facts, or to a number of facts so bound together by resemblance as virtually to constitute one fact. Any such fact that appears important should receive a name. This will insure to it a correspondingly important place in our thought and will serve there as a nucleus around which further information that we obtain may be systematically organized. As soon, however, as a considerable number of objects in a given field, and their relations, have been observed, arbitrary proper names will no longer serve. We must have names essentially relational, which will not only serve to identify the objects but will at the same time indicate their relations to others of the given field; so that "the direct relation between the name and the object shall materially assist the solution of the problem '*given the one, to determine the other.*'"[13] Now, "any one may give an arbitrary name to a thing, merely to be able to talk of it; but, to give a name which shall at once refer it to a place in a system, we must know its properties; and we must *have* a system, large enough, and regular enough, to receive it in a place which belongs to it and to no other."[14] Nomenclature cannot therefore usefully at any time be given much more system and precision than our knowledge itself possesses; else we run the risk of mistaking the means for the end by sacrificing convenience and distinctness to a rage for arrangement.

2. THE UNIFORMITY OF NATURE

Attention must now be called to a character which Herschel, in the fourth chapter, states facts must have if they are to possess any scientific value, and which is of a very different order from the others already mentioned. Herschel's words on the point are as follows: "The only facts which can ever become useful as grounds of physical enquiry are those which happen uniformly and invariably under the same circumstances . . . for if they have not this character they cannot be included in laws."[15] Whether a given fact is anomalous only in appearance, owing to our having failed to detect some difference in the circumstances; or whether it is really anomalous, owing to caprice, i.e., owing to "the arbitrary intervention of mental agency"—in either case all we can do with anomalous facts is to record them as curiosities or as problems awaiting explanation, but "we can make no use of them in scientific enquiry."[16]

This character of being regular is obviously not one that is a possible matter of observation in any individual instance, and it cannot therefore be used as a guide in the selection of instances. The practical consequence Herschel draws from the above remarks is only the one already mentioned, namely, that not merely facts themselves, but also the circumstances under which they occurred must be carefully observed and recorded. But the broader import of his remarks as above quoted concerns the status of the so-called principle of the uniformity of nature. They present uniformity, in effect, as a postulate definitive of the scope of science. Herschel does not say that we know nature to be wholly or partially ruled by law; neither does he directly say that we have to assume it to be so for practical purposes. The import of his words is merely that the realm of science extends so far and so far only into nature as it may happen to have regularity. This follows directly from his view that science essentially consists of a body of general propositions. Science is the study of such phenomena as are governed by laws.

3. THE ANALYSIS OF PHENOMENA

We now come to what Herschel, in the second chapter of Part II of the *Discourse*, calls "the analysis of phenomena." A large portion of the chapter is given to the discussion of some of the terms in which his methodological doctrine is stated, and of some of the assumptions that underlie it. An examination of these terms and assumptions, as considered by Herschel at this and other places in the *Discourse*, is indispensable to an understanding of his doctrine and in particular has to be more or less closely combined with any critical exposition of the part of it that concerns the analysis of phenomena.

The term *phenomenon* is explicitly defined by Herschel as follows: "Phenomena . . . are the sensible results of processes and operations carried on among external objects, or their constituent principles, of which they are only signals." [17] He explains what he means by this in the following words:

> As the mind exists not in the place of sensible objects, and is not brought into immediate relation with them, we can only regard sensible impressions as signals conveyed from them by an . . . inexplicable mechanism to our minds, which receives and reviews them, and by habit and association, connects them with corresponding qualities or affections in the objects.[18]

This situation Herschel compares to that of a person who, having constantly observed that a certain telegraphic signal was sure to be followed by the announcement of the arrival of a ship, would connect the two facts by a link of the same nature as that which connects the notion of a large floating building filled with sailors with the impression of the outline of a ship on the retina of a spectator on the beach.[19]

Herschel here obviously deals in summary fashion with the classic philosophical problem of our knowledge of the "external world." Since he was writing as a methodologist of science, it was of course not incumbent upon him to undertake a full discussion of that problem; but, since he nevertheless touches upon it and attempts to state his methodological doctrines in terms of the answer to it that he sketches, it is necessary at least to make as explicit as possible the import of his words on the subject.

Herschel intends to use the terms "sensible impression" and "phenomenon" synonymously; also, the terms "objects" and "sensible objects" as synonymous. "Sensible object" seems to mean in Herschel's usage an object *qua* causing or susceptible of causing an impression, i.e., a sensation: "We know nothing of the objects themselves which compose the universe, except through the medium of the impressions they excite in us, which impressions are the results of certain actions and processes in which sensible objects and the material parts of ourselves are directly concerned." [20] The sensation or impression, then, constitutes the "signal" of the existence and of the nature of the object, in the sense in which any effect observed may be described as a signal for us of its cause. The objects themselves are declared by Herschel not to be known by us otherwise than through the sensations they cause in us. The nature of the objects, as distinguished from that of the sensations that signify them, is known to us, he seems to hold, because the qualities and affections of the objects "correspond" to the sensations in us. But these "corresponding" qualities in the object are, so far as can be gathered from Herschel's words, merely such qualities as in our minds have become habitually *associated* with given sensations. The question then arises, whence our ideas of these *objective* qualities if our sensations are all we ever directly experience. The status of associate of some sensation of ours could be acquired by an objective quality only *after* we had had experience of the quality itself, and this he believes we never have, since he holds that all we ever are directly acquainted with are our own sensations. It would seem then that, if this is the situation, the nature of the objects must forever remain a mere X to us; i.e., all we can know of them is that they are the sorts of entities, if any, of which our sensations are signs.

Herschel's illustration of the telegraph and the ship does not clarify the situation, for the analogy between it and the relation of "phenomena" to "objects," as the latter is abstractly described by him, does not hold. In the case of the telegraphic signal, *both* the sight of the motion of the telegraph and the sight of the arrival of the ship have been "sensible impressions." We have thus here two terms first *given*, and *then* a relation of association established between them. In Herschel's abstract description, however, we are supposed to *construct* the second term purely out of a relation (habitual association) of it to the first, notwithstanding that this relation can come to exist only *after* the experience of the second independently of the relation.

With regard to the other illustration—the connection between the "notion" of a large floating building filled with sailors and the impression of the outline of a ship on the retina of the spectator on the beach—the following may be said. If Herschel means by "the outline on the retina" just what he says, viz., the same sort of physical fact in the eye as occurs in any other physical camera and not the *sensation* this fact in the eye causes in the mind, then the "connection" between the outline on the retina and the "notion" of a large floating building is something very different from the sort of connection called association of ideas. It is, namely, connection between the stimulation of nerve endings and the occurrence of mental states. If, on the other hand, Herschel means by "the outline on the retina" the sensation caused in the mind by the presence of this outline on the back of the eye, then that sensation can become associated with the "notion" of a large floating building only if this notion has already been acquired; and the question then remains *how* it was acquired. Obviously, association between the sensation and the notion is not a possible answer to it. Herschel's statements quoted above thus wholly fail to provide an intelligible account of the relation of phenomena to objects.

When, however, we pass to actual cases of the "analysis of phenomena" as offered by Herschel, we find another story, which must be taken to represent his true thought much more than does the short and precarious venture into epistemology we have just examined. The "analysis of phenomena" in these concrete cases turns out to be not as first announced a "rendering sensible" of objective processes that the mind was assumed incapable of sensing, but merely an analysis of complex objects or objective processes into simpler ones. These, however, are referred to by Herschel at a number of places in the chapter as themselves "phenomena"—inconsistently, let it be noted, with his initial definition of phenomena as the sensible *results in the mind* of processes and operations among external objects. For example, in section 79, the propagation of motion from one body to another is termed a "phenomenon." At most places, indeed, Herschel seems to use the word "phenomenon" to mean any objective event susceptible of being somehow "observed." Moreover, the relation between a complex phenomenon (e.g., the sensation of sound) and the simpler "phenomena" into which it is analyzed by Herschel (e.g., the rapid vibratory motion of the parts of a bell, and its communication to the air, and by the air to our eardrum, ossicles, etc., and ultimately to the auditory nerve—this then exciting sensation) turns out to be not of the nature of an association of our ideas resulting from habit, but a causal relation among external objects, where one process causes another *to occur*, not merely to be thought of.

Discussing the "analysis" of the phenomenon of sound just given, Herschel goes on to say that "two other phenomena, of a simpler, or, it would be more correct to say, of a more general or elementary order,[21] into which the complex phenomenon of sound resolves itself" are brought to light: (1) the excitement and propagation of motion, and (2) the production of sensation. Again, the communication of motion from body to body is resolvable into several other "phenomena": the original setting in motion of a material body; the behavior of a moving particle when it comes into contact with another; the behavior of that other—the last two in turn pointing to another phenomenon that must also be considered, namely, the connection of the parts of material masses, by which the parts influence each other's motion.

Thus . . . an analysis of the phenomenon of sound leads to the enquiry, first, of two *causes*, viz., the cause of motion, and the cause of sensation, these being phenomena which (at least as human knowledge stands at present) we are unable to analyze further; and therefore we set them down as simple, elementary, and referable . . . to the immediate action of their causes.

And, second, to "several questions relating to the connection between the motion of material bodies and its cause, such as, *What will happen* when a moving body is surrounded on all sides by others not in motion? *What will happen* when a body not in motion is advanced upon by a moving one?"[22] The answer to such questions consists in *laws of motion,* in the sense attributed by Herschel to a law of nature, viz., "a statement in words of what will happen in such and such general contingencies." Last, we are led to the two other general phenomena of the cohesion and the elasticity of matter, which, until means are found to analyze them further, must be regarded "as *ultimate phenomena,* and referable to the direct action of causes, viz., an attractive and a repulsive *force.*"[23]

The above concrete picture of the "analysis of a phenomenon" makes clearer what Herschel means by the phrase than his abstract description does. But it makes clear also that what is then analyzed is not a "phenomenon" in the sense he has specified for this term. To complete the account of his views on this particular part of the task of science, we may add his declaration, in section 88, that no general rules can be given for the analysis of phenomena "any more than . . . general rules can be laid down by the chemist for the analysis of substances of which all the ingredients are unknown." As to the utility of the analysis of phenomena in the development of science, Herschel goes on to say that it principally

> . . . enables us to recognize, and mark for special investigation, those [phenomena] which appear to us simple; to set methodically about determining their laws, and thus to facilitate the work of raising up general axioms, or forms of words, which shall include the whole of them;[24] which shall, as it were, transplant them out of the external into the intellectual world, render them creatures of pure thought, and enable us to reason them out *a priori,*

this often leading to the discovery, by deduction, of previously unsuspected particular facts.

Herschel's use of the terms "cause," "force," and "law" now calls for separate examination. There seem to be some four distinguishable senses in which the term "cause" is employed by Herschel in the *Discourse;* but, as he does not appear to have realized that he was in fact using the term in more than *two* different senses, the four, in spite of their distinctness from one another, can hardly be exhibited by means of wholly separate quotations from his work. There will accordingly be a certain amount of overlapping in our remarks concerning each of them.

The first of the four is that in which a cause is conceived as an *act of will*. According to Herschel there is one case and only one where we have an "immediate consciousness" of an "act of direct *causation*,"[25] and this is "the production of motion by the exertion of force"[26]—"we feel within ourselves a *direct* power to produce" motion of our limbs. But our *original* impression of the nature of force is independent of any perception of motion; it is received from "our own effort and sense of fatigue," for when, for instance, "we press our two hands violently together, so as just to oppose each other's effort, we still perceive, by the fatigue and exhaustion, and by the impossibility of maintaining the effort long, that something is going on within us, of which the mind is the agent, and the will the determining cause."[27] This internal consciousness, Herschel claims, gives us "a complete idea of *force*." That force as so understood is the cause of motion, however, and that motion can therefore be regarded as the *signal* of the exertion of force, is something we come to know only by finding that the same action of the mind, which, when our limbs are encased in plaster enables us to fatigue and exhaust ourselves by the tension of our muscles, puts it into our power, when our limbs are free, to move ourselves and other bodies.[28] The conclusion Herschel draws from these remarks is that, since the *process* by which the exertion of force causes motion of our limbs remains obscure to us even in the case where the force is exerted by ourselves, there is very little prospect that, "in our investigation of nature, we shall ever be able to arrive at a knowledge of ultimate causes." This "will teach us to limit our views to that of *laws*, and to the analysis of complex phenomena by which they are resolved into simpler ones, which appearing to us incapable of further analysis, we must consent to regard as causes."[29]

Here, then, we find Herschel speaking of "the exertion of force" as the cause of motion of our limbs through an intervening process; also, of the *will* as the "determining cause" of the fact that something (viz., muscular tension) is going on within our body. In a passage to be quoted below, moreover, Herschel clearly identifies the exertion of force with the act of volition by stating that, when a nerve is cut, the exertion of force occurs without any muscular tension resulting.

We come now to a second sense of "cause," that in which a cause is conceived as an *ultimately simple phenomenon*. In the passage quoted above Herschel uses the term "ultimate causes" to mean, apparently, the elements that would be revealed by a complete analysis of the process through which the exertion of force (act of will) causes tension or motion of our muscles, if such analysis were possible. An ultimate cause, as so conceived, is thus an ultimate phenomenon—a phenomenon that does not merely seem to us unanalyzable and simple, but really is so. Moreover, the exertion of force (act of will) of which Herschel speaks above is not the cause in *this* sense of what takes place in the muscle, since what he regards as obscure is not the nature of force as exerted by us (of which we have "a complete idea"), but the *process* through which it causes the change in the muscle. We have then already two different senses of the word "cause" in the passages quoted: in one, "cause" means the exertion of force or act of will; and, in the other, "cause" means the simple, unanalyzable phenomena of which the process intervening between the act of will and the resulting motion ultimately consists.

That we have no knowledge of causes in the second of these two senses (ultimate causes), however, is a thesis which assumes that our act of will is not the immediate cause of what occurs in the muscle, but causes it only through an intervening mechanism. And the ground of this assumption in Herschel's mind appears to be that, "when we put any limb in motion, the seat of the exertion seems to us to be *in* the limb, whereas it is demonstrably no such thing, but either in the brain or in the spinal marrow; the proof of which is that if . . . a nerve, which forms a communication between the limb and the brain, or spine, be divided in any part of its course, however we make the effort, the limb will not move."[30] The intervening mechanism of which Herschel thinks can then only be the nerve which connects the place of motion, viz., the muscle, with the place of the exertion of force, viz., supposedly the brain or spinal cord. But Herschel is here on very precarious ground, as will appear from the following considerations.

That the act of will or effort, in Herschel's sense of fact of consciousness directly known to introspection, is *located in* the brain or spinal cord is an assumption the correctness of which is in no way evidenced by the experiment of cutting the motor nerve ending, say, in the muscles of the arm. For one thing, it seems impossible without absurdity to take the relation "being in" or "being the seat of" literally in any case where both its terms are not supposed to be material, space-occupying objects, but, on the contrary, one a material object (the brain) and the other a state of consciousness directly known to introspection (not the hypothetical molecular brain process which "corresponds" to it). The absurdity of taking "being in" literally is of the same sort as that in speaking, say, of transfixing on a pin the feeling of dizziness. It must, then, be taken figuratively and loosely, to mean the relation (parallelism, or interaction, or whatever it may be) that obtains between the psychical act of will and its constant bodily correlate. But this correlate may well be neither the contraction of the muscle (say the biceps) that would occur if the effort "succeeded" nor the molecular processes in the nervous connections of this muscle, but very possibly, on the contrary, either a contraction in some *other* muscle [31] or the molecular process in the nerve connections of this other muscle, left untouched, of course, by the severing of the motor nerve leading to the biceps.

In any case, however, Herschel's volitional analysis of cause would seem to have no applicability to the case of inorganic nature, where no nervous system at all is present in the production of motion. The "exertion of force," let it be remembered, has been defined by Herschel only in terms of volition as directly introspected. Therefore, unless he means to assume that when one billiard ball pushes another it "exerts force," in the sense of having the state of consciousness that consists in willing that the other ball move, the occurrence of motion in inanimate nature cannot be taken as the signal of the "exertion of force" *as defined*. But, if not, force then remains undefined so far as inanimate nature is concerned; and therefore so does the "cause" of motion in inanimate nature if that cause is meant to be something not observed but hidden and inferred.

Herschel declares *both* that the cause of motion in inanimate nature is force, *and* that the forces of nature are never observable, but are only inferred from their effects by means of an analogy to the human will as fact of consciousness. This analogy, however, if it is to hold, presupposes some sort of panpsychism. Herschel seems to have had some awareness of this, but it is not clear whether he accepts it or not. His most explicit statement bearing on the point is this:

Of force, as counterbalanced by opposing force, we have . . . an internal consciousness; and though it may seem strange to us that matter should be capable of exerting on matter the same kind of effort, which, judging alone from this consciousness, we might be led to regard as a mental one; yet we cannot refuse the direct evidence of our senses, which shows us that when we keep a spring stretched with one hand, we feel our effort opposed exactly in the same way as if we had ourselves opposed it with the other hand [italics ours].[32]

If Herschel is not willing to say that the spring *wills* what it then does as we will to stretch it, then "force" as exerted by the spring constitutes a third sense in which he uses the term "cause." Attention to the examples Herschel gives suggests that "force," in the case of a "force of nature," is thought of by him not much more clearly or significantly than as a *deus ex machina.* I have already quoted the end of section 80, where he speaks of the cohesion and elasticity of matter as, so far as we now know, "*ultimate phenomena* and referable to the direct action of causes, viz., an attractive and a repulsive *force.*" From this, the meaning he really attaches to "ultimate cause" can easily be construed: an ultimate phenomenon is one that cannot be further analyzed, and an ultimate cause is the cause of an ultimate phenomenon in the sense of "cause" in which an attractive or cohesive force is the cause of cohesion—which sense, let us add, is also that in which the famous *vis dormitiva* of opium is the cause of sleep! In other words, when we find anything T behaving in a particular manner M under given conditions C, and cannot answer the question why T so behaves by pointing to an intermediary I which is caused by C and itself in turn causes M (that is, by giving an analysis of the phenomenon), we answer by saying: T behaves in manner M because there exists an X such that X causes T to do M in the presence of C. And a *deus ex machina*—an X of this sort defined *merely* as such a cause as would produce an effect such as we suppose it to cause—seems to be all that Herschel, without realizing it, really means by a force in external nature. This is a third sense in which he uses the term "cause," since it is here not as before the ultimately simple phenomena that are spoken of as causes but the nonphenomenal forces to the "direct action" of which these phenomena are to be "referred."

We come now to the fourth and most fruitful of the several conceptions of cause to be found in the *Discourse,* viz., that of *proximate cause,* where that which causes is some phenomenon antecedent in time to its effect. Section 83, in which "cause" in this sense is introduced, opens with these words: "Dismissing, then, as beyond our reach, the enquiry into causes, we must be content at present to concentrate our attention on the laws which prevail among phenomena, and which seem to be their immediate results." Toward the end of the same section Herschel writes as follows:

Thus, in a modified and relative sense, we may still continue to speak of causes, not intending thereby those ultimate principles of action on whose exertion the whole frame of nature depends, but of those proximate links which connect phenomena with others of a simpler, higher, and more general or elementary kind. [For example,] we may regard the vibration of a musical string as the proximate cause of the sound it yields, receiving it, so far, as an ultimate fact, and waving [*sic*] or deferring enquiry into the cause of vibrations, which is of a higher and more general nature.

When cause is so conceived, the two terms of the causal relation are both then phenomena and are one temporally sequent to the other. For example, in section 138 Herschel says:

Experience having shown us the manner in which one phenomenon depends on another in a great variety of cases, we find ourselves provided, as science extends, with a continually increasing stock of such antecedent phenomena, or causes (meaning at present merely proximate causes) competent under different modifications, to the production of a great multitude of effects, besides those which originally led to a knowledge of them.

He goes on to tell us that he has in mind what Newton called *verae causae*, which he takes to mean phenomena "which experience has shown us to exist" and "to be efficacious in producing similar phenomena" (to those they are called upon to explain).[33] He then gives the following three illustrations: the elevation of the bottom of the sea till it becomes dry land, as cause of the presence of shells in rocks at a great height above the sea; the sinking of old continents and the elevation of new, as a possible cause of great changes in the general climate of large tracts of the globe; the increase of the minor axis of the ellipse that the earth describes about the sun, as cause of the diminution of the mean temperature of the surface of the earth.

We may now, for purposes of contrast, recapitulate briefly the four meanings of the term "cause" that we have discerned in Herschel's text. In the first sense examined, the relation of a "cause" to its "effect" is the sort of relation that subsists between an act of will (as known in introspection) and the tension or motion of a muscle that accompanies it. In the second sense, "causes" are the genuinely simple phenomena into which a complex phenomenon is theoretically analyzable. In the third sense, the relation of "cause" to "effect" is that of a postulated noumenal "force" to its phenomenal manifestation, e.g., of "cohesive force" to cohesion as observed, or of "soporific power" to the phenomenon of sleep. Such "causes" are always "forces" postulated *ad hoc;* they are never directly observed but supposedly only "inferred" from their so-called effects. In the fourth sense, the relation of a "cause" to its "effect" is that of an event to another event sequent to it in time and following it "necessarily." What "necessarily" means here, Herschel explains in section 145 when he enumerates "the characters of that relation which we intend by cause and effect." What these are we shall examine later in detail. Suffice it to say here that "invariable connection" is the character most nearly inclusive of its various features. This fourth sense of causation is relevant to Herschel's formulation of the principles which, following J. S. Mill, are called the methods of agreement and of difference, and their derivatives.[34]

Before turning to an examination of them, we must first say a few words concerning the use of the terms "law" and "induction" in Herschel's *Discourse.* A number of passages in which the term "law" occurs have already been quoted. We may recall that the answer to such a question as "What will happen when a body not in motion is advanced upon by a moving one?" constitutes, Herschel says, a "law of motion." Again, in section 89, he says:

A law of nature, being the statement of what will happen in certain general contingencies, may be regarded as the announcement in the same words, of a whole group or class of phenomena. Whenever, therefore, we perceive that two or more phenomena agree in so many or so remarkable points, as to lead us to regard them as forming a class or group, if we lay out of consideration, or *abstract,* all the circumstances in which they disagree, and retain in our minds those only in which they agree, and then, under this kind of mental convention, frame a definition or statement of one of them, in such words that it shall apply equally to them all, such statement will appear in the form of a general proposition, having so far at least the character of a law of nature," [for instance] double refracting substances exhibit periodical colors by exposure to polarized light.

A law of nature, such as this proposition constitutes, may, Herschel goes on, be regarded in one of three ways:

1. "As a general proposition, announcing, in abstract terms, a whole group of particular facts relating to the behavior of natural agents in proposed circumstances."[35] For instance, the law stated above includes "among others the particular facts that rock crystal and saltpeter exhibit periodical colours."

2. "As a proposition announcing that a whole class of individuals agreeing in one character agree also in another"; for instance, here, as declaring the constant association between double refraction and exhibition of periodical colors.

3. "As a proposition asserting the mutual connection, or in some cases the entire identity of two classes of individuals"; for instance, here,

> . . . if observation had enabled us to establish the existence of a class of bodies possessing the property of double refraction, and observations of another kind had, independently of the former, led us to recognize a class possessing that of the exhibition of periodical colours in polarized light, a mere comparison of lists would at once demonstrate the identity of the two classes, or enable us to ascertain whether one was or was not included in the other.[36]

Laws may thus in Herschel's view briefly be characterized as *general facts*, as distinguished from particular facts. Such general facts are themselves susceptible of being grouped into classes, i.e., "included in laws which, as they dispose of groups, not individuals, have a far superior degree of generality, till at length, by continuing the process, we arrive at *axioms* of the highest degree of generality of which science is capable."[37]

As to the relation between laws and causes, Herschel is not very definite. We have seen that cause, in the last of the senses distinguished above, appears to mean for him a fact antecedent to another that follows it "necessarily," and that by "necessarily" he appears to mean, broadly, *according to a law.* Note also the end of the following passage, which seems to regard the notion of cause as bound up in that of Law:

> Every law is a provision for cases which *may* occur, and has relation to an infinite number of cases that never have occurred, and never will. Now, it is this provision, *a priori*, for contingencies, this contemplation of possible occurrences, and

predisposal of what shall happen, that impresses us with the notion of a *law* and a *cause*.[38]

As definite a statement as can be found in the *Discourse* as to the relation between cause and law is the following: "Whenever two phenomena are observed to be invariably connected together, we conclude them to be related to each other, either as cause and effect, or as common effects of a single cause." [39] In this statement the words "we conclude" probably are not intended by Herschel to mean that the invariability is *evidence of* something other than itself, called causation. Rather, he probably means to adhere to Hume's view, according to which invariability *constitutes* the relation called causation.

If so, however, *all* laws could then be spoken of as causal laws. They would all be such that the pairs of events that can be subsumed under them would *eo ipso* be related either as one the cause of the other, or both as effects of some antecedent third. This would have to be so even in the case of "empirical laws" as defined by Herschel, that is, "laws . . . derived, by the direct process of including in mathematical formulae the results of a greater or less number of measurements." [40] He describes such laws, it is true, as "unverified inductions" and says that until they have been verified "no confidence can ever be placed in them beyond the limits of the data from which they are derived." But what he has in mind for them is verification "theoretically by a deductive process," not, as one could fairly expect, verification by observation of the "invariability" that he seems to equate with causation. Like J. S. Mill after him, and like anyone else who accepts Hume's identification of causation with regularity of conjunction, Herschel can be confronted here with the objection that according to this view we should have to class as causal certain relations that he like anyone else would, as a matter of common sense, refuse so to class. These would include not only such a relation as Ohm's law describes between electrical current, potential, and resistance, but also such a case as that of the close covariation over a good many years, which Morris R. Cohen mentions in his Carus lectures,[41] between the death rate in the state of Hyderabad, India, and the membership in the International (American) Machinists Union.

But, whatever may be Herschel's view of the relation between law and cause, he declares in section 95 that the process of formulating laws "is what we mean by induction." Induction may, he holds, be carried on in two different ways: either by noting the agreements and disagreements of ascertained *classes;* or by considering the *individuals* of a class and seeking what character they have in common besides that which constitutes them one class. The latter employs the division of labor and is better adapted to the infancy of science; the former, which is more suitable to the maturity of science, "mainly relies on individual penetration, and requires a union of many branches of knowledge in one person." [42]

4. THE FIRST STAGE OF INDUCTION

In the sixth and seventh chapters, Herschel passes to a direct enunciation of the methodological precepts that are to govern induction. He distinguishes two broad stages in induction: at the first, science is concerned with the discovery of "proximate causes" and of laws of the lowest degree of generality, and with the verification of these laws. At the second stage, the inductions of science have for their material no longer individual facts, but, on the contrary, general facts, viz., the laws themselves and the causes, which, in the first stage, were obtained from the examination of the individual facts. The results of this second stage of induction consist in laws of a higher generality, to which he gives the name of theories, and which must, like the others, be verified.

In our examination of what Herschel has to say concerning the first stage, it is worth while to note first of all that the problem he has directly in mind when he approaches the formulation of his "rules of philosophizing," and throughout the exposition of them, is that of the *discovery* of causes. The problem of *proof* of causal connections, on the other hand, is for him that of verifying—by deduction of predictions and comparison of the predictions with observed facts—the *causal hypotheses* that resulted from the use of the method of discovery:

. . . when the cause of a phenomenon neither presents itself obviously on the consideration of the phenomenon itself, nor is as it were forced on our attention by a case of strong analogy . . . we have then no resource but in a deliberate assemblage of all the parallel instances we can muster; that is, to the formation of a class of facts, having the phenomenon in question for a head of classification; and to search among the individuals of this class for some other common points of agreement, among which the cause will of necessity be found.[43]

If more than one such point of agreement appears, we must then devise "crucial" experiments where—some of these points of agreement being absent and the phenomenon nevertheless still present—we find evidence adequate to the rejection of such absent points from the cause.

According to this broad prefatory description of the course of scientific investigation, the role assigned by Herschel to experimentation appears to be that of eliminating from a number of possible hypotheses as to the cause of a phenomenon those which are incorrect; and, eventually, that of devising more and more severe verifications of the adequacy of the hypothesis which has survived this process of elimination. That Herschel is directly concerned with the discovery of causal hypotheses is further evidenced by the following words, which immediately introduce his detailed presentation of principles of method: "When we would lay down general rules for guiding and facilitating our *search,* among a great mass of assembled facts, *for their common cause,* we must have regard to the characters of that relation which we intend by cause and effect" (italics ours).[44]

Let us now examine these defining characters of the causal relation, and the rules of practice Herschel bases upon them. These characters are:

"1st. Invariable connection, and, in particular, invariable antecedence of the cause and consequence of the effect, unless prevented by some counteracting cause." To this are appended some remarks concerning the difficulty of deciding which one, of two phenomena, precedes the other, when the cause and the effect either are gradual changes—so that a late stage of the cause coexists with an early stage of the effect—or else occur in almost instantaneous succession.

2nd. Invariable negation of the effect with absence of the cause, unless some other cause be capable of producing the same effect.

3rd. Increase or diminution of the effect, with the increased or diminished intensity of the cause, in cases which admit of increase and diminution.

4th. Proportionality of the effect to its cause in all cases of *direct unimpeded* action.

5th. Reversal of the effect with that of the cause.

On the basis of these characters of the causal relation, Herschel then formulates ten observations which may be considered as "rules of philosophizing." [45] Thus, he says, we conclude:

1st. That if in our group of facts there be one in which any assigned peculiarity, or attendant circumstance, is wanting or opposite, such peculiarity cannot be the cause we seek.

2nd. That any circumstance in which all the facts without exception agree, *may* be the cause in question, or, if not, at least a collateral effect of the same cause: if there be but one such point of agreement, this possibility becomes a certainty; and on the other hand if there be more than one, they may be concurrent causes.

The third observation is an injunction against a priori rejection of a cause in favor of which we have a unanimous agreement of strong analogies, merely because we do not see how such a cause can produce the effect, or even can exist under the circumstances of the case.

4th. That contrary or opposing facts are equally instructive for the discovery of causes with favourable ones.

5th. That causes will very frequently become obvious, by a mere arrangement of our facts in the order of intensity in which some peculiar quality subsists; though not of necessity because counteracting or modifying causes may be at the same time in action.

For example, the rapidity of vibration of a medium and the pitch of the note heard are judged to be causally connected owing to the correspondence between the series of frequencies and the series of pitches.

6th. That such counteracting or modifying causes may subsist unperceived, and annul the effects of the cause we seek, in instances which, but for their action, would have come into our class of favorable facts; and that, therefore, exceptions may often be made to disappear by removing or allowing for such counteracting causes.

7th. If we can either find produced by nature, or produce designedly for ourselves, two instances which agree *exactly* in all but one particular, and differ in that one, its influence in producing the phenomenon, if it have any, *must* thereby be rendered sensible. If that particular be present in one instance and wanting altogether in the other, the production or non-production of the phenomenon will decide whether it be or be not the only cause: still more evidently, if it be present *contrariwise* in the two cases, and the effect be thereby reversed. But if its total presence or absence only produces a change in the *degree* or intensity of the phenomenon, we can then only conclude that it acts as a concurrent cause or condition with some other to be sought elsewhere.

Herschel adds here that although such cases of single difference are rare in nature, they are easily devised in experimentation, which becomes the more valuable as it more closely approximates the requirement of having exact agreement in all its circumstances but one.

8th. If we cannot obtain a complete negative or opposition of the circumstance whose influence we would ascertain, we must endeavor to find cases where it varies considerably in degree.

9th. Complicated phenomena, in which several causes concurring, opposing, or quite independent of each other, operate at once, so as to produce a compound effect, may be simplified by subducting the effect of all the known causes, as well as the nature of the case permits, either by deductive reasoning or by appeal to experience, and thus leaving as it were, a *residual phenomenon* to be explained.

Herschel concludes these observations by noting that:

10th. The detection of a *possible* cause, by the comparison of assembled cases, *must* lead to one of two things: either, 1st, the detection of a real cause, and of its manner of acting, so as to furnish a complete explanation of the facts; or 2ndly, the establishment of an abstract law of nature, pointing out two phenomena of a general kind as invariably connected; and asserting, that where one is, there the other will always be found.

The application of these rules is then illustrated by a number of examples—the investigations of Wells into the cause of dew, particularly, being discussed at length.

In the remainder of the sixth chapter, Herschel deals mainly with the verification of inductions of the first stage. The manner in which we seek for possible causes or laws, although important in practice, is theoretically of no moment: "provided only we verify them carefully when once detected, we must be content to seize them wherever they are to be found."[46] In practically every case, however, the statement of a law of nature that we frame goes beyond the cases actually examined; and we cannot rely on its thus "enabling us to extend our views beyond the circle of instances from which it was obtained, unless we have already had experience of its power to do so; unless it actually *has* enabled us before trial to say what will take place in cases analogous to those originally contemplated." [47] To examine whether it thus actually enables us to predict is to *verify* it. Thus "the successful process of scientific enquiry demands continually the alternate use of both the *inductive* and *deductive* method"; [48] and the confidence which we may justifiedly place in the universality of a law is proportionate to the severity of the verifications made, extreme cases affording particularly convincing tests.[49] But

> . . . the surest and best characteristic of a well-founded and extensive induction . . . is when verifications of it spring up, as it were, spontaneously, into notice, from quarters where they might be least expected, or even among instances of that very kind which were at first considered hostile to them. Evidence of this kind is irresistible and compels assent with a weight which scarcely any other possesses.[50]

Empirical laws, "derived by the direct process of including in mathematical formulae the results of a greater or less number of measurements," cannot be trusted beyond the limits of the data from which they are derived, and even within those limits must be carefully scrutinized, to ascertain whether the differences between their results and actual facts may fairly be attributed to errors of observation.[51]

Herschel then remarks that, "in forming inductions, it will most commonly happen that we are led to our conclusions by the especial force of some two or three strongly impressive facts, rather than by affording the whole mass of cases a regular consideration," [52] and he devotes the remainder of the chapter to concrete illustrations from modern science of some of the chief kinds of such "prerogative instances" described by Bacon. But he observes that, although much is usually made of this part of Bacon's work, the classification of the instances under various headings, however just, is not of much practical help, for the misfortune is that the choice of those that would be most useful does not rest with us: "We must take the instances as nature presents them." [53]

5. THE SECOND STAGE OF INDUCTION

The seventh and last chapter is devoted by Herschel to the higher degrees of inductive generalization, and to the formation and verification of theories. "As particular inductions and laws of the first degree of generality are obtained from the consideration of individual facts," he tells us, "so Theories result from a consideration of these laws, and of the proximate causes brought into view in the previous process, regarded all together as constituting a new set of phenomena. . . . The ultimate objects we pursue in the highest theories are the same as those of the lowest inductions," and the means are closely analogous in both cases.[54]

An important part of the knowledge we seek of the hidden processes of nature depends on the discovery of the actual structure of the universe and its parts, and the agents concerned in these processes. But "the mechanism of nature is for the most part either on too large or too small a scale to be immediately cognizable by our senses, and her agents in like manner elude direct observation, and become known to us only by their effects."[55] Are we then in such cases "to be deterred from framing hypotheses and constructing theories" because we find it difficult to decide between rival theories or because we "find ourselves frequently beyond our depth"? "Undoubtedly not," Herschel answers. "Hypotheses, with respect to theories, are what presumed proximate causes are with respect to particular inductions: they afford us motives for searching into analogies," and if well constructed frequently lead to additional steps in generalization. In certain cases, even,

> . . . such a weight of analogy and probability may become accumulated on the side of an hypothesis, that we are compelled to admit one of two things; either that it is an actual statement of what really passes in nature, or that the reality, whatever it be, must run so close a parallel with it, as to admit of some mode of expression common to both, at least in so far as the phenomena actually known are concerned.[56]

Herschel then goes on to discuss the process of constructing theories:

> In framing a theory which shall render a rational account of any natural phenomenon, we have *first* to consider the agents on which it depends, or the

causes to which we regard it as ultimately referable. These agents are not to be arbitrarily assumed; they must be such as we have good inductive grounds to believe do exist in nature, and do perform a part in phenomena analogous to those we would render an account of; or such, whose presence in the actual case can be demonstrated by unequivocal signs. They must be *verae causae*, in short, which we can not only show to exist and to act, but the laws of whose action we can derive independently, by direct induction, from experiments purposely instituted; or at least make such suppositions respecting them as shall not be contrary to our experience.

As an instance of an agent that is such a *vera causa*, Herschel mentions "force, or mechanical power" in the theory of gravitation.[57] This is hardly a happy illustration since the description of a "force" always turns out, on examination, to be a statement that objects of a certain kind are such that circumstances of certain sorts cause them to behave in a certain manner. That is, a force is essentially a causal law; and a law—even a causal law—never *itself* functions as a cause. A true example of an agent that is a *vera causa* in the sense specified by Herschel would be the *atom*.

But, second, we have to consider "the laws which regulate the action of these our primary agents; and these we can only arrive at in three ways: 1st, By inductive reasoning," that is, by examining all the particular cases, piecing together the results of our observations, and generalizing from them;

2dly, By forming at once a bold hypothesis, particularizing the law, and trying the truth of it by following out its consequences and comparing them with facts; or 3dly, By a process partaking of both of these . . . viz., by assuming indeed the laws we would discover, but so generally expressed that they shall include an unlimited variety of particular laws;—following out the consequences of this assumption . . . comparing them in succession with all the particular cases within our knowledge; and lastly, *on this comparison*, so modifying and restricting the general enunciation of our laws as to *make the results agree*.[58]

Third, in cases where the laws that regulate the actions of our ultimate causes do not apply at once to the materials and directly produce the result (as in the instance of gravitation), we have to consider "a system of mechanism, or a structure of parts through the intervention of which [the effects of our ultimate causes] become sensible to us."[59]

As to the estimation of the value of a theory, i.e., the "verification" of a theory, what is important to know "is whether our theory truly represent *all* the facts, and include *all* the laws to which observation and induction lead." [60] While theories are best arrived at by the consideration of general laws, they are "most securely verified by comparing them with particular facts, because this serves as a verification of the whole train of induction, from the lowest term to the highest"; [61] but these particular facts must be widely diversified and include extreme cases if reasonable probability of detecting error is to be afforded. "When two theories run parallel to each other, and each explains a great many facts in common with the other, any experiment which affords a crucial instance to decide between them . . . is of great importance." [62]

Except in point of generality, the inductions of the second stage are thus not conceived by Herschel to differ markedly from those of the first stage. They are still, ultimately, *obtained from* experience, as well as verified by appeal to experience. Even when guessed at by a bold stroke of genius, the guess is still one that springs from examination of the facts, rather than from the spontaneous demand of man for intellectual tools adapted to his logical needs. That is, for Herschel it is an act of *discovery* rather than of invention.[63] Whewell, on the other hand, urges a few years later that the task of science as it actually presents itself in many cases is to be described much rather as that of inventing a way of stating the facts observed that shall satisfy our logical demands.

6. CONCLUSION

In the light of this examination of Herschel's *Discourse*, what can we say now of his place in the history of the philosophy of science? For one thing, Herschel—unlike his contemporary Whewell whose *Novum organon renovatum* represents the influence of Kant in this field—clearly belongs in the line of British empiricists that may be considered to begin with Francis Bacon and to include as principal figures Locke, Berkeley, Hume, and John Stuart Mill.[64]

Herschel, of course, was not like Locke, Berkeley, and Hume concerned to formulate an epistemology, but the theory of scientific method is so intimately connected with epistemology that, as we have seen, he finds himself led at times into epistemological excursions. These are improvised on empiricistic and associationist principles, but they remain amateurish and contribute nothing to the solution of the epistemological problems concerned. Herschel's definition of "phenomena," for instance, is as we have seen at variance with his own actual use of the term; and his account of the relation of phenomena to objects is unacceptable. Again, he mistakenly believes that a force can be a cause; that in acts of will the ultimate nature of force is evident to our observation; and that the forces of nature are essentially akin to volitions. Again, as already pointed out, Herschel, like Hume before him and Mill after, never distinguishes clearly between the two notions of cause and of law. On this account, many of the statements in which he uses the terms have the sort of obscurity confusion breeds. His conception of theories as simply wider laws, having to narrower ones a relation analogous to that of the latter to the particular facts they generalize, is inherited from Bacon and is unsound. However, the question of the nature of theories and of their relations to laws is a difficult one. Whewell, whose *Philosophy of the Inductive Sciences* was published ten years after Herschel's *Discourse,* exhibits a sounder insight into it; but it is not until comparatively recent years that a clear answer can be said to have become available.[65]

On the credit side are to be placed Herschel's statements on the subjects of Observation, Experiment, Classification, Hypothesis, and Verification; not, however, because they are particularly original since even in Bacon's writings rather similar remarks are to be found, but rather because of the clear light thrown on these scientific procedures by the true concrete illustrations of them Herschel furnishes. His remarks on terminology, and the relation to the scope of science which he assigns to the postulate of the uniformity of nature, are also sound and valuable. His greatest contribution, however, is undoubtedly to be found in the ten "rules of philosophizing" already quoted. It is in them and in the remarks that accompany them that we find, for the first time both distinctly enunciated and amply illustrated, the famous four methods of agreement, difference, concomitant variations, and residues. These names, by which they are now universally known, are due to John Stuart Mill; but Mill himself declares that in Herschel's *Discourse* alone, "of all the books which I have met with, the four methods of induction are distinctly recognized." [66] And at least one logician, W. S. Jevons, has expressed his preference for Herschel's discussion of them over that of Mill.[67] These methods, nowadays often referred to as "Mill's Methods," would, if they are to be tied to any man's name, therefore seem better described as "Herschel's Methods." The history of at least some of them, however, can be traced back as far as Aristotle. Bacon had discerned the first three (which are the most significant) and was fully aware of their great power, although the material to which he proposed to apply them was inappropriate and barren.[68] A very clear statement—the best prior to Herschel's—of the first three is to be found in Hume's *Treatise of Human Nature* (1739); [69] although the conception of cause implicit in the rules Hume there gives is, without his being conscious of it, different from that of cause as event observed to have been regularly followed by a certain other, which is the conception he explicitly adopts.

The clear and abundantly illustrated discussion of these four methods given by Herschel would be sufficient to assure him a permanent place in the history of the philosophy of science; but this place is further assured by the fact that his *Discourse* contains much the best and most comprehensive formulation up to its date of the methods of scientific investigation, and strongly stimulated or influenced the labors in this field of his contemporaries Whewell and J. S. Mill.

CHAPTER NINE

William Whewell's Philosophy of Scientific Discovery

WHEWELL's views as to the nature of scientific inquiry were first published shortly after those of Herschel (1830) and immediately before those of Mill (1841). Several later editions were printed with various additions, including comments on some of Mill's opinions. The great popularity quickly attained by Mill's *System of Logic* not only then tended to eclipse in some measure the importance of Herschel's *Discourse;* it also stood in the way of general recognition of the merits of Whewell's theory of the nature of scientific knowledge and of the process of discovery. Disregard of its merits was the easier because of its sharp break with the traditions of the British empiricists and its alliance instead with the Kantian point of view. In the light of subsequent developments in the philosophy of science, an examination of Whewell's contributions has considerable interest.

The German metaphysicians, Whewell writes, "saw at once that ideas and things, the subjective and the objective elements of our knowledge, were, by Kant's system, brought into opposition and correlation, as equally real and equally indispensable"; [1] and Whewell makes the equal reality and equal indispensability of these two elements the very core of his own theory of knowledge. He acknowledges freely his general sympathy with Kant's fundamental position and states that he has "adopted some of Kant's views, or at least some of his arguments," even to the extent that "the chapters on the Ideas of Space and Time in the *Philosophy of the Inductive Sciences* were almost literal translations of chapters in the *Kritik der Reinen Vernunft.*" Yet Whewell—whose views had been regarded by Mansel [2] as misrepresenting the Kantian philosophy and by G. H.

Lewes,[3] on the contrary, as identical with the views of Kant—protests that he "does not profess himself a Kantian," and that he regards his main views as "very different from Kant's."[4] He criticizes Kant, in respect to the fundamental antithesis between ideas and things, for having fixed his attention almost entirely upon ideas[5] and indicates in other passages[6] how he conceives himself further to depart from Kant. He "agrees with Kant in placing in the mind certain sources of necessary truth," which he, however, calls "Fundamental Ideas"; but he believes that there are many others of these beside those Kant admitted. It is only as regards space, time, and, to some extent, causality that he considers himself in close agreement with Kant. He does not appear to have been particularly interested in or impressed by Kant's argument for the exhaustiveness and a priori validity of the table of categories given in the *Critique*. Indeed, he rather suggests that an account of the fundamental ideas ought, in his view, to be constructed a posteriori, for, after declaring that (except as regards space, time, and possibly causation) his doctrine of Fundamental Ideas has "no resemblance to any doctrines of Kant or his school," he immediately goes on to say that "the nature and character of the other Scientific Ideas which [he, Whewell, has] examined . . . have been established by an analysis of the history of the several Sciences to which those Ideas are essential." What Whewell borrows from Kant is the general conception of knowledge as essentially involving both a subjective, "necessary" element, and an objective, empirical one. But he interprets this general conception in his own fashion, following Kant only to the limited extent indicated.

1. THE FUNDAMENTAL ANTITHESIS INVOLVED IN ALL KNOWLEDGE, AND THE NATURE OF "FUNDAMENTAL IDEAS"

Whewell's views on the general nature of knowledge are most fully stated in Book I of the *History of Scientific Ideas*, but many statements on the subject are to be found in the *Philosophy of Discovery* also, notably in chapters xxiv, xxviii, xxix, and Appendix E. The *Novum organon renovatum* contains not so much a general doctrine of the nature of knowledge, as an account, on the basis of such a doctrine, of the processes by which science is constructed. The detailed examination of these processes will occupy us later.

The fundamental contention of Whewell's theory of knowledge is that all knowledge essentially involves the antithesis of two elements. One of them is given to us by pure observation, and the other is superimposed by ourselves upon what we observe. Only when the two elements are united do we have knowledge properly so-called.

1. Many pairs of terms exist which all refer to one or another aspect of that fundamental antithesis. One such pair is *thoughts* and *things*—"in all human knowledge both Thoughts and Things are concerned." Thus, in the knowledge that a solar year consists of 365 days, there are involved on the one hand the sun as given, and on the other the mental act of counting. "Without Thoughts, there could be no connexion; without Things, there could be no reality." [7] Again, we are familiar with the contrast of *necessary* and *experiential* truths: "Necessary Truths are derived from our own Thoughts, Experiential Truths are derived from our observation of Things about us." [8]

2. The opposition of *deduction* and *induction* constitutes another aspect of the same fundamental antithesis: "The term *Deduction* is specially applied to . . . a course of demonstration of truths from definitions and axioms" supplied by our own thoughts; in *Induction*, however, "truths are obtained by beginning from observation of external things and by finding some notion with which the Things, as observed, agree." [9]

3. Another antithesis, which involves that of thoughts and things, but is not identical with it, is the antithesis of *theory* and *fact*. A theory is a general experiential truth, and facts are the particular observations from which theories are inductively obtained. This implies the antithesis of thoughts and things, for a (true) theory

. . . may be described as a Thought which is contemplated distinct from Things and seen to agree with them; while a Fact is a combination of our Thoughts with Things in so complete agreement that we do not regard them as separate . . . for we know Facts only by thinking about them. The Fact that the year consists of 365 days . . . cannot be known to us, except we have the Thoughts of Time, Number and Recurrence. But these Thoughts are so familiar, that we have the Fact in our mind as a simple Thing without attending to the Thought which it involves.[10]

4. Of the various pairs of terms that express or refer to the fundamental antithesis of thoughts and things, the one that appears to separate the members of the antithesis most distinctly is, Whewell tells us, *ideas* and *sensations*.

We see and hear and touch external Things, and perceive them by our senses; but in perceiving them, we connect the impressions of sense according to relations of space, time, number, likeness, cause, etc. Now some at least of these kinds of connexion, as space, time, number, may be contemplated distinct from the Things to which they are applied; and so contemplated, I call them *Ideas*. And the other element, the impressions upon our senses which they connect, are called *Sensations*. I term space, time, cause, etc., *Ideas*, because they are general relations among our sensations, apprehended by an act of the mind, not by the senses simply. These relations involve something beyond what the senses alone could furnish.[11] . . . We use the word *Ideas* . . . to express that element, supplied by the mind itself, which must be combined with Sensation in order to produce knowledge. For us, Ideas are not objects of Thought (as they are according to Locke's use of the term), but rather Laws of Thought. Ideas are not synonymous with Notions; they are Principles which give to our Notions whatever they contain of truth.[12]

The term "idea" plays so important a part in Whewell's theory of knowledge that, to elucidate still further what he means by it, a number of other statements concerning it may profitably be quoted here. He writes, for instance, that the ideas themselves "cannot be fixed in words," that "our elementary Definitions and Axioms, even taken all together, express the Idea incompletely" for, in addition to them, other axioms, independent of them, can be stated. They express it, however, sufficiently for the purposes of science.[13] Again,

. . . the ground of the axioms belonging to each science is the *Idea* which the axiom involves. The ground of the Axioms of Geometry is the *Idea of Space;* the ground of the Axioms of Mechanics is the *Idea of Force*, of *Action* and *Reaction*, and the like. . . . It is, not the logical, but the philosophical, not the formal but the real foundation of necessary truth, which we are seeking,

and it is in the ideas that we find the ground "not only of demonstration but of truth." [14] Elsewhere he says: "By speaking of Space as an Idea I intend to imply . . . that the apprehension of objects as existing in Space, and of the relations of position, etc., prevailing among them is, not a consequence of experience, but a result of a peculiar constitution and activity of the mind."[15] And, similarly, "the Idea of cause . . . is a part of the active powers of the mind. The relation of cause and effect is a relation or condition under which events are apprehended, which relation is not given by observation, but supplied by the mind itself."[16] Again, "by the word Idea (or Fundamental Idea) used in a peculiar sense, I mean certain wide and general fields of intelligible relation, such as Space, Number, Cause, Likeness."[17] Among the fundamental ideas that are the basis of the material sciences, Whewell mentions and discusses, beside space, time, and causation, also number, force, matter, outness of objects, media of perception of secondary qualities, polarity, chemical composition and affinity, substance, likeness, means and ends (whence the notion of organization), symmetry, vital powers.[18]

5. The fundamental antithesis of philosophy has yet other aspects. In modern German philosophy, Whewell goes on to say, it has been indicated by the terms "subjective" and "objective."

6. But another and one of the most ancient ways of referring to the antithesis is "that which speaks of Sensations as the Matter, and Ideas as the Form, of our knowledge," this comparison having the advantage of "showing that two elements of an antithesis which cannot be separated in fact, may yet be advantageously separated in our reasonings."[19] In terms of the metaphor of matter and form, the erroneous opinion that "every Idea is a transformed Sensation" may be corrected, while retaining the metaphor, "by saying, that ideas are not *transformed* but *informed* sensations . . . our sensations, from their first reception, have their form not *changed*, but *given* by our Ideas."[20]

7. Again, nature has been opposed to man, the latter being spoken of as the interpreter of nature, and science as the right interpretation. "The facts of the external world are marks, in which man discovers a meaning, and so reads them," and this illustration serves to explain

. . . the very different degrees in which, in different cases, we are conscious of the mental act by which our sensations are converted into knowledge. For the same difference occurs in reading an inscription. If the inscription were entire and plain, in a language with which we were familiar, we should be unconscious of any mental act in reading it. We should seem to collect its meaning by the sight alone. But if we had to decipher an ancient inscription, of which only imperfect marks remained, with a few entire letters among them, we should probably make several suppositions as to the mode of reading it, before we found any mode which was quite successful; and thus, our guesses, being separate from the observed facts, and at first not fully in agreement with them, we should be clearly aware that the conjectured meaning, on the one hand, and the observed marks on the other, were distinct things, though these two things would become united as elements of one act of knowledge when we had hit upon the right conjecture.[21]

This enables us to sharpen the distinction already made between theory and fact:

In Theory the Ideas are considered as distinct from the Facts: in Facts, though Ideas may be involved, they are not, in our apprehension, separated from the sensations. In a Fact, the Ideas are applied so readily and familiarly, and incorporated with the sensations so entirely, that we do not see *them*, we see *through them* . . . thus a True Theory is a Fact, a Fact is a familiar Theory.[22] . . . Theories become Facts by becoming certain and familiar.[23]

In the second chapter of the *History of Scientific Ideas*, on technical terms, Whewell points out that, when "some ideal conception, which gives unity and connexion to multiplied and separate perceptions," has been found and has become thoroughly incorporated with them in our minds, a definite step in the pursuit of knowledge has then been made. Such successful ideal conceptions are then always "recorded, fixed, and made available by some peculiar form of words" called technical terms, in which in every case an induction thus lies packed, even though the term be commonly said only to denote a fact. Examples of such technical terms would be accelerating force, attraction, neutral salts, affinity, anode, cathode.

2. CONTROVERSY WITH MILL AND MANSEL CONCERNING NECESSARY TRUTHS

The remainder of Book I of the *History of Scientific Ideas* is devoted to a more detailed comparison of necessary with experiential truths and to a discussion of the grounds of the former. That there are necessary truths, i.e., truths in which we not only learn that the proposition *is* true, but see that it *must be* true, cannot be doubted,

Whewell declares. That 3 plus 2 make 5 is an instance of such a truth, which, he insists, is not a definition (the definition of 5 being 4 plus 1), but a proposition the truth of which can be demonstrated from definitions and axioms.[24] One way of expressing the difference between necessary truths and truths of experience is by saying that the former are "those of which we cannot distinctly conceive the contrary," while in the case of the latter we find no difficulty in doing so.[25] The qualification introduced by the word "distinctly" in this account of necessary truths has the effect of paving the way for one of the most characteristic points of Whewell's doctrine, namely, that *necessary truths are to be found not only in mathematics, but also in other subjects* such as mechanics, hydrostatics, and chemistry—although, Whewell grants, "the discipline of thought which is requisite to perceive them distinctly may not be so usual among men . . . as it is with regard to the sciences of geometry and arithmetic."[26] Whewell's view of the ground of necessary truths met with considerable opposition, notably on the part of Mill and Mansel, and the main contentions set forth in the course of the controversy will here briefly be passed in review.

Mill contends[27] that the definitions to which Whewell refers are not necessary truths but merely some of our simplest and earliest generalizations from experience. Strictly speaking, they are not even exactly true in any individual case, but only "so nearly true that no error of any importance in practice will be incurred by feigning [them] to be exactly true." Axioms—which both Mill and Whewell agree are needed by mathematics in addition to definitions—are, Mill says, not approximately but exactly true. They are, however, also inductive generalizations from observed facts.

To Whewell's contention that experience, although it suggests the axioms, cannot prove them,[28] Mill replies that on the contrary—and whether or not they be evident independently of experience—they are also evident from experience.[29] Thus, for the axiom that two straight lines cannot enclose a space,

> . . . experimental proof crowds in upon us in such endless profusion, and without one instance in which there can be even a suspicion of an exception to the rule, that we should soon have stronger ground for believing the axiom, even as an experimental truth, than we have for almost any of the general truths which we confessedly learn from the evidence of our senses.

Why then seek any other origin for the axiom?

It is urged, indeed, that we do not need to observe straight lines to become convinced of the truth of the axiom, that it is seen to be true by merely thinking of them. Moreover, sense observation cannot make the experiment, since it cannot follow the lines to infinity. But to these objections Mill replies that we do not believe the axiom "on the ground of the imaginary intuition simply, but because we know that the imaginary lines exactly resemble real ones." And because of this we can also, by transporting ourselves in imagination to the point, however remote, at which the lines may be supposed to begin to converge, declare that, did they indeed converge, they would be imagined—and therefore if perceivable also perceived—not as "straight" but as "bent."

Again, it is urged that axioms are conceived not only as true, but also as universally and necessarily true, and that sense experience cannot possibly give us propositions of this character but only general propositions.[30] To this Mill replies as follows: by a necessary proposition, Whewell means one the negation of which is not only false but inconceivable. But there is "ample experience to show, that our capacity or incapacity of conceiving a thing has very little to do with the possibility of the thing in itself; but . . . depends on the past history and habits of our own minds."[31] Many things that have been pronounced inconceivable have afterward been proved to be true. In the case of the axiom mentioned, to conceive it false would demand going contrary to a lifelong habit of our minds, and no analogy is to be derived from any other branch of our knowledge to help us in doing this. The psychological laws of association, under such circumstances, make it impossible that the falsity of the axiom should appear otherwise than inconceivable to us, and Whewell's assertions on the subject afford a striking instance of the working of these laws in producing the illusion of the necessity of axioms.

In chapter xxviii of the *Philosophy of Discovery*, Whewell endeavors to meet this criticism of his view that the truth of axioms is based upon the inconceivability of their contrary. To this end he formulates his doctrine of "*progressive intuition*." "The special and characteristic property of all Fundamental Ideas," he tells us, "is . . . that they are the mental sources of necessary and universal scientific truths . . . and the way in which those ideas become the foundation of science is, that *when they are clearly and distinctly entertained in the mind, they give rise to inevitable convictions or intuitions* which may be expressed as *axioms*." [32] The difficulty, he goes on, is then this: "The test of Axioms is that the contrary of them is inconceivable; and yet persons, till they have in some measure studied the subject, do not see the inconceivableness. Hence our Axioms must be evident only to a small number of thinkers; and seem not to deserve the name of self-evident or necessary truths." [33] Whewell's answer to the objection that he has thus stated consists in stressing the distinction between two meanings of "self-evidence" or "intuition." What he means by self-evidence is not self-evidence to men in general, irrespective of their mental training, but only self-evidence to such men as have achieved a thoroughly *clear and distinct* conception of the ideas upon which the axioms rest; and he readily acknowledges that, with regard to some ideas at least, the degree of clearness and distinctness in the conceiving of them that is necessary to render the corresponding axioms intuitively evident may well be a rare and difficult attainment.[34] Thus with regard to the two principles of chemical science—that combinations are definite in kind and in quantity —what he holds is not only that they are in fact true, but also that, "*if* we could conceive the composition of bodies *distinctly*, we might be able to see that it is necessary that the modes of this composition should be definite." He declares, however, that even in his own mind "the thought of such a necessity was rather an anticipation of what the intuitions of philosophical chemists in another generation would be, than an assertion of what they now are or ought to be." [35] "*There are scientific truths which are seen by intuition, but this intuition is progressive.*" [36]

At this point, however, the question recurs whether experience, as Mill argued, is not to be regarded as the source of these axioms, since the discovery of them and the apprehension of their character as necessary truths do not precede but follow experience. But, says Whewell, the case is the same here as with light, which "reveals to us at the same time the existence of external objects and our own power of seeing," [37] but does not confer upon us that power. Thus, while experience does indeed suggest universal truths, i.e., furnishes the occasion for thinking of them, it never does and never can yield them in the sense of furnishing adequate (logical) evidence for them.[38] The most that it can yield are general, but not necessary and universal, truths. The general truths yielded by experience—the definitions as well as the axioms—are not known to be universal and necessary (if they really are so) until they have been intuitively perceived to flow from our corresponding fundamental ideas. "*What* should happen universally, experience might be needed to show: but that what happened should happen *universally*, was implied in the nature of knowledge" (as dependent upon fundamental ideas).[39] That, for instance, laws of motion existed, i.e., "universal formulae connecting the causes and effects when motion takes place," was from the beginning regarded as certain. "The rule was not accepted as particular at the outset, and afterwards generalized more and more widely; but from the very first, the universality of the rule was assumed, and the question was, how it should be understood so as to be universally true." [40]

Mill's reply to Whewell's doctrine of "progressive intuition" consists in giving a different account of the progress of the scientific mind. The habit of conceiving facts in terms of the laws that regulate them comes, Mill says, only by degrees. In the case of newly discovered laws, so long as the habit is not thoroughly formed, no necessary character is ascribed to the new truth. However, when the habit of conceiving facts according to a new (true) theory is thoroughly formed in man,

> . . . any other mode in which he tries, or in which he was formerly accustomed, to represent the phenomena, will be seen by him to be inconsistent with the facts that suggested the new theory—facts which now form a part of his mental picture of nature. And since a contradiction is always inconceivable, his imagination rejects false theories, and declares itself incapable of conceiving them. Their inconceivableness to him does not, however, result from anything in the theories themselves, intrinsically and *a priori* repugnant to the human faculties; it results from the repugnance between them and a portion of the facts; which facts as long as he did not know, and did not distinctly realize in his mental representations, the false theory did not appear otherwise than conceivable; it becomes inconceivable, merely from the fact that contradictory elements cannot be combined in the same conception. Although, then, his real reason for rejecting theories at variance with the true one, is no other than that they clash with his experience, he easily falls into the belief, that he rejects them because they are inconceivable, and that he adopts the true theory because it is self-evident, and does not need the evidence of experience at all.[41]

But Whewell's contention was not that true theories are self-evident, nor that false theories are inconceivable owing to their containing an explicit or implicit contradiction in terms. A theory is always a theory *about certain facts,* reference to which is an intrinsic part of the theory, since the theory is a proposal to conceive those facts in a certain manner. The contradiction from which arises the inconceivability Whewell has in mind is a contradiction between a statement of those facts, already known to be true, and the statement the theory makes *about* those facts. For instance, with regard to the supposition of chemical combinations otherwise than as definite in kind and quantity, Whewell does not claim that it is a self-contradictory supposition, i.e., to use Mill's terms, that it is "intrinsically and *a priori* repugnant to the human faculties." What Whewell says is that if that supposition were true there would be no fixed kinds of bodies; while on the contrary we *know* "that the world consists of bodies distinguishable from each other by definite differences capable of being classified and named, and of having general propositions asserted concerning them." [42] His further assertion, that "we cannot conceive a world in which this should not be the case," is more questionable. The basis for it could only be the contention that, when one speaks of "a world," one means a cosmos, and that a chaos therefore could not be called a world. Mill's criticism misses the essential point of Whewell's contention; but Whewell apparently does not perceive this clearly, for, instead of meeting Mill's criticism in the manner just suggested, his reply proceeds on the assumption that what Mill questions is the self-evidence of the assertion that, if combinations were otherwise than definite in kind and quantity, there would be no fixed kinds of bodies (i.e., there would be chaos). And he explains that he does not claim this to be as yet fully self-evident to anyone, and that what he meant to do was "to throw out an opinion, that *if* we could conceive the composition of bodies *distinctly,* we might be able to see that it is necessary that the modes of this composition should be definite." [43] And, as to this, Whewell was right. In the light of the knowledge of the composition of bodies that modern physics possesses, it has become evident that combinations must be definite and why they must be so. This is evident to us today, because, Whewell would say, the investigations of recent years have sufficiently *clarified* for us the idea of substance. Whewell thus is very far from contending, as Mill in the passage quoted seems to think, that in the formulation of theories the scientist "does not need the evidence of experience at all."

The ultimate difference between Whewell and Mill then appears to be as follows: according to Whewell, an intimate knowledge of facts is necessary to make our "Fundamental Ideas" clear; then, and then only, the axioms are perceived to be necessary, i.e., their negatives inconceivable as clashing with the "Ideas" that experience has *clarified for us*. Mill, on the other hand, would say that the axioms are perceived to be necessary in that their negatives clash with our experience. Both would admit that additional experience—whether by clarifying our ideas further or by furnishing us with further facts—might force us to revise our axioms; and the difference between the two doctrines thus turns out to be largely verbal. Whewell, it may be added, does not adequately establish the need of supposing latent in us the "Fundamental Ideas" of which he speaks. Their function, in his theory of knowledge, is only to certify the universality and necessity of certain propositions, which experience by itself could reveal to us only as general. But this universality and necessity is never really certified by the "Fundamental Ideas" since we are never sure that we have as yet an apprehension of them that is completely clear. Whewell fails to perceive that some sort of a priori deduction of them, such as Kant attempted, was needed if they were to be capable of the function he assigned to them. Whewell's attempt to break down any sharp distinction between mathematics and the natural sciences, although of quite a different sort from Mill's, is no more convincing. Both assume that mathematics cannot be a "science respecting nonentities." But that geometry, for instance, which is their favorite example, is a science that studies the nature of "real space," would hardly be contended today, at least by mathematicians. In both Mill and Whewell, moreover, the lack of an explicitly formulated theory of necessity, distinguishing the various senses of the term, and agreed upon by both, as a background for the various statements they make about necessary truths, is apparent throughout.[44]

Mansel disagrees with both Whewell and Mill. He thinks that Mill appears to prove the case for empiricism only because Whewell "has not more accurately observed Kant's distinction between the necessary laws under which all men think, and the contingent laws under which certain men think of certain things."[45] The difference between a priori principles and empirical generalizations, he urges, "is not one of degree but of kind." The increased clearness and distinctness of a conception, upon which Whewell lays such stress, "may enable us to multiply to any extent our analytical judgments, but cannot add a single synthetical one." Without something more than this, viz., without a psychological deduction such as given by Kant in the instance of space and time, "the philosopher has failed to meet the touchstone of the Kantian question: *How are synthetical judgments a priori possible?*"

Whewell, however, replies that Mansel and the Kantians admit the existence of necessary truths based upon other fundamental ideas than those of time and space, for instance those of substance and causality. To these, they ascribe not a "mathematical," but a "metaphysical," necessity. Mansel recognizes also "logical" necessity; and, Whewell says, there seems no reason why, on his own principles, he should not recognize yet others, as indeed he appears to him to do. We may then well speak, in general, of grounds of scientific necessity, and these, Whewell says, are precisely what he means by fundamental ideas. He might well have added that the "transcendental deduction" that Kant gave of his categories would, for whatever it may be worth, apply automatically to any other categories that Kant might have overlooked, and that the question really at issue between him and Mansel is whether or not Kant's attempted "metaphysical deduction" of his table of categories, based as it is upon a notoriously empirical table of judgments, has yielded none but, *and all of*, the true categories of thought (fundamental ideas). It has, moreover, been urged by Paulsen[46] that it is very difficult to see why experience should have to be appealed to for the knowledge of some syntheses if, as Kant maintains, *every* synthesis is solely an act of the understanding, and some syntheses can be known without such appeal. The drawing of a line between those that can and those that cannot be so known seems wholly arbitrary, and the only consistent procedure the adoption of a pure rationalism or of a pure empiricism.

3. THE PROCESSES OF SCIENTIFIC DISCOVERY

We now pass to the examination of the particular processes upon which the work of scientific discovery relies according to Whewell. He discusses them chiefly in the *Novum organon renovatum.*

A. *Outline of the essential steps in the inductive process*

Our knowledge, Whewell reminds us, consists in applying ideas to facts;

> . . . and . . . the conditions of real knowledge are that the ideas be distinct and appropriate, and exactly applied to clear and certain facts. The steps by which our knowledge is advanced are those by which one or the other of those two processes is rendered more complete;—by which *Conceptions* are *made more clear* in themselves, or by which the Conceptions more strictly *bind together the Facts.*[47]

These two processes he calls respectively the *explication of conceptions* and the *colligation of facts.* They involve various special processes, the nature and mutual relations of which may be indicated in a general way as follows, before each is examined in detail.

The task of inductive inquiry is the *colligation of facts,* that is, the *binding together of a set of facts by the invention and the introduction among them of an exact and appropriate conception, expressing them all at once.* To this task, two things are requisite: the possession of *conceptions* and of *facts.* Both the conceptions and the facts will be serviceable in proportion as they are clearly analyzed; and, therefore, preliminary to the effecting of the colligation of facts, there must occur a process of analysis and clarification. This process of clarification, on the idea side, Whewell calls the *explication* of *conceptions,* and on the facts side the *decomposition of facts.* These two operations constitute the first step in the progress of knowledge. In the second step, namely, in the process of effecting the *colligation* of a set of facts, three stages representing progressive definiteness in this colligation can be distinguished: *selection of the idea, construction of the conception,* and *determination of the magnitudes.* The third and final step in the induction of knowledge is the *verification* of the colligating character of the conception that was applied to the facts. This verification is made through the application of three tests: the possibility of *predicting* facts by means of the conception; the *consilience* of separately induced conceptions; and the progressive *simplification of the conception* as it is extended to new cases. This summary statement of the elements of the inductive process as conceived by Whewell may be presented in tabular form as follows:

First step: *Clarification of the Elements of Knowledge by Analysis:*
Explication of Conceptions
Decomposition of Facts

Second step: *Colligation of Facts by Means of a Conception:*
Selection of the Idea
Construction of the Conception
Determination of the Magnitudes

Third step: *Verification of the Colligation by:*
Prediction
Consilience
Simplification

It is not to be inferred from the above, however, that the first and the second steps, at all events, take place in strict temporal succession, the first being completed before the second is begun. Rather, "they are inseparably connected with each other" [48] and go on simultaneously.

B. *The explication of conceptions*

Each of the elements of the inductive process appearing in the foregoing outline will be examined in detail. We shall begin with the explication of conceptions.

What Whewell means by conceptions, and what relation he regards them as having to ideas, is made clear by him in the following passage:

> We have given the appellation of *Ideas*, to certain comprehensive forms of thought, —as space, number, cause, composition, resemblance,—which we apply to the phenomena which we contemplate. But the special modifications of these ideas which are exemplified in particular facts, we have termed *Conceptions;* as a circle, a square number, an accelerating force, a neutral combination of elements, a genus.[49]

By the explication of conceptions as so characterized, Whewell means "their clear development from Fundamental Ideas in the discoverer's mind, as well as their precise expression in the form of Definitions or Axioms when that can be done." [50] The explication of conceptions, he also says, "is the process by which we bring the Clearness of our Ideas to bear upon the formation of our knowledge"; [51] and the criterion of clearness of an idea is "that the person shall *see* the necessity of the Axioms belonging to each Idea;—shall accept them in such a manner as to perceive the cogency of the reasonings founded upon them." [52] The explication of a conception, he further says,

> . . . is effected not generally nor principally by laying down a definition of it, but rather by acquiring such a possession of it in our minds as enables, indeed compels us, to admit, along with the conception, all the axioms and principles which it necessarily implies, and by which it produces its effects upon our reasonings.[53]

This acquisition, in practice, has resulted mainly from controversies between men of science, which have indeed often assumed the form of battles of definitions. But, Whewell insists,

> . . . these controversies have never been questions of insulated and *arbitrary* Definitions . . . in all cases there is a tacit assumption of some proposition which is to be expressed by means of the definition, and which gives it its importance. The dispute concerning the definition thus acquires a real value, and becomes a question concerning true and false. The question really is, how the conception shall be understood and defined in order that the proposition may be true.[54]

> Definition and Proposition are the two handles of the instrument by which we apprehend truth; the former is of no use without the latter. . . . To unfold our conceptions by means of Definitions, has never been serviceable to science, except when it has been associated with an immediate *use* of the Definitions.[55]

Definitions, thus, are not the initial, but rather the terminal, point of the progress of knowledge, and the omission of them at the beginning or during the course of an inquiry is not to be regarded as a neglect which involves some blame: "The business of Definition is part of the business of discovery . . . the Definition as well as the discovery, supposes a decided step in our knowledge to have been made." [56] Definitions, moreover, can often be dispensed with, and axioms substituted for them without loss of clearness; for example, "the Idea of cause which forms the basis of the science of Mechanics, makes its appearance in our elementary mechanical reasonings, not as a Definition, but by means of the axioms that 'causes are measured by their effects,' and that 'Reaction is equal and opposite to action.' " [57]

Whewell goes on to say that, for any real advance in the discovery of truth, our ideas must be not only clear, but also *appropriate*. Thus, "Aristotle and his followers endeavored in vain to account for the mechanical relation of forces in the lever by applying the *inappropriate* geometrical conceptions of the properties of the circle." [58] The injunction that our ideas must be appropriate cannot be made use of in practice except to turn us away from labors certain to be fruitless. But, although "no maxims can be given which inevitably lead to discovery," on the other hand, "no *scientific discovery* can, with any justice, be considered *due to accident*," [59] for discovery depends upon the presence in the observer's mind of a clear and appropriate conception by which the facts observed may be analyzed and connected, and this is never the case with the minds of common men.

The explication of conceptions "does not admit of being much assisted by methods, although something may be done by Education and Discussion." [60] "The authors and asserters of the new opinions, in order to make them defensible, have been compelled to make them consistent. . . . The opinions discussed have been, in their main features, the same throughout the debate; but they have at first been dimly, and at last clearly apprehended," and this as the very result of the debate. Such debates are thus anything but idle controversies. They are, on the contrary, the very method (if they may be called a method) "by which the Explication of Conceptions is carried to the requisite point among philosophers." [61]

C. *The decomposition of facts*

"When we inquire," Whewell says, "what Facts are to be made the materials of Science, perhaps the answer which we should most commonly receive would be, that they must be *True Facts*, as distinguished from any mere inferences or opinions of our own." [62] There is, however, he maintains, no such thing as a fact in which ideas do not enter as an essential element: "We cannot obtain a sure basis of Facts, by rejecting all inferences and judgments of our own, for such inferences and judgments form an unavoidable element in all Facts. We cannot exclude our Ideas from our Perceptions, for our Perceptions involve our Ideas."[63] The solution of this difficulty is to be found, therefore, not in seeking to exclude ideas from facts, but rather in trying "to discern, with perfect distinctness, the Ideas which we include."[64] This will mean that "Facts when used as the materials of physical science, must be *referred to conceptions of the intellect only*, all emotions of fear, admiration, and the like, being rejected or subdued."[65] The facts, that is to say, are "to be observed, as far as possible, *with reference to place, figure, number, motion,* and the like conceptions; which, depending upon the Ideas of Space and Time, are the most Universal, exact and simple of our conceptions."[66] Moreover, "we are not to confine ourselves to these; but are to consider the phenomena *with reference to other conceptions also*," e.g., in the case of sound, to such conceptions as concord and discord.

This process of exclusion of emotional elements, and of inclusion, on the other hand, of definite relations of time, space, cause—or other ideas equally clear—into our recorded observations of facts is what Whewell means by the *decomposition* of the complex facts of experience into their elementary facts. And although the possession of such elementary facts, "clearly understood and surely ascertained," is not sufficient to the discovery of the laws of nature, yet it is a step that must necessarily precede all such discovery. The decomposition of complex facts into simple generally leads to the introduction of technical terms by which the simple facts are described, such as light *rays, reflection, refraction, force, momentum,* and *inertia.*

In chapter ii of the third book of *Novum organon renovatum,* Whewell gives an account of some of the methods most useful in observing facts with reference to the ideas involved in them, particularly those of number, space, and time. Such a discussion of methods and instruments of measurement, having no essential connection with his theory of induction, need not detain us here.

D. *The colligation of facts; induction*

Whewell's doctrine of the colligation of facts constitutes the most important and most original part of his contribution to the theory of induction. We shall therefore examine it in some detail.

Colligation of facts, he writes, is a term that can be applied "to every case in which, by an act of the intellect, we establish a precise connection among the phenomena which are presented to our senses." [67] With colligation so characterized, Whewell then defines induction as follows: "*Induction* is a term applied to describe the *process* of a true Colligation of Facts by means of an exact and appropriate conception. *An Induction* is also employed to denote the *proposition* which results from this process." [68] Again, when the two processes called "the Explication of the Conceptions of our own minds, and the Colligation of observed Facts by the aid of such Conceptions" are

> . . . united, and employed in collecting knowledge from the phenomena which the world presents to us, they constitute the mental process of *induction;* which is usually and justly spoken of as the genuine source of all our *real general knowledge* respecting the external world . . . *real* because it arises from the combination of Real Facts . . . *general,* because it implies the possession of general ideas.[69]

> In each inference made by induction there is introduced some General Conception, which is given, not by the phenomena, but by the mind. The conclusion is not contained in the premises, but includes them by the introduction of a New Generality. In order to obtain our inference, we travel beyond the cases which we have before us; we consider them as mere exemplification of some Ideal Case in which the relations are complete and intelligible. We take a Standard and measure the facts by it; and this Standard is constructed by us, not offered by Nature.[70]

Hence, Whewell goes on,

> . . . in every inference by Induction, there is some conception *superinduced* upon the Facts: and we may henceforth conceive this to be the peculiar import of the term *Induction.* I am not to be understood as asserting that the term was originally or anciently employed with this notion of its meaning; for the peculiar feature just pointed out in Induction has generally been overlooked. [Aristotle, for instance,] turns his attention entirely to the *evidence* of the inference; and overlooks a step which is of far more importance to our knowledge, namely, the invention of the second extreme term.[71]

It is, for instance, Whewell says elsewhere but still referring to Aristotle,

> . . . no adequate example of induction to say, "Mercury describes an elliptical path, so does Venus, so do the Earth, Mars, Jupiter, Saturn, Uranus; therefore all the planets describe elliptical paths." This is as we have seen, the mode of stating the *evidence* when the proposition is once suggested, but the Inductive step consists in the *suggestion* of a conception not before apparent,[72]

which constitutes the "Second Extreme Term" of Aristotle's inductive syllogism.

When we have once invented this "Second Extreme Term," we may, or may not, be satisfied with the evidence of the syllogism; we may, or may not, be convinced that, so far as this property goes, the extremes are coextensive with the middle term; but the *statement* of the syllogism is the important step in science. We know how long Kepler labored, how hard he fought, how many devices he tried before he hit upon this *Term*, the Elliptical Motion. . . . When he had established his premises, that "Mars does describe an ellipse about the Sun," he does not hesitate to *guess* at least that . . . he might . . . assert that "All the planets do what Mars does." But the main business was, the inventing and verifying the proposition respecting the Ellipse. The Invention of the Conception was the great step in the *discovery;* the Verification of the Proposition was the great step in the *proof* of the discovery.[73]

As to this, it might be objected that Kepler's original contribution did not in the least consist in the invention of the conception of an ellipse—which had been formulated much earlier—but in the thinking of it in connection with the endeavor to formulate a general description of the motion of Mars; and therefore that Whewell's terming the essence of the inductive step "the invention of a conception" is unfortunate. Although in many cases of discovery a new conception is invented by the discoverer, that is not so in every case. But Whewell's statements elsewhere indicate that he would freely admit this.[74] His real meaning is abundantly clear. In every induction, "there is a Conception supplied by the mind and superinduced upon the Facts," and although the conception may indeed be *suggested* by the facts, it is no more to be described as itself an additional fact observed like the rest, than is the key to a cryptogram to be described as itself observed in the cryptogram in the same way as the letters of it. The letters are *perceived*, the key is *thought of*, even though both the letters and the particular order in which they make sense are facts of the cryptogram.

Whewell goes on to point out that the contributions of the inventive intellect to our knowledge, to the extent that they are successful, cease correspondingly soon to be remembered as having had their origin in the intellect.

The pearls once strung, they seem to form a chain by their nature. Induction has given them a unity which is so far from costing us an effort to preserve, that it requires an effort to imagine it dissolved. For instance, we usually represent to ourselves the Earth as *round* . . . we hardly can understand how it could cost the Greeks . . . so much pains and trouble to arrive at a view which to us is so familiar . . . for in what other manner (we ask in our thoughts) could we represent the facts to ourselves? . . . Eclipses arrive in mysterious confusion: the motion of a *Cycle* dispels the mystery. The planets perform a tangled and mazy dance; but *Epicycles* reduce the maze to order. The Epicycles themselves run into confusion; the conception of an Ellipse makes all clear and simple. And thus from stage to stage, new elements of intelligible order are introduced. . . . Men ask whether eclipses follow a cycle; whether the planets describe Ellipses; and they imagine that so long as they do not *answer* such questions rashly, they take nothing for granted. They do not recollect how much they assume in *asking* the question;—how far the conceptions of Cycles and Ellipses are beyond the visible surface of the celestial phenomena. . . . And thus they treat the subject . . . as if it were a question, not of invention, but of proof . . . as if the main thing were not *what* we assert, but *how* we assert it. . . . In collecting scientific truths by induction we often find . . . a Definition and a Proposition established at the same time,—introduced together, and mutually dependent upon each other. The combination of the two constitutes the inductive act; and we may consider the Definition as representing the superinduced Conception, and the Proposition as exhibiting the Colligation of Facts.[75]

We might, Whewell says elsewhere, ask whether induction does not have a typical *formula* such as the syllogism constitutes for deduction. As such a formula, it would not be enough to say that all known particulars of a given kind are exactly included in a certain general proposition, for this brings out only the *evidence* for the induction but not the inductive step itself at all, which "consists in the suggestion of a conception not before apparent."[76] The inductive formula might then, he says, be something like the following: "These particulars, and all known particulars of the same kind, are exactly expressed by adopting the Conceptions and Statement of the following Proposition." Or, better still, if the clarification of the conception has resulted in a definition: "These facts are completely and distinctly expressed by adopting the following Definition and Proposition."[77] Moreover the conviction is present in the most perfect examples of induction, that the conception and assertion used in stating it are not only *sufficient* to express all the facts, but *also necessary*. The tabular arrangement of induced propositions in order of increasing generality is regarded by Whewell as of so much value that he calls such tables the "criterion of truth" for the doctrine they include. They are, he goes so far as to say, "the Criterion of Inductive Truth, in the same sense in which Syllogistic Demonstration is the Criterion of Necessary Truth." [78]

While the framing of a hypothesis is the inductive act itself, the complete process of *induction* is that of *hypothesis* and *verification*, or process of *trial and error*, for success in the use of which no positive rules can be given that will be of much practical value. Correct hypotheses, Whewell writes, are obtained "never without labour, never without preparation;—yet with no constant dependence upon preparation, or upon labour, or even upon personal endowments." [79] Whewell, however, probably having in mind Newton's much quoted "*Hypotheses non fingo*," insists that "a facility in devising hypotheses . . . is so far from being a fault in the intellectual character of a discoverer, that it is, in truth, a faculty indispensable to his task." [80] Such hypotheses or guesses will, of course, often or even usually be wrong, but "to try wrong guesses is, with most persons, the only way to hit upon right ones." Our theorizing faculties "obtain something, by aiming at much more. . . . The character of the true philosopher is, not that he never conjectures hazardously, but that his conjectures are clearly conceived, and brought into rigid contact with facts." [81] True philosophers, moreover, must possess a "pure love of truth and comparative indifference to the maintenance of their own inventions," [82] so that, if the facts are found to clash with their hypotheses, the latter are abandoned by them without hesitation.

E. *Rules of procedure and stages in the process of colligation*

"The Colligation of ascertained facts into General Propositions," Whewell writes, "may be considered as containing three steps, which I shall term the *Selection of the Idea*, the *Construction of the Conception*, and the *Determination of the Magnitudes*." [83] These are not so much three separate operations as three stages of definiteness in the process of colligating the facts which, as a preliminary, have been "decomposed." And, although the all-important requisite for the performance of successful inductions is the possession of a fertile, sagacious, ingenious, and honest mind, certain rules and methods of procedure useful in various degrees may be formulated in connection with the three stages in the colligation of facts mentioned.

i. *Selection of the idea.* This, in practice, amounts to the *proposing* of some scientific problem, to the *formulating* of some question concerning some set of known facts.[84] Concerning the selection of the idea, Whewell mentions, for what little it may be worth, the rule that "*the idea and the facts must be homogeneous* . . . thus if facts have been observed and measured by reference to space, they must be bound together by the idea of space." [85] Again, "*the idea must be tested by the facts*," although this, he adds, can hardly be called a rule.

ii. *Construction of the conception and determination of the magnitudes.* Whewell discusses the methods applicable to the construction of the conception under two headings, according to whether facts of *quantity* or facts of *resemblance* are in question. As special methods of induction applicable to *quantity*, Whewell mentions the methods of curves, of means, of least squares, and of residues. These, of course, do not in any sense constitute original contributions by Whewell to the theory of induction. Therefore, the only point of immediate interest here is the manner in which Whewell regards them as connecting themselves with his theory of induction, and the place at which they enter it. A very brief exposition of his account of them will suffice to indicate this.

a. *The method of curves.* "This consists in drawing a curve, of which the observed quantities are the Ordinates, the quantity on which the change of these quantities depends being the Abscissa. The efficacy of this method depends upon the faculty which the eye possesses, of readily detecting regularity and irregularity in forms," [86] while on the contrary to detect the relations of number considered directly as number is not easy. This method enables us not only "to obtain laws of nature from *good* observations, but also, in great degree, from observations which are very *imperfect*," [87] and even to obtain "data which are *more true than* the individual *facts themselves*." [88] This possibility of correcting the observations by the curve, so far as it really exists, depends on the fact that, though the observations may appear irregular, "the correct facts which they imperfectly represent, are really regular." [89] Two obstacles, however, impede the application of the method of curves: our ignorance of the nature of the quantity upon which the changes we are studying depend; and the complication of several laws with one another.[90]

b. *The method of means.* The correction of our observations by the method of curves consists, virtually, in taking the *mean* of the observations. But this may be done, without curves, arithmetically. The efficacy of the method of means depends upon this: "in cases in which observed quantities are affected by other inequalities, besides that of which we wish to determine the law, the excesses *above* and defects *below* the quantities which the law in question would produce, will, in a collection of *many* observations, *balance* each other." [91] However, knowledge of the quantity upon which the changes that we are studying depend (their "argument") is more indispensably necessary in the method of means than in the method of curves; for, to detect the law that holds within a collection of numbers, the subcollections *within* it that constitute the dependent term of the law must be established by *a selection of our own;* while in the case of a curve these subcollections are represented by the sinuosities of the curve, which the eye may be able readily to detect.

c. *The method of least squares.* This is a method of means the object of which is to discover the *best mean,* or *most probable law,* of a number of quantities obtained from imperfect observations. It assumes that small errors are more probable than large ones and defines the *best mean* as that which makes the sum of the *squares* of the errors the least possible.

d. *The method of residues.* When, in a series of changes of a variable quantity, *one* law has been detected by such methods as already mentioned, but that law does not fully account for the observed changes, thus leaving a *residue* of unexplained fact, that residue is itself in turn studied as the original set of observations was studied, until another law governing the quantities that constitute the residue has been detected. If this second law in turn leaves a residue, that residue is studied, and so on until the facts are completely accounted for. Whewell mentions Herschel's discussion of this method, stating that, as described by Herschel, its application is not limited to quantitative data.

As inductive methods depending upon *resemblance,* Whewell mentions the method that appeals to the law of continuity, the method of gradation, and the method of natural classification.

e. *The law of continuity.* "A quantity cannot pass from one amount to another by any change of conditions, without passing through all intermediate degrees of magnitude according to the intermediate conditions." [92] Although this law is primarily applicable to quantities, Whewell treats of it under the heading of resemblance because "its inferences are made by a transition from one degree to another among contiguous cases." The law of continuity "is a test of truth, rather than an instrument of discovery"; [93] it enables us to detect the falsity of inductions that would involve a violation of it. The evidence of the law of continuity itself is of the same intuitive nature as in the case of other fundamental ideas.[94]

f. *The method of gradation.* When classes of things and properties have been established in virtue of comparisons in respect of agreements and differences, "it may still be doubtful whether these classes are separated by distinctions of opposites, or by differences of degree." To settle such questions the method of gradation is used. It consists in "taking intermediate stages of the properties in question, so as to ascertain by experiment whether, in the transition from one class to another, we have to leap over a manifest gap, or to follow a continuous road." [95]

g. *The method of natural classification.* This consists "in grouping together objects, not according to any selected properties, but according to their most important resemblances; and in combining such grouping with the assignation of certain marks of the classes thus formed." [96] What Whewell means by speaking of a resemblance as "important" and as giving rise to a classification that is "true" or "natural" is that the resemblance is one on the basis of which *general propositions become possible.* This, he says, is "the criterion of a true classification," [97] and distinguishes natural classifications from artificial classifications based upon *arbitrary* definitions. In natural classification, we "find our classes in Nature, and do not make them by marks of our own imposition," to which exceptions constantly would be found.

F. *The notion of cause*

Whewell concludes the discussion of the methods of induction applicable to quantity and resemblance with a statement that they "commonly lead us to *Laws of Phenomena* only." He adds, however, that "Inductions founded upon other Ideas, those of Substance and Cause for example, appear to conduct us somewhat further into a knowledge of the essential nature and real connexions of things." [98] In proceeding to such inductions, however, "we can no longer lay down any Special Methods by which our procedure may be directed." [99]

On the part of a writer thoroughly acquainted, as Whewell was, with Herschel's *Discourse*, and therefore with the "methods" to which Mill later gave the names of methods of agreement, of difference, and of concomitant variations, the last statement quoted is rather startling. The explanation of it is to be found in the fact that Whewell is here essentially concerned with *discovery* and that he does not admit these methods to be methods of discovery: "They take for granted the very thing which is most difficult to discover, the reduction of phenomena to formulae" such as Mill sets forth in terms of A, B, C, and a, b, c, etc., to represent the elements of antecedent and sequent phenomena. But where, Whewell asks, "are we to look for our A, B, C, and a, b, c? Nature does not present to us the cases in this form." [100] The characters of agreement, difference, and concomitant variation may well belong to the causal relation, but they then constitute the *object* of discovery, not the method. And he quotes with approval Herschel's comment upon Bacon's "Prerogative Instances," of which they are cases, to the effect that the difficulty is to find them, not to perceive their force when found.

It is true that Herschel nevertheless presents his five statements on the characters of the causal relation and his ten "rules of philosophizing" based upon them as useful for guiding and facilitating our *search* for causes, but he is well aware of the applicability to them of his own remarks concerning Bacon's "Prerogative Instances." And he conceives his rules as methods of discovery mainly in the negative sense: they are principles of *criticism* of the hypotheses concerning causes, which we frame. The method of discovery is, for him no less than for Whewell, essentially the method of hypothesis, i.e., the method of trial and error. In the case of Mill's discussion of the "methods," it is, as usual in his writings, difficult if not impossible to determine which of his contradictory statements represents his true view. In spite of the distinction between proof and discovery which is made so much of by him and his followers,[101] the very first line of the famous chapter of the *System of Logic* on the "Methods of Experimental Inquiry" refers to them as "modes of *singling out*" (our italics) the causes or effects of a phenomenon from among its antecedents or sequents, and on the next page he speaks still more explicitly of "the mode of discovering and proving laws of nature" which he has been discussing.

But, even granting Whewell's objection to the "methods" as described by Herschel and Mill, it may yet well be maintained that they are often of a good deal of importance at least in *suggesting* causal hypotheses—certainly of no less importance than the methods Whewell acknowledges as of value to that end—and therefore that they should have been given by him a place at the point of his exposition that we are now considering. At any rate, they are of commanding importance as methods *of verification of causal hypotheses,* and Whewell's neglect of them under that heading also constitutes a grave fault. Probably, it is due in no small measure to the haziness of his notion of cause, complicated by the preoccupation evident throughout in his discussion that that notion shall be so conceived as to enable us to speak of God as the supreme cause of the universe.[102] This, it may be noted, furnishes a clear, even if an unfortunate, example of what he means when he contends that definitions are ever so framed as to make some contemplated proposition true. Brief mention of some of his characteristic statements in regard to causation will make evident the complete negligibility of his attempt to analyze the notion.

By cause, Whewell declares, "we mean some quality, power or efficacy, by which a state of things produces a succeeding state. Thus the motion of bodies from rest is produced by a cause which we call *Force:* and in the particular case in which bodies fall to the earth, this force is termed *Gravity.*"[103] And later he writes: "When one event gives rise to another, the first *event* is, in common language, often called the cause, and the second the effect,"[104] as when one billiard ball, striking another, turns out of its path. For the purpose of expressing general and abstract truths concerning cause and effect, however, the term cause must not be applied to such occurrences,

> . . . but to a certain conception, *force,* abstracted from all such special events, and considered as a quality or property by which one body affects the motion of the other. And in like manner in other cases, cause is to be conceived as some abstract quality, power, or efficacy, by which change is produced; a quality, not identical with events, but disclosed by means of them.

That cause as so conceived is not an idea derived from experience is proved, Whewell believes, by the fact that we know the proposition, "Every event must have a cause," to be true "not only probably, and generally, and as far as we can see; but we cannot suppose it to be false in any single instance." [105] And in further evidence of the non-empirical origin of the idea of cause, he adduces the fact that we apply that idea to infer the existence of a First Cause of a nature different from the events themselves, which would be unwarranted if experience were the source of the idea of cause!

It is perhaps superfluous to remark that the legitimacy of such an application of the idea is not in the least established by the fact that some people do so apply the idea. Moreover, the proposition, "Every event must have a cause," not only *can* be supposed false, for aught Whewell shows to the contrary, but in fact has been regarded as false or doubtful by many persons since Hume. And, with regard to the characterizing of "cause" as "force," or in general as "power or efficacy," it is obvious that when this is done, and at the same time forces or powers are declared (as by Whewell no less than Herschel) never to be observable but only to be "inferred" from their "effects," all that has been effected is a purely verbal substitution, viz., that of the word "force" for the word "cause," which does not explain or clarify anything. It is but a case of the time-worn practice of giving a name to that for which one is inquiring and calling that name the answer.

Whewell, it may further be remarked, wholly fails to support his assertion that, for the purpose of expressing general and abstract truths concerning cause and effect, the word "cause" must not be applied to events or occurrences. Since, as he himself observes, the word *is* so applied in common language, some show of reasons for departing from current usage would certainly appear to have been in order. But his only reason for the assertion seems to have been that already mentioned, viz., that God is not an event or an occurrence, and that cause ought to be so defined as to make true the proposition that God is the cause of the universe. To Comte's injunction that "science must study only the laws of phenomena, and never the mode of production," Whewell replies that

> . . . the laws of phenomena, in many cases, cannot be even expressed or understood without some hypothesis respecting their mode of production. How could

the phenomena of polarization have been conceived or reasoned upon, except by imagining a polar arrangement of particles, or transverse vibrations, or some equivalent hypothesis? [106]

But, while the justice of the reply may be conceded, it must be noted once more that what it thus appears necessary to suppose is not the existence of a "force" or "forces," but, on the contrary, of such entities as *particles,* and of such *events* as the transverse vibration of those particles.

Whewell, however, is prone to regard axioms as of more value than definitions in the clarification of conceptions, and we therefore quote the three axioms relating to the idea of cause which he gives,[107] and for the self-evidence of which he refers us to our own thoughts.[108] The first is: "*Nothing can take place without a cause.*" To this, Whewell says, we irresistibly assent as soon as it is understood. The second axiom is: "*Effects are proportional to their Causes, and Causes are measured by their effects*"; and the third: "*Reaction is equal and opposite to action.*" This means that cause and effect mutually determine each other, and Whewell erroneously regards this as implying that in direct causation cause and effect are not successive, as is generally asserted, but strictly simultaneous. The current maxim that cause precedes effect is true, he thinks, only when (as usually) *cumulative* effect, and not *direct* effect, is considered. Thus if a ball A strikes a ball B, and B a ball C, the motion of C can be spoken of as the effect, though not the direct effect, of the impact of A, this illustration being quoted by Whewell from Herschel's review of his book in the *Quarterly Review,* already cited.

G. *The process of verification*

Verification, in Whewell's view, is a most important element of the inductive process. Three different tests of hypotheses are discussed by him.

The first is that of *adequacy.* The hypotheses that we entertain should be sufficient to explain all the phenomena that we have observed. Whewell, in this connection, remarks that, even if a hypothesis only *approximates* to this desideratum, it may yet be of value: "hypotheses may often be of service to science, when they involve a certain portion of incompleteness, and even of errour," [109] for "if our scheme has so much of truth in it as to conjoin what is really connected, we may afterwards duly correct or limit the mechanism of this connexion." Thus, he says, it is after all true that "nature *does* abhor a vacuum and does all she can to avoid it," even though her power to avoid it is not unlimited, and her exertions to do so are determined by the pressure of the circumambient air. However, in spite of the value that partially adequate hypotheses may have, "we are never to rest in our labours or acquiesce in our results, till we have found some view of the subject which *is* consistent with *all* the observed facts." [110]

The second test of a hypothesis consists in its capacity "to *foretell* phenomena which have not yet been observed; at least all phenomena of the same kind as those which the hypothesis was invented to explain." [111] "The prediction of results, even of the same kind as those which have been observed, in new cases, is a proof of real success in our inductive processes." [112]

The third test Whewell mentions is the capacity of a hypothesis to explain and predict cases of a different kind from those which were contemplated in the formation of the hypothesis. When this takes place, we have what he terms a *"Consilience of Inductions"*; that is, two laws obtained by independent inductions and concerning apparently heterogeneous classes of phenomena turn out to be, both of them, deducible from one and the same hypothesis. This constitutes the most striking and convincing test of the correctness of a hypothesis. Such a hypothesis may equally be described as *rendering consilient* separate inductions, or as *tending to simplicity* and harmony, or again as constituting an ascent to a higher degree of *generality*. This third test of Whewell's is thus in its essence identical with that constituted by the principle that Sir William Hamilton has called the "principle of parsimony" [113] and with the well-known maxim connected with the name of William of Occam.

Considering the test constituted by the "simplicity" of a hypothesis, Whewell says that the various elements of a hypothesis are often not thought of all at once, but are, on the contrary, added one after another, on the occasion of different researches. In the case of true theories,

> . . . all the additional suppositions *tend to simplicity* and harmony; the new suppositions resolve themselves into the old ones, or at least require only some easy

modification of the hypothesis first assumed: the system becomes more coherent as it is further extended. The elements which we require for explaining a new class of facts are already contained in our system. Different members of the theory run together, and we have thus a constant convergence to unity. In false theories, the contrary is the case. The new suppositions are something altogether additional;—not suggested by the original scheme; perhaps difficult to reconcile with it. Every such addition adds to the complexity of the hypothetical system, which at last becomes unmanageable, and is compelled to surrender its place to some simpler explanation.[114]

As examples may be mentioned the ancient astronomical doctrine of eccentrics and epicycles, the Cartesian system of vortexes, and the doctrine of phlogiston.

4. CONTROVERSY WITH MILL OVER THE NATURE OF INDUCTION

In the second chapter of the third book of his *System of Logic,* Mill discusses at length Whewell's use of the term induction, which he deems improper as "confounding a mere description, by general terms, of a set of phenomena, with an induction from them." When, Mill says, a navigator sails along the shore of an unknown land and eventually (without having changed the course of his ship with reference to the shore) returns to the point from which he started, he concludes that the land in question is an *island.* This is not a generalization of his observations; it is not an inference, in the sense of an assertion going in any way beyond those observations; it is simply a *summary description of his observations.* And the matter stands likewise, Mill says, with regard to Kepler's proposition that the path of the planet Mars is an *ellipse.* To apply the conception of an ellipse to the observed places of Mars was but to give a summary description of them; it was not an induction. "The only real induction concerned in the case, consisted in inferring that because the observed places of Mars were correctly represented by points in an imaginary ellipse, therefore Mars would continue to revolve in that same ellipse," and also in inferring that the positions of the planet between observations were on the same regular curve as the observed positions. But, Mill says, "these inferences were so far from being a part of Kepler's philosophical operation, that they had been drawn long before he was born.

. . . He merely applied his new conception to the facts inferred, as he did to the facts observed." And he did not "(which is the true test of a general truth) add anything to the power of prediction already possessed."

That a navigator's statement that the land he has kept in sight is an island is *more* than a summary description of his observations, however, is evident as soon as we reflect that, so far as the observations supposed go, the land in question might be not an island but a continent containing an inland sea, which the ship has circumnavigated. To say that an assertion concerning some observations goes beyond them need not necessarily mean that it generalizes them. It may go beyond them in the sense of asserting that with the facts observed go certain other facts which were not observed and which are not mere extrapolations or intrapolations, but facts of a different sort. For instance, to call the land in sight of which one has kept until one returned to one's starting point an island is to assert something concerning the relative sizes of the continental and water areas, or concerning the right or left curvature of the shore with regard to the ship's course, since, on a globe, it would be in some such terms that the distinction between an island and a continent containing an inland sea would have to be defined. Mill does not perceive that increase in our power of prediction, which he requires, occurs not only when we discover a new law, but also when a concretely present fact is discovered to be a case of a kind of whose properties we already have knowledge. In other words, the minor instead of the major premise of a syllogism may be what was as yet unknown, and with knowledge of it comes power of prediction concerning the subject of that minor.

Whewell would acknowledge that the idea of ellipticity as applied to the orbit of Mars does summarize our observations in the sense that it states them all at once, but not in the sense that it was to be obtained from them by a mere process of summation. Moreover, as Venn pointedly remarks,[115] the orbit of Mars, as a bare matter of fact, is *not* an ellipse. From the purely empirical standpoint that was Kepler's, the observations of positions out of the elliptical path were data of precisely the same importance as any others. They were not "deviations" until a "normal" orbit had been conceived. Kepler could not have *seen* the ellipse in the facts, first because it is not there, and second because the path is so nearly circular that, if it were displayed, no ordinary eye could detect that it was not a circle.

What Kepler did was thus neither to "see" the ellipse in the facts, nor to "sum them up" and find the result an ellipse. Rather, as Whewell maintains, Kepler did something exactly analogous to what is done when the key of a cipher is being discovered. He thought of a hypothesis from which all the known facts, and others as yet unobserved, could be logically deduced. And there is a vast difference between stating what one perceives (describing), and stating that what one perceives is such as it would be if a certain conjecture were true (diagnosing). Whewell is undoubtedly right in maintaining that it is through such acts of discovery that science progresses.

Whether such acts are to be called "inductions," however, is another question, which is essentially one of terms. Whewell, as we have seen, acknowledges that this is not the traditional meaning of the term. But, if the term induction is to be used, as it generally is, to denote the process by which science grows, then certainly the discovering of a correct hypothesis (whether of fact or of law) has a better claim to the name of induction than the more or less mechanical act of extrapolation or intrapolation by which generalizations are obtained. And etymologically induction, i.e., leading in, would describe the leading in of the colligating conception no less accurately than it describes the leading in of the particulars to be generalized.

Mill objects to Whewell's view of induction also on the ground that "induction is proof," [116] and that Whewell's theory passes over altogether the question of proof. But, of course, this is not so. Whewell on the contrary emphasizes the need of *verifying* our hypotheses; and the tests he indicates to that end have been considered above. Whewell, as we have seen, contends that the method of discovery is that of trial and error. But that very method consists in *proving* true or false each hypothesis thought of, in its turn. That is precisely what differentiates a discovery, properly so called, from a mere guess never tested.

Moreover, Mill's emphatic assertion that induction is proof should not be taken too seriously. For one thing, he does not himself stick to it very strictly. In an earlier passage he defines induction as "the operation of *discovering and* proving general propositions" (italics ours).[117] Also, to the assertion that induction is proof he adds immediately that "it is inferring something unobserved from something observed"—a statement that certainly describes discovery more nearly than proof; for proof consists in adducing evidence in support of a proposition already thought of, not in the inferring of something that one had not hitherto thought of. However, the probability is that when Mill said that induction is proof what he was thinking of were the "Four Methods" that he adopted from Herschel's *Discourse*.

5. ESTIMATE OF WHEWELL'S IMPORTANCE

Of Whewell's view as a whole, it may be said that it defines induction as the discovery of hypotheses adequately accounting for known facts. For him, hypotheses concerning causes would constitute only one species among others. Adequate hypotheses are discovered by the method of trial and error, each hypothesis thought of being tested in turn against the facts until a hypothesis is found which the facts verify. That other aspect of Whewell's view, according to which the correct hypotheses are "necessary truths" apprehended by (progressive) intuition, is reconcilable with the doctrine just described only if the process of comparing hypotheses with facts is conceived as essentially a device for the clarification of the meaning of our hypotheses. It can be so conceived, of course, although perhaps rather artificially. Whewell's doctrine on the subject of necessary truths, however, is at least in part rooted in misapprehensions concerning the meaning of "necessary." As he uses the term, it evidently means "absolutely necessary." He fails to see that whatever is "necessary" is *necessitated by* something or other, i.e., that one can never say that anything is necessary simply, but only that it is necessary *relatively to* something else. Absolute, i.e., nonrelative, necessity is therefore no less of a contradiction in terms than would be a childless father. The problem that is important for the theory of knowledge is not whether there are necessary truths—which may be granted—but what exactly are the grounds of their necessity. This problem Whewell never discusses adequately. Concerning it, we find in his writings only echoes of Kant's proposed solution, than which they are still more unsatisfactory.

Whewell's importance in the history of theories of induction is owing mainly to three things. One is his insistence on the thesis that proposition and definition are the two handles of the instrument by which knowledge grows, and the admirable way in which he makes clear the relation of definitions to propositions. The second is that Whewell's doctrine is the most conspicuous example, up to his time, of an attempt to base a detailed and concrete philosophy of science upon essentially Kantian epistemological premises. But the third and most significant is that Whewell is the first to formulate a comprehensive and systematic theory of induction throughout in terms of the so-called Newtonian method of Hypothesis—Deduction—Verification.

CHAPTER TEN

John Stuart Mill's System of Logic

I. THE MEANING OF "INDUCTION"

WITH REGARD to the meaning Mill attaches to the word "induction," several clear-cut statements are to be found in Book III of his *A System of Logic:* "For the purposes of the present inquiry, Induction may be defined, the operation of discovering and proving general propositions." [1] Again: "We shall fall into no error, then, if in treating of induction, we limit our attention to the establishment of general propositions. The principles and rules of induction as directed to this end, are the principles and rules of all induction." [2] And in the next chapter: "Induction is that operation of the mind, by which we infer that what we know to be true in a particular case or cases, will be true in all cases which resemble the former in certain assignable respects." [3] In another section of the same chapter, he refers to "that transition from known cases to unknown, which constitutes Induction in the original and acknowledged meaning of the term." [4] And in another chapter we find:

> Induction properly so-called may be summarily defined as Generalization from Experience. It consists in inferring from some individual instances in which a phenomenon is observed to occur, that it occurs in all instances of a certain class; namely, in all which *resemble* the former in what are regarded as the material circumstances.[5]

This definition of induction as generalization, however, is somewhat qualified by Mill in the chapter in which he sets forth his well-known doctrine that all induction is really from particulars to particulars, and that the formulation of a universal proposition in between is not essential. The reasons on which he bases this view appear in the following passage, which must be quoted before his new definition of induction can become intelligible:

The mortality of John, Thomas, and others is, after all, the whole evidence we have for the mortality of the Duke of Wellington. Not one iota is added to the proof by interpolating a general proposition [viz., that all men are mortal]. Since the individual cases are all the evidence we can possess, evidence which no logical form into which we choose to throw it can make greater than it is; and since that evidence is either sufficient in itself, or, if insufficient for the one purpose, cannot be sufficient for the other; I am unable to see why we should be forbidden to take the shortest cut from these sufficient premises to the conclusion.[6]

These considerations, he believes, enable us to discern the essence of all reasoning:

We have thus obtained what we were seeking, a universal type of the reasoning process. We find it resolvable in all cases into the following elements: Certain individuals have a given attribute; an individual or individuals resemble the former in certain other attributes; therefore they resemble them also in the given attribute.[7]

And with regard to the use of the term induction, he now says:

Although, therefore, all processes of thought in which the ultimate premises are particulars, whether we conclude from particulars to a general formula, or from particulars to other particulars, according to that formula, are equally induction; we shall yet, conformably to usage, consider the name induction as more peculiarly belonging to the process of establishing the general propositions, and the remaining operation, which is substantially that of interpreting the general proposition, we shall call by its usual name, Deduction. And we shall consider every process by which anything is inferred respecting an unobserved case, as consisting of an Induction followed by a Deduction; because, although the process needs not necessarily be carried on in this form, it is always susceptible of the form, and must be thrown into it when assurance of scientific accuracy is needed and desired.[8]

2. IS THE INDIRECT ASCERTAINMENT OF INDIVIDUAL FACTS INDUCTION?

The preceding quotation represents Mill's most explicit and careful statement of the meaning in which he wishes to use the term induction. There is a question, however—that of the indirect ascertainment of individual facts—the discussion of which leads him into inconsistencies so as to render it impossible to say what after all he does, or does not, mean by induction. To his expressions on the subject we may now advert.

Mill, in several places, points out that individual facts are in many cases obtained not by observation alone, but as a result of inference from observation. The question then arises in connection with the definition of the term induction, whether the process of inference by which such individual facts are established is to be called inductive in the sense that he has placed on the term. On this point, Mill expresses himself as follows:

> It is true that the process of indirectly ascertaining individual facts is as truly inductive as that by which we establish general truths. But it is not a different kind of induction; it is a form of the very same process . . . whenever the evidence which we derive from observation of known cases justifies us in drawing an inference respecting even one unknown case, we should on the same evidence be justified in drawing a similar inference with respect to a whole class of cases.[9]

As an instance of what he means, Mill offers the problem of ascertaining the distance of the moon from the earth by trigonometry, on the basis of simultaneous observations of the zenith distances of the moon taken from two far distant points on the earth's surface. In the process, he points out, the share of direct observation is limited to the ascertainment of these zenith distances; all the rest, he asserts, is induction. At each step in the demonstration, "a new induction is taken in, represented in the aggregate of its results by a general proposition." The process whereby the individual fact of the distance of the moon was ascertained is therefore exactly similar, he says, to that by which science establishes its general truths; in fact, "a general proposition might have been concluded instead of a single fact," namely, "a theorem respecting the distance, not of the moon in particular, but of any inaccessible object; showing in what relation that distance stands to certain other quantities."[10]

Mill's thought in the discussion of this example, however, appears to be confused. First, he forgets that not the theorem but the individual fact of the distance of the moon was in question. The theorem, *in the instance,* could *not* have been concluded instead of the single fact, because the theorem (or, if Mill should prefer, the evidence on which the theorem could be established) constitutes a necessary *premise* of the conclusion as to the individual fact of the distance of the moon —a premise, one may add, neither obtained from nor confirmed by the observation of the moon. As well say that, when I sharpened my pencil, the knife wherewith I did so might equally well have been used to sharpen itself! And, again, because the process by which the distance of *our individual moon* was ascertained does include the observed fact of *our moon's* zenith distances—this constituting the minor premise of the argument—*that* process, in its entirety, could *not* have yielded the distance of any other object but just *that* moon. Either, therefore, the process of indirect ascertainment of individual facts is induction, and then, in spite of Mill's assertion, induction is not always capable of yielding general propositions; or else induction is always capable of yielding general propositions, and then the process of indirect ascertainment of individual facts is not induction.

Mill, in the passage quoted, apparently fails to differentiate between two different processes. It is indeed true that, *before* one can establish indirectly an individual fact, one must have (direct) evidence adequate to establish a general proposition concerning facts that resemble in certain respects the one given. But the process by which the individual fact is established—once that evidence is known as adequate to the general proposition—and the process whereby that evidence itself becomes known as adequate to the general proposition, are *quite distinct and different in kind*, and the word induction cannot be used to designate either one indifferently.

But the most illuminating comment on the passage just discussed is really provided by comparing it with other statements of Mill himself. Mill contends that what Whewell calls induction is merely description. He points out that the statement of an observation always involves the assertion of a resemblance, so that "there is always something introduced which was not included in the observation itself; some conception common to the phenomenon with other phenomena to which it is compared," and thus,

> . . . it is impossible to express in words any result of observation, without performing an act possessing what Dr. Whewell considers to be characteristic of Induction. . . . But this identification of an object—this recognition of it as possessing certain known characteristics—has never been confounded with Induction. . . . It is a perception of resemblances, obtained by comparison.

Moreover, Mill goes on, and here we reach the point of interest, "these resemblances are not always apprehended directly . . . they are often ascertained through intermediate marks, that is, deductively." [11] As an example of this, he mentions the process by which the shape of the earth is ascertained through inference from such marks as that its shadow thrown upon the moon is circular, or that, on the sea, the horizon is always a circle.

Now, it should be carefully noted that both the fact of the shape of the earth and the process of inference by which Mill represents it as being ascertained are of precisely the same sort as the fact of the distance of the moon and the process by which that distance is obtained. Both are inferences of an individual unobserved fact, on the basis of the observation of more or less direct marks of it. If the distance of the moon was an induction, as he explicitly declared it to be,[12] the shape of the earth must also be an induction (and, incidentally, Whewell's view of induction is then vindicated). Mill, in fact, in Book III, chapter i, section 2 (as Whewell himself indeed notes [13]), mentions them *both* as instances of individual facts that have to be ascertained indirectly, and he refers to *both* as inductions. But in Book IV, chapter i, section 3, Mill denies with equal explicitness that such facts are inductions:

> We could not without impropriety call either of these assertions [that the earth is a globe, and that it is an oblate spheroid] an induction from facts respecting the earth. They are not general propositions collected from particular facts, but particular facts deduced from general propositions. They are conclusions obtained deductively from premises originating in induction: but of these premises some were not obtained by observation of the earth, nor had any peculiar reference to it. If, then, the truth respecting the figure of the earth is not an induction, why should the truth respecting the figure of the earth's orbit [as Whewell contends] be so? [14]

And, we may go on, why should the truth respecting the distance of the moon from the earth, ascertained by precisely the same sort of process, and *called by Mill* an induction, be so?

One may further point out that the ascertainment of the shape of the earth through the process described by Mill is as exact a case as one might wish to find of what he characterizes in the following passage, already quoted, as the type to which all reasoning can be reduced: "Certain individuals have a given attribute; an individual or individuals resemble the former in certain other attributes; therefore they resemble them also in the given attribute." For we may paraphrase thus: Certain individuals have a given attribute, globularity; an individual, the earth, resembles the former in certain other attributes, e.g., casting a circular shadow, etc.; therefore it resembles them also in the given attribute, globularity. Are we not then entitled to conclude that the truth regarding the figure of the earth is not merely "a perception of resemblances," but an inference and, indeed, an inductive inference?

Mill's various utterances now before us, concerning what is and what is not induction, contain such large inconsistencies that it is impossible to say to what, exactly, he proposes to limit the use of the term. That the process by which general propositions are established is induction, is clear; but beyond this little can be said that could not be infirmed as definitely as confirmed by reference to Mill's own words.[15]

3. THE UNIFORMITY OF NATURE

Let us now turn to the question, on what ground, according to Mill, it is possible to pass from particular observations to a general proposition—a question to which he gives considerable attention. He deals with it in the third chapter of Book III, entitled "Of the Ground of Induction," and in the twenty-first chapter of the same book, entitled "Of the Evidence of the Law of Universal Causation."

Induction, he tells us in a passage already quoted, may be "summarily defined as Generalization from Experience. It consists in inferring from some individual instances in which a phenomenon is observed to occur, that it occurs in all instances of a certain class." But, he observes:

> . . . there is a principle implied in the very statement of what Induction is; an assumption with regard to the course of nature and the order of the universe; namely, that there are such things in nature as parallel cases; that what happens once will, under a sufficient degree of similarity of circumstances, happen again, and not only again, but as often as the same circumstances recur. This, I say, is an assumption involved in every case of induction.[16]

What now, according to Mill, is our warrant for assuming this uniformity of nature? His answer is that the proposition that the course of nature is uniform is itself an induction from experience. "It would yet be a great error," he tells us, "to offer this large generalization as any explanation of the inductive process." On the contrary:

I hold it to be itself an instance of induction, and induction by no means of the most obvious kind. . . . The truth is, that this great generalization is itself founded on prior generalizations. The obscurer laws of nature were discovered by means of it, but the more obvious ones must have been understood and assented to as general truths before it was ever heard of. . . . In what sense, then, can a principle which is so far from being our earliest induction be regarded as our warrant for all the others?

This "ultimate major premise of all inductions" can be so regarded, he answers, only in the one sense in which the major premise of a syllogism ever really is a warrant of the conclusion, viz., "not contributing at all to prove it, but being a necessary condition of its being proved; since no conclusion is proved, for which there cannot be found a true major premise." [17]

Now, both this particular application of Mill's peculiar view of the relation of the major premise of a syllogism to its conclusion, and the doctrine itself in general, seem to rest on a confusion in Mill's mind between psychological and logical considerations. That the formulation of the principle of the uniformity of nature is, as an event in the history of the psychological processes of the race, a late fact, is indeed true; but this has no bearing whatever on the question of the *logical* priority of that principle to all inductive generalization. For there is no necessary parallelism between the temporal order of our thoughts as psychological facts and the order of logical implication among the meanings of those thoughts. As a bare matter of fact, indeed, the logically prior considerations are usually thought of later than the truths that presuppose them.

But let us deal directly with Mill's position concerning the function of the major premises of syllogisms in general. We may grant that, *as a matter of psychological fact,* we often pass from the assertion of some particular proposition known to be true to that of another particular proposition, without passing through a generalization (though it is very questionable whether such merely associative processes have better claim to the name of reasoning than any other instance of mere habit).[18] That we in fact do so, however, and that the second proposition often turns out to be true, throws no light whatsoever on the question whether the first particular proposition *proves* the second, or whether on the contrary a certain general proposition alone does so; for proof is a matter of logical evidence, not of psychological sequence. And if, as we would assert, the general proposition *is* logically necessary to the proof of the second, we might grant further that that general proposition is itself logically evidenced only by a number of particular observations, and yet maintain that the statement that the second is an inference of a particular from particulars is false, or at least entirely misleading. For it is not from these particulars *as particulars* that the second is inferred, but from these particulars *as bound to one another by a specific relation* (such as contemplated in the canons of Mill's Four Methods of Experimental Inquiry), which makes them, *as one whole so bound, logically equivalent* to the general proposition. And with reference to the function that Mill does assign to the major premise of a syllogism, we may remark that, if we were told that the air does not contribute at all to support a bird in flight, but is a necessary condition of its being supported in flight, we would say that the distinction attempted between "contributing to" and "being a necessary condition of" is a distinction without an adequate difference, and that the assertion is therefore a plain contradiction. Now, when Mill tells us that the major premise in a syllogism does not contribute at all to prove the conclusion, but is a necessary condition of its being proved, we must reply that, if this is not an equally plain contradiction, it can be only because the word "prove" is used therein ambiguously; first to refer to a psychological process and next to refer to a logical relation. That is, the statement, if true, can mean only that the correct conclusion is often arrived at, as a bare matter of psychological fact, without the *thought* of the major premise entering into our psychological processes, although the *truth* of the major premise is logically necessary evidence of the truth of the conclusion (with the minor as datum). But then we must insist that, unless we have *actually noted* the truth of the major, or noted that we have sufficient evidence to establish it,[19] we may indeed have reached the true conclusion, but we do not know that we have done so, i.e., we have not proved it.

Mill's reference to the principle of the uniformity of nature as having to all inductions the same relation that the major premise of a syllogism has to the conclusion, therefore, in no wise disposes of the difficulty that confronts him when he tells us both that the principle "is an assumption involved in every case of induction" and that it is itself an instance of induction founded on prior generalizations. Mill's own statement of the difficulty, which in Book III, chapter xxi, section 2, he tries to meet otherwise, is so clear as to deserve quoting:

> We arrive at this universal law by generalization from many laws of inferior generality. . . . As, however, all rigorous processes of induction presuppose the general uniformity, our knowledge of the particular uniformities from which it was first inferred was not, of course, derived from rigorous induction, but from the loose and uncertain mode of induction *per enumerationem simplicem:* and the law of universal causation, being collected from results so obtained, cannot itself rest on any better foundation. . . . Is there not then an inconsistency in contrasting the looseness of one method with the rigidity of another, when that other is indebted to the looser method for its own foundation? [20]

Mill endeavors to escape this inconsistency by telling us:

> . . . the precariousness of the method of simple enumeration is in an inverse ratio to the largeness of the generalization. The process is delusive and insufficient, exactly in proportion as the subject-matter of the observation is special and limited in extent. As the sphere widens, this unscientific method becomes less and less liable to mislead; and the most universal class of truths, the law of causation, for instance, and the principles of number and of geometry are duly and satisfactorily proved by that method alone, nor are they susceptible of any other proof.[21]

The truth of these remarks may be admitted, but the important question is, *on what assumption is their truth admissible?* Obviously only on the assumption that the nature of the observed—in proportion to the extensity of the observations, if at all, perhaps—is evidence of the nature of the unobserved. But this is the principle of the uniformity of nature itself; if it is not assumed to start with, Mill's statement that the precariousness of induction by simple enumeration is in inverse ratio to the largeness of the generalization is therefore entirely groundless and nothing but a dogmatic assertion. The same *petitio principii* is involved in the following statement: "I fully admit that if the law of causation were unknown, generalization in the more obvious cases of uniformity in phenomena would nevertheless be possible, and though in all cases more or less precarious, and in some extremely so, would suffice to constitute a certain measure of probability." [22] Why would it, we ask, except on the assumption that the unobserved cases resemble, or tend to resemble, the observed?

Mill's position on the relation of the principle of the uniformity of nature to induction is thus seen to be thoroughly untenable. The implications of strict empiricism in respect to this question were realized by Hume, in spite of his own numerous inconsistencies, with far greater clearness than by Mill, whose discussion of the uniformity of nature obscures the issues without solving any of the difficulties.

4. CAUSATION AND METHODS OF INDUCTION

When we turn to the consideration of the methods of induction formulated by Mill, we find the notion of causation assuming in this connection a prominence that it did not have in most of the references to induction that we have so far had occasion to quote from his work. The conception of induction that governs his discussion of these methods is stated by him as follows: "To ascertain therefore, what are the laws of causation which exist in nature; to determine the effect of every cause, and the causes of all effects is the main business of Induction; and to point out how this is done is the chief object of Inductive Logic." [23] The reason for this is that, according to Mill, "all the uniformities which exist in the succession of phenomena, and most of the uniformities in their coexistence, are either . . . themselves laws of causation, or consequences resulting from, and corollaries capable of being deduced from, such laws," [24] so that the notion of cause is "the root of the whole theory of Induction." [25]

This leads us, first of all, to a consideration of Mill's notion of cause. "I premise, then," we find him telling us, "that when, in the course of this inquiry I speak of the cause of any phenomenon, I do not mean a cause which is not itself a phenomenon"; [26] and the terms cause and effect he defines by saying, "Between the phenomena . . . which exist at any instant, and the phenomena which exist at the succeeding instant, there is an invariable order of succession. . . . The invariable antecedent is termed the cause; the invariable consequent, the effect." [27] As the antecedents and consequents, however, are seldom simple we must say that "the cause, then, philosophically speaking, is the sum total of the conditions, positive and negative taken together; the whole of the contingencies of every description, which being realized, the consequent invariably follows."[28] But Mill observes:

When we define the cause of anything to be "the antecedent which it invariably follows," we do not use this phrase as exactly synonymous with "the antecedent which it invariably *has* followed in our past experience" . . . it is necessary to our using the word cause, that we should believe not only that the antecedent always *has* been followed by the consequent, but that, as long as the present constitution of things endures, it always *will* be so.[29]

In other words, the regularity of the sequence is not merely actual, but *necessary*, and this means unconditional:

If there be any meaning which confessedly belongs to the term necessity, it is *unconditionalness*. That which is necessary, that which *must* be, means that which will be, whatever supposition we may make in regard to all other things. . . . Invariable sequence, therefore, is not synonymous with causation unless the sequence, besides being invariable, is unconditional . . . we may define therefore, the cause of a phenomenon, to be the antecedent, or the concurrence of antecedents, on which it is invariably and *unconditionally* consequent.[30]

The burden of the distinction thus insisted upon by Mill is to meet the possible objection that, according to his first definition of cause, night would be the cause of day, and day the cause of night.

The cause of day, he tells us, is the union of the following conditions: that the sun, or some such luminous body, be above the horizon; that his light be not extinct; that no opaque body intervene between us and him; and that no change take place in the properties of matter. If this combination of antecedents prevails day will be present "quite independently of night as a previous condition," i.e., day is then *necessarily* consequent. But "the succession of day and night," he goes on, "evidently is not necessary in this sense. It is conditional on the occurrence of other antecedents. That which will be followed by a given consequent when, and only when, some third circumstance also exists, is not the cause. . . ."[31] Furthermore, "it is experience itself which teaches us that one uniformity of sequence is conditional, and another unconditional."[32] Sunset and sunrise constitute *experimenta crucis* that the cause of day is the sun. We have thus "an experimental knowledge of the sun which justifies us on experimental grounds in concluding, that if the sun were always above the horizon, there would be day, though there had been no night. . . . We thus know from experience that the succession of night and day is not unconditional." [33]

This is all very clearly stated, and, as far as it goes, very obvious. Unfortunately for Mill's argument, however, it is largely irrelevant to the point really at issue, which is whether, for a strict empiricist such as Mill claims to be, the addition of the requirement of unconditionalness to the definition of cause really modifies it or is not rather utterly empty of the virtue that he needs.

That the latter is the case can easily be shown: *If* we know that the sun (or, more accurately, the combination of circumstances elsewhere enumerated) is the cause, i.e., the "unconditional" antecedent of the day, in the sense that the day not only has succeeded, *but will succeed* invariably the presence of that combination of circumstances, we certainly *do not know it from experience*, which, as Hume showed, is of the past only. If we know it, we know it as a consequence of the principle that in the same circumstances the same sequences *will* recur, or, as Mill himself approximately phrases it in this connection (thus, apparently unawares, begging the question), we know it "as long as the present constitution of things endures." But his principle of the uniformity of nature, no matter how worded, is, as Hume, again, clearly showed, not itself obtained from experience, but only assumed. Therefore, qua empiricists, we do not know that the sun is the cause, i.e., the "unconditional" antecedent of the day, in the sense desired by Mill, but *only* in the sense that the day *has* invariably followed the appearance of the sun. The addition of the word "unconditional" to the definition of cause, so far as we remain strict empiricists, thus in no way changes its meaning in the direction needed—that of making it provide a guarantee of the future—and is thus entirely idle. And, to complete the refutation, one may further point out that, again on strictly empiricistic premises, we do not even know that the night is not the unconditional antecedent of the day, in the sense that if the night does not precede, the day *will* nevertheless perhaps be present. We know it *only so far* as we assume that the interruption of the sequence of night and day, observed to have been, under certain circumstances, a characteristic of situations in other respects similar (e.g., the illumination of an orange), *would* also be a characteristic of the case of the earth, as yet unobserved, under analogous circumstances. We know it, that is to say, not from experience, but again only under the assumption "that the present constitution of things endures."

Mill, thus, is here the victim of his own fatal plausibility. The root of his difficulties in connection with the concept of causation is after all the fundamental inconsistency, already demonstrated at length in our examination of Hume, between a radical empiricism, such as his and Mill's, and the use of the methods of agreement and difference.

A few words may now be said in connection with the methods of induction propounded by Mill. The methods described in his chapter entitled "Four Methods of Experimental Inquiry," viz., the methods of agreement, of difference, of concomitant variations, and of residues, are usually regarded as the most important part of his contribution to the theory of induction. That these methods were in no sense discovered by Mill has appeared from our examination of the writers before him. In Herschel's *Discourse,* especially (to which, indeed, Mill fully acknowledges his debt), we found them stated with great clearness and illustrated at length; and some logicians have expressed their preference for his treatment of them to that of Mill.[34]

Whatever needed to be said in connection with the relation of these methods to strict empiricism has already been said in our discussion of Hume, and the nature of the methods themselves sufficiently indicated there and in the examination of Herschel's *Discourse.* Since Mill's treatment of them, so far as it is a contribution to the theory of induction, is so not owing to its setting forth anything new, but rather to the detailed and exhaustive manner of it, and is moreover so widely known, we may pass over it without further discussion.

Aside from these four methods, however, Mill devotes a chapter to the consideration of what he terms the "Deductive Method," which has to be used in cases where the complexity of the phenomena renders the direct methods of observation and experiment inapplicable. This mode of investigation, he tells us, "consists of three operations: the first, one of direct induction; the second, of ratiocination; the third, of verification." [35] His exposition of it is as follows: "The problem of the Deductive Method," he tells us, "is to find the law of an effect, from the laws of the different tendencies of which it is the joint result. . . . To ascertain, then, the laws of each separate cause which takes a share in producing the effect, is the first desideratum of the Deductive Method." [36] The second part of the process consists in "determining from the laws of the causes what effect any given combination of those causes will produce." [37] This is a matter of ratiocination, and: "By such ratiocination from the separate laws of the causes, we may, to a certain extent, succeed in answering either of the following questions: Given a certain combination of causes, what effect will follow? and What combination of causes, if it existed, would produce a given effect?" [38] The test of the validity of results so obtained is verification; they must, that is, "be found, on careful comparison to accord with the results of direct observation wherever it can be had." [39]

It is often possible, Mill notes in a later chapter, to dispense with the first of these three steps, viz., the induction, replacing it by a hypothesis, "the law which is reasoned from being assumed instead of proved." [40] This enables us to apply the deductive method earlier to phenomena and is justified when the verification can be carried out with rigor sufficient to establish not only that the hypothesis used accounts for the actual facts, but also that no other could do so. To this deductive method, Mill declares, "the human mind is indebted for its most conspicuous triumphs in the investigation of nature."

In the second of the two forms mentioned by Mill, namely, where the first step is a hypothesis, this deductive method is identical with the method described by Newton, Herschel, and Whewell. It constitutes what Whewell conceives the inductive process as a whole to be. His unfavorable comments upon "Mr. Mill's Hope from Deduction" [41] are really directed to the word "deduction" and what it usually implies, and have little if any bearing on the method that Mill, unfortunately perhaps, designated by that word. Mill does not propose the substitution of what Whewell means by deduction for what Whewell means by induction. The net import of Mill's discussion of the "Deductive Method" is, much rather, an acknowledgment of the superiority of the method that *Whewell* calls induction, to that which *Mill* calls induction (viz., the direct application to phenomena of his Four Methods of Experimental Inquiry).

The results of our examination of Mill's *System of Logic* in respect of the three questions with which we are especially concerned may be summed up briefly. With reference to the meaning of the term induction, Mill's statements are of interest primarily by way of contrast with Whewell's. By induction, Whewell, as we have seen, means essentially *discovery,* while Mill, in spite of the fact that in one place he defines it as "the operation of discovering and proving general propositions," means essentially *proof* [42] and contends that the introduction of a new conception is not, as Whewell claims, essential to induction.

The important question is also raised for the first time, in connection with Mill's treatment, whether or not the proof of individual facts is induction—although, as we have seen, no consistent answer can be extracted from his statements on the question. His reference to inductive logic as the logic of truth—deductive logic being the logic of consistency—since it makes no distinction between the truth of general and the truth of particular propositions, is of interest in this connection.

Concerning the basis of the possibility of generalization—the uniformity of nature—Mill's views are of value perhaps chiefly as exhibiting the insuperable obstacles confronting a radical empiricism in that direction. Otherwise, he adds nothing to Hume's discussion of the matter.

And on the question of methods of induction, his discussion, as already noted, is of value on account of its fullness, and of the wide attention that it has attracted, rather than of its novelty.

CHAPTER ELEVEN

W. S. Jevons on Induction and Probability

MORE THAN most writers on inductive theory, W. S. Jevons tried to relate in a systematic manner his views on the nature of induction, probability, and hypothesis. I shall examine chiefly his characterization of the "inverse" nature of induction and his explications of the classical theory of probability and the nature of hypothetical inference, suggest some difficulties along the way, and, finally, point out how his influential work is a bridge from nineteenth-century philosophy of science to current discussions in this field.

I. INDUCTION

In deduction, Jevons writes, we are engaged in developing the consequences of a law; we learn the meaning, contents, results, or inferences that attach to any given proposition.[1] Induction, however, is exactly the inverse process. Given certain results or consequences, we are required to discover the general law from which they flow.[2] (It may be mentioned, in connection with Jevons' reference to induction as the *inverse* of deduction, that Herschel had already noted, though without dwelling on the fact, that the relation of the two processes could be so characterized. After pointing out that when we arrive at a general law of nature in any branch of science it usually enables us to extend our knowledge of others, Herschel says: "This remark rather belongs to the inverse or *deductive* process, by which we pursue laws into their remote consequences." [3])

Consider, Jevons writes,[4] this general *deductive* problem: if we are given as premises, A = AB and B = BC, what can we conclude? The possibilities defined by the three terms A, B, C, are the following (the small letters denoting the contradictories of the capitals):

(1) ABC	(5) aBC
(2) ABc	(6) aBc
(3) AbC	(7) abC
(4) Abc	(8) abc

The assertion that whatever is A is also B, given as premise, cancels the possibilities of A's being bC (3) and bc (4). And the other premise, to the effect that whatever is B is also C, cancels the possibilities of B's being Ac (2) and ac (6), leaving as the conclusion:

Either A is B and C
or a is B and C
or a is b and C
or a is b and c.

Now, Jevons tells us, the inverse or inductive problem corresponding to the present deductive problem would be stated as follows: "Having given certain combinations of terms [i.e., that either ABC, or aBC, or abC, or abc is true] we need to ascertain the propositions with which the combinations are consistent and from which they may have proceeded,"[5] viz., in this case, that A = AB, and B = BC. Inverse processes, Jevons notes, are commonly far more difficult to perform than the corresponding direct processes. "Given any two numbers," for instance, "we may by a simple and infallible process obtain their product; but when a large number is given it is quite another matter to determine its factors."[6] "Compare again the difficulty of decyphering with that of cyphering. . . . Induction is the decyphering of the hidden meaning of natural phenomena. Given events which happen in certain definite combinations, we are required to point out the laws which govern these combinations."[7]

The type of induction examined thus far Jevons calls *perfect* induction, because "all the objects or events which can possibly come under the class treated have been examined."[8] But in most actual cases such exhaustive examination is impossible, because the number of instances is practically infinite, or some of the instances are not accessible to us. "In all such cases, induction is *imperfect*, and is affected by more or less uncertainty." [9] In reply to the commonly made assertion that perfect induction, though certain, is useless because it does not increase our knowledge and is merely a summary of it, Jevons says that "mere abbreviation of mental labour is one of the most important aids we can enjoy in the acquisition of knowledge. The powers of the human mind are so limited that multiplicity of detail is alone sufficient to prevent its progress in many directions. Thought would be practically impossible if every separate fact had to be separately thought and treated." [10] Moreover,

> . . . perfect induction is a process absolutely requisite . . . in the performance of imperfect induction, [for] if I can draw any inference at all concerning objects not examined, it must be done on the data afforded by the objects which have been examined. . . . Adams and Leverrier, for instance, must have inferred that the undiscovered planet Neptune would obey Bode's law, because *all the planets known at that time obeyed it.*[11]

Imperfect induction, Jevons constantly claims, is always uncertain; and one reason for this uncertainty lies in the assumption of the uniformity of nature, which, while it is a postulate of inductive inquiry (induction is not justifiable without it), nevertheless can never be proved. Jevons avers that we ever hang upon the will of the Creator for maintaining the framework of the world unchanged from moment to moment.[12] He amplifies his uniformity concept in this way:

> All predictions, all inferences which reach beyond their data, are purely hypothetical, and proceed on the assumption that new events will conform to the conditions detected in our observation of past events. [But] we cannot be sure . . . that our observations have not escaped some fact, which will cause the future to be apparently different from the past; nor can we be sure that the future really will be the outcome of the past. We proceed then in all our inferences to unexamined objects and times on the assumptions—
>
> 1. That our past observation gives us a complete knowledge of what exists.
> 2. That the conditions of things which did exist will continue to be the conditions which will exist.[13]

The character of our knowledge of nature is, according to Jevons, best illustrated by the simile of a ballot box:

> Nature is to us like an infinite ballot-box, the contents of which are being continually drawn, ball after ball, and exhibited to us. Science is but the careful observation

of the succession in which balls of various character present themselves; we register the combinations, notice those which seem to be excluded from occurrence, and from the proportional frequency of those which appear we infer the probable character of future drawings. But under such circumstances certainty of prediction depends on two conditions:—

1. That we acquire a perfect knowledge of the comparative numbers of balls of each kind within the box.
2. That the contents of the ballot-box remain unchanged.[14]

These two requirements are not, however, separate, as Jevons thinks they are. The second assumption is a form of the Millian postulate of the uniformity of nature; and the effect of the first is usually allowed for when this uniformity is postulated, since the uniformity is assumed to be present only so far as the same conditions are present. Thus, if the uniformity assumption itself is granted, and yet the future happens to be different from the past, our ignorance of some relevant change in the conditions present is postulated.

Jevons' strong belief in the uncertainty of "imperfect" induction made him suspicious of Mill's analysis of cause, which, Jevons thought, mistakenly it may be, implies "that when once we pass within the circle of causation we deal with certainties."[15] If a cause is defined as the necessary or invariable antecedent of an event, then, Jevons writes, we know that an effect will certainly happen if we know its cause. But nothing is more unquestionable, he thought, than that finite experience can never give us certain knowledge of the future, so he defined "cause" in a strictly Humean manner without any modal term like "necessary condition."

If a cause is an invariable and necessary condition of an event, we can never know certainly whether the cause exists or not. To us, then, a cause is not to be distinguished from the group of positive or negative conditions which, with more or less probability, precede an event. In this sense, there is no particular difference between knowledge of causes and our general knowledge of the succession of combinations, in which the phenomena of nature are presented to us, or found to occur in experimental inquiry.[16]

That Jevons was not clearly conscious of the implications of this view, however, is shown by his making statements like the following, among others: "Chance then exists not in nature, and cannot coexist with knowledge; it is merely an expression, as Laplace remarked, for our ignorance of the causes in action. . . . In nature the happening of an event has been pre-determined from the first fashioning of the universe." [17] For here the notion of causation is obviously taken in its usual dynamic, intensive, objectively necessitating sense (which notion Jevons apparently thought sensible but unknowable); [18] but if Jevons' definition of cause were introduced where the terms "causes in action" and "pre-determined" occur, the statement as a whole would lose all meaning. In another place,[19] where Jevons is discussing process laws, he observes that in a solid universe, in at least approximate equilibrium, a not inconceivable state, the relation of cause and effect would be no more than the relation of before and after, which again indicates his implicit dynamic interpretation of process laws.

2. INDUCTION AND PROBABILITY

Only "perfect" knowledge, whether deductive or inductive, can give certainty; but, Jevons insists, since perfect knowledge of nature is beyond our power we must content ourselves with the partial knowledge of imperfect induction—"knowledge mingled with ignorance, producing doubt" [20]—and the measure of such knowledge, he was convinced, was the Laplacian theory of probability. He expounds the theory formally and later shows that, in its inverse form, it is the essence of imperfect induction.

Jevons states the classical interpretation of probability thus:

> Calculate the number of events which may happen independently of each other, and which, as far as is known, are equally probable. Make this number the denominator of a fraction, and take for the numerator the number of such events as imply or constitute the happening of the event, whose probability is required. Thus, if the letters of the word *Roma* be thrown down casually in a row, what is the probability that they will form a significant Latin word? The possible arrangements of four letters are 4 x 3 x 2 x 1, or 24 in number, and if all the arrangements be examined, seven of these will be found to have meaning, namely *Roma*, *ramo*, *oram*, *mora*, *maro*, *armo*, and *amor*. Hence the probability of a significant result is 7/24.[21]

Events are "independent," Jevons explains, when the happening of one does not make the other either more or less probable than it was before. "Thus the death of a person is neither more nor less probable because the planet Mars happens to be visible." [22] He never really comes to grips with the notion of equiprobability (not really clearly distinguishing it from equipossibility) but introduces it through the usual examples of coins and dice. His positive contribution is his attempt to justify the notion of equiprobability on the basis of his Substitution of Similars Principle:

> The calculation of probabilities is really founded, as I conceive, upon the principle of reasoning set forth in preceding chapters. We must treat equals equally, and what we know of one case may be affirmed of every case resembling it in the necessary circumstances. . . . Throw a penny into the air, and consider what we know with regard to its way of falling. We know that it will certainly fall upon a side, so that either head or tail will be uppermost; but as to whether it will be head or tail, our knowledge is equally divided. Whatever we know concerning head, we know also concerning tail, so that we have no reason for expecting one more than the other. The least predominance of belief to either side would be irrational; it would consist in treating unequally things of which our knowledge is equal.[23]

Jevons correctly draws several corollaries from these basic elements of the classical theory of probability. First, the probability of any statement for which we have no evidence whatsoever is ½; or, in Jevons' words, "even when we have no other information we must not consider a statement as devoid of all probability" for "the true expression of complete doubt is a ratio of equality between the chances in favour of and against it, and this ratio is expressed in the probability ½."[24] Second, probability measures do not refer to any property of events themselves but belong "wholly to the mind," wholly "to our mental condition, to the light in which we regard events, the occurrence or non-occurrence of which is certain in themselves."[25] Jevons believes this view is proved by the fact that different men may and do ascribe to the same event at the same time widely different degrees of probability. A steamship, for instance, is missing and some people think she has sunk in mid-ocean, others not. Now the probability of the event's having occurred will vary from day to day, and from mind to mind, Jevons says, depending upon the slightest information regarding vessels met at sea, prevailing weather, condition of vessel, signs of wreck, and so forth, which may turn up.[26] But in the event itself, of course, there is no uncertainty; the ship has either sunk or not, "and no subsequent discussion of the probable nature of the event can alter the fact." Jevons, however, does not accept De Morgan's dictum that by degree of probability we really mean, or ought to mean, degree of belief, but asserts that by degree of probability we really mean what our degree of belief ought to be.[27]

The inverse application of the classical theory of probability, Jevons averred, *is the very essence of imperfect inductive inference.* This idea of the inverse application of the rules of probability Jevons took over from Laplace, who had said essentially that, if any event can be produced by any one of a certain number of different causes, all equally probable a priori, the probabilities of the existence of these causes as inferred from the event are proportional to the probabilities of the event as derived from these causes.[28] Jevons formulates the inverse method of probability and induction as follows:

Suppose, to fix our ideas clearly, that E is the event, and C_1 C_2 C_3 are the three only conceivable causes. If C exists, the probability is p_1 that E would follow; if C_2 or C_3 exist, the like probabilities are respectively p_2 and p_3. Then as p_1 is to p_2, so is the probability of C_1 being the actual cause to the probability of C_2 being it; and, similarly, as p_2 is to p_3, so is the probability of C_2 being the actual cause to the probability of C_3 being it. By a simple mathematical process we arrive at the conclusion that the actual probability of C_1 being the cause is

$$\frac{p_1}{p_1 + p_2 + p_3};$$

and the similar probabilities of the existence of C_2 and C_3 are,

$$\frac{p_2}{p_1 + p_2 + p_3} \text{ and } \frac{p_3}{p_1 + p_2 + p_3}.$$

The sum of these three fractions amounts to unity, which correctly expresses the certainty that one cause or other must be in operation.

We may thus state the result in general language. *If it is certain that one or other of the supposed causes exists, the probability that any one does exist is the probability that if it exists the event happens, divided by the sum of all the similar probabilities.* . . . No one can possibly understand the principles of inductive reasoning, unless he will take the trouble to master the meaning of this rule, by which we recede from an event to the probability of each of its possible causes.[29]

A fundamental problem, however, remains for Jevons, namely, to account for the possibility and nature of inductive extrapolation or prediction within this framework of inverse probability calculations. He formulates this "general inverse problem" in this way: "*An event having happened a certain number of times, and failed a certain number of times, required the probability that it will happen any given number of times in the future under the same circumstances.*" [30] Using the familiar example of balls drawn from an urn, he tries to clarify the way a Laplacian solves this general problem.

Suppose we know that an urn contains four balls, some of which are black and some white, but the exact proportion is unknown. We draw a ball from the urn four times, each time replacing the ball drawn, and find that three times the ball is white and once black. Now the problem is to discover what the probability is that a white one will appear on the next draw. To calculate this probability, of course, we need to know the contents of the urn. Only the following hypotheses about the contents are possible ones: (1) 4 white and 0 black; (2) 3 white and 1 black; (3) 2 white and 2 black; (4) 1 white and 3 black; (5) 0 white and 4 black. However, since we know from our four drawings that at least one is white and at least one is black, only three of these hypotheses—(2), (3), (4)—could be correct.

Now consider for each hypothesis—(2), (3), (4)—what the probability is of getting the results obtained on our four draws, namely, 3 white and 1 black. If the urn contains 3 white and 1 black, the probability of getting a white one on each draw is $\frac{3}{4}$ and of getting a black $\frac{1}{4}$. The probability of the compound event observed —the 3 white and 1 black of our four draws—is $\frac{3}{4} \times \frac{3}{4} \times \frac{3}{4} \times \frac{1}{4}$; and, since the balls could occur in any one of four orders (the black one might have been first, second, third, or fourth), we must multiply this result by 4, giving the final result of $\frac{27}{64}$. If the urn contains 2 white and 2 black, following the same simple mathematical procedure, we arrive at the probability $\frac{16}{64}$; and, if the urn contains 1 white and 3 black, we arrive at the probability $\frac{3}{64}$. Since one or the other of these hypotheses must be correct, Jevons continues, the joint probability of the observed event, according to these hypotheses, should be unity or 1. In order to obtain this "absolute probability" value, we simply alter these fractions in a uniform ratio so that their sum is 1, the expression of certainty. Since $27 + 16 + 3 = 46$, the required alteration is achieved by dividing each fraction by 46, and multiplying by 64. Thus, the probability of the second, third, and fourth hypothesis is, respectively, $\frac{27}{46}$, $\frac{16}{46}$, and $\frac{3}{46}$.

Now, Jevons writes, we are in a position, finally, to infer the probability that our next draw, the fifth one, will yield a white ball —the original problem. If the urn actually contains 3 white and 1 black (hypothesis 2), the probability of drawing a white one next is clearly $\frac{3}{4}$; and, since the probability of the urn's being so constituted is $\frac{27}{46}$, the compound probability of the urn's being so filled and of a white ball's appearing on the next draw is $\frac{27}{46} \times \frac{3}{4}$ or $\frac{81}{184}$. If the urn actually contains 2 white and 2 black (hypothesis 3), following the same procedure, we arrive at the probability $\frac{32}{184}$; and, if 1 white and 3 black (hypothesis 4), the probability $\frac{3}{184}$. "Since one and not more than one hypothesis can be true, we may add together these separate probabilities, and we find that $\frac{81}{184} + \frac{32}{184} + \frac{3}{184}$ or $\frac{116}{184}$ is the complete probability that a white ball will be next drawn under the conditions and data supposed." [31]

The three rules of predictive inference which are implicit in Jevons' present example, and later made explicit by him, are these: [32] (1) To find the probability that an event which has never been observed to fail will happen once more, divide the number of times the event has been observed increased by one, by the same number increased by two. (2) To find the probability that an event which has never failed will not fail for a certain number of new occasions, divide the number of times the event has occurred increased by one, by the same number increased by one and the number of times it is to happen. (3) If an event has occurred and failed to occur a certain number of times, then to find the probability that it will happen the next time, divide the number of times the event has occurred increased by one, by the whole number of times the event has occurred or failed increased by two.

Another fundamental problem, however, arises for Jevons, of

which, again, he is thoroughly aware. Is it possible to accommodate, within this inverse framework, not only games of chance and artificial systems of balls and ballot boxes, where a priori interpretations are plausible, but also ordinary enumerative inference and scientific inductive inference? Jevons realized that, while his example of drawing balls from ballot boxes might bring enumerative inference into the picture, still he did not by any means have a paradigm of inductive inference as far as empirical science is concerned. In science one cannot operate on the simplifying assumptions implied by a few balls drawn from discrete ballot boxes. Moreover, since the scientist has no a priori information about the phenomena in question, there is no limit to the variety of hypotheses that could be suggested. So, Jevons writes,

> . . . mathematicians have therefore had recourse to the most extensive suppositions which can be made, namely, that the ballot-box contains an infinite number of balls; they have then varied the proportion of white to black balls continuously, from the smallest to the greatest possible proportion, and estimated the aggregate probability which results from this comprehensive supposition.[33]

While the integral calculus is used as the method of summing infinitely numerous results, Jevons concludes, one in no way abandons the principles of combining simple into complex probabilities set forth in the traditional "calculus" of probabilities. Unfortunately, however, for aught these comments show, the way that Jevons gets from artificial systems and games of chance, where simplifying conditions exist, to ordinary and scientific inductive inference remains as obscure as before.

Jevons' version of the classical theory of probability and its inverse inductive form is, on the whole, clearer and more straightforward than most, and its merits do not require pointing out. There are, however, numerous difficulties that he did not meet successfully. First, his assertion that ignorance is denoted by the probability ½, and therefore that in the absence of all knowledge the antecedent probability of an event should be assumed to be ½, rests, I believe, as Venn and Peirce claimed, on a confusion between the subjective and the objective meanings of the word probability. The probability expressed under such circumstances by the fraction ½ is not the probability *of the event;* it is the probability *of our answering* yes rather than no if, remaining all the while in complete ignorance, we are asked to reply a number of times with yes or no to the question: Will the event occur? C. S. Peirce adumbrates this view in his "Theory of Probable Inference":

There is a sense in which it is true that the probability of a perfectly unknown event is one half; namely, the assertion of its occurrence is the answer to a possible question answerable by "yes" or "no," and of all such questions just half the possible answers are true. But if attention be paid to the denominators of the fractions, it will be found that this value of 1/2 is one of which no possible use can be made in the calculation of probabilities.[34]

Jevons, it is true, is right in insisting, against De Morgan, that by "degree of probability" one must mean "what degree of belief ought to be," not simply "degree of belief," for beliefs do not have the property of additivity, and so it is impossible to estimate "strength" of beliefs; but, on the other hand, Jevons himself gives no analysis at all of "what degree of belief ought to be" or no explanation of the sources of the imperative.

The difficulties centering around the notion of equipossibility are notorious, and Jevons does not resolve them. On the classical theory of probability, a probability statement can be made only if the case is analyzable into a group of "equipossible" alternatives, but in many prima facie probability statements such an analysis is not possible. Assume a biased coin is assigned the probability of .63 that it will turn up a head when tossed; it is difficult to see how this number can be interpreted as the ratio of equipossible alternatives. The difficulty is perhaps even more evident, Ernest Nagel writes, for statements like "the probability that a thirty-year-old man will live at least another year is .945."

It is absurd to interpret such a statement as meaning that there are a thousand possible eventuations to a man's career, 945 of which are favorable to his surviving at least another year. Moreover, the Laplacian definition requires a probability coefficient to be a rational number. But irrational numbers frequently occur as values for such coefficients, and there is no way of interpreting them as ratios of a number of alternatives.[35]

One should notice, next, that Jevons' rules 1 and 2 lead to a peculiar paradox, which Jevons does not resolve but in fact appears to regard as a virtue.[36] If one assumes the sun has risen one thousand million times, then the probability that it will rise next morning, based simply on this information, is, on rule 1, $\frac{1{,}000{,}000{,}000 + 1}{1{,}000{,}000{,}000 + 1 + 1}$. However, the probability that it will continue to rise in the future for exactly the same period is, on rule 2, only $\frac{1{,}000{,}000{,}000 + 1}{2{,}000{,}000{,}000 + 1}$, which is almost exactly ½. Certainly it seems queer that empirical evidence, even when it is induction by enumeration, should lend one extrapolation a certain probability but each subsequent one less. If the generalization is casual, then no extrapolation is legitimate; but if the generalization is not casual, then any number of extrapolations is as good as any other.

Jevons, finally, in some places, seems aware that the ratio of the number of favorable alternatives to the whole number of equipossible ones is not, on this theory, any legitimate clue to the relative frequency of the occurrence of the favorable event. It is not impossible, he says, that a person should always throw a coin head uppermost and appear unable to get a tail by chance. The classical theory would not be falsified, he continues, "because it contemplates the possibility of the most extreme runs of luck. Our actual experience might be counter to all that is probable; the whole course of events might seem to be in complete contradiction to what we should expect, and yet a casual conjunction of events might be the real explanation." [37] Jevons, of course, wishes to minimize this paradox; "coincidences" of this sort are so unlikely, he writes, that the whole duration of history does not give any appreciable probability of their being encountered:

> At the same time, the probability that any extreme runs of luck will occur is so excessively slight, that it would be absurd seriously to expect their occurrence. It is almost impossible, for instance, that any whist player should have played in any two games where the distribution of the cards was exactly the same, by pure accident (p. 191). Such a thing as a person always losing at a game of pure chance, is wholly unknown. Coincidences of this kind are not impossible, as I have said, but they are so unlikely that the lifetime of any person, or indeed the whole duration of history, does not give any appreciable probability of their being encountered. Whenever we make any extensive series of trials of chance results, as in throwing a die or coin, the probability is great that the results will agree nearly with the predictions yielded by theory.[38]

Unfortunately, however, Jevons' use of "unlikely" and "probability" in this minimizing of the paradox cannot be given a classical interpretation and appears to be a poorly concealed concession to the fundamental role of empirical frequencies and a consequent abandonment of pure apriorism.

3. THE NATURE OF HYPOTHESIS

Jevons, like Whewell, believed that the invention and successive trial of hypotheses constitute the very essence of the inductive process—and this, incidentally, leads him to note that, without the use of deduction, whereby to draw from hypotheses the consequences the comparison of which with facts constitutes the "trial" of a hypothesis, all induction would be impossible.[39] Jevons' description of the concrete process of induction follows:

> Being in possession of certain particular facts or events expressed in propositions, we imagine some more general proposition expressing the existence of a law or cause; and, deducing the particular results of that supposed general proposition, we observe whether they agree with the facts in question. Hypothesis is thus always employed, consciously or unconsciously.[40]

There are only three steps in this process of induction: (1) framing some hypothesis as to the character of the general law; (2) deducing consequences from that law; (3) observing whether the consequences agree with the particular facts under consideration.

This procedure, however, is the hypothetico-deductive process at its simplest, viz., when the nature of the subject under investigation admits of *certain* deductions from the hypothesis; for, "of several mutually inconsistent hypotheses, the results of which can be certainly compared with facts, but one hypothesis can ultimately be entertained." But, when deduction is only probable, it is not possible to adopt one hypothesis to the exclusion of the others. Then one must entertain at the same time all conceivable hypotheses and regard each with the degree of esteem proportionate to its probability. We go through the same steps as before: (1) we frame a hypothesis; (2) we deduce the *probability* of various series of possible consequences; (3) we compare the consequences with the particular facts and observe the *probability* that such facts would happen under the hypothesis. This process must be performed for every conceivable hypothesis, and then the absolute probability of each will be yielded by the principle of the inverse method,[41] viz., that the most probable cause of an event which has happened is the one which would most probably lead to the event, supposing the cause to exist. In other words, "as in the case of certainty we accept that hypothesis which certainly gives the required results, so now we accept as most probable that hypothesis which most probably gives the results. . . ." [42]

The mere agreement of the hypothesis with already observed facts, however, is not sufficient. "When once we have obtained a probable hypothesis, we must not rest until we have verified it by comparison with new facts. We must endeavour by deductive reasoning to anticipate such phenomena, especially those of a singular and exceptional nature, as would happen if the hypothesis be true." [43] And such agreement with facts, both those already observed and those yet to be observed, "is the sole and sufficient test of a true hypothesis." But this, Jevons declares, may be said to involve three constituent conditions: (1) that it allow of the application of deductive reasoning and the inference of consequences capable of comparison with the results of observation; (2) that it not conflict with any laws of nature, or of mind, which we hold to be true; (3) that the consequences inferred agree with facts of observation.[44]

Jevons states that the conception of hypothesis now outlined was suggested to him by the study of the inverse method of probability, but he notes that it "also bears much resemblance to the so-called Deductive Method" of Mill, which he regards as probably the most original and valuable part of Mill's treatise. That this method is not original with Mill, but was fully described before him by Whewell, and before Whewell by Herschel and Newton, we have already seen. Indeed Jevons himself elsewhere refers to it as the Newtonian Method.[45]

Finally, Jevons' analysis of "hypothesis" includes a point that is much emphasized in present-day discussions of the logic of science. He saw clearly, and presented in a fresh light, how different the hypothetico-deductive method of science is from the enumerative induction of the logic texts.[46] Science is not a series of enumerations—what some contemporary analysts call "natural history generalizations"—which are subject to upset by a single negative instance, but is rather a system of interconnected and verified hypotheses, where what is "hypothesized" is either a cause or a law. The particular significance of Jevons' hypothetico-deductive interpretation of science lies in his generalization of the procedure to fit all scientific inference. The method does not apply only to the discovery of *causes*, which, since they frequently involve the assumption of theoretical entities, clearly qualify as hypotheses, but also refers to the discovery of *laws*. In formulating a law, Jevons wrote, we replace, in essence, an empirical curve with a rational function; and, while we may achieve this replacement, perchance, by trial and error or by trying functions that give a similar form of variation, we generally anticipate the form of the law from previous knowledge—"theory and analogical reasoning, in short, must be our guides." [47]

Jevons, it will be noted, carefully interwove his hypothetico-deductive interpretation of science and his classical theory of probability. The measure of the probability of a hypothesis, he thought, can only be taken by the Laplacian inverse method—a point that nicely bridges nineteenth-century and contemporary probability theory. Some contemporary authors [48] write that a hypothesis is more or less probable in terms of different evidence at different times, or that one hypothesis is more or less probable than another in terms of the same evidence, and so forth—in which usage of "probable" favorable evidence *increases* the probability of the hypothesis while unfavorable evidence *decreases* it. Within certain bounds, these relationships between hypothesis and evidence are exhibited in Jevons' classical theory of probability; and, since these relationships seem indispensable to the nature of "hypothesis" and apparently cannot be accounted for by the frequency theory, Jevons' classical theory, in spite of its shortcomings, continues to fascinate some logicians of science. These logicians, trying to avoid the difficulties of the classical theory and yet retain its evidential analysis, have, in turn, devised various "logical" theories of probability that are either all-inclusive interpretations of "probability" or else interpretations of only one of the senses of "probable."

CHAPTER TWELVE

Charles Sanders Peirce's Search for a Method

CHARLES Sanders Peirce's interpretations and analyses of induction, probability, and hypothesis, which he frequently called abduction, are thoroughly penetrating and, for the most part, clear; but, as one finds in all areas of Peirce's thought, the relations among his key concepts shift in the course of his career and are not always made explicit or systematically worked out. "What he has left us," Thomas Goudge points out, "are piecemeal analyses in which profound insights and unresolved problems exist side by side." [1]

I. THE NATURE OF INDUCTION

In his article entitled "A Theory of Probable Inference," which first appeared in the Johns Hopkins *Studies in Logic* (1883), Peirce characterized induction as the *inversion* of a statistical deduction. A statistical deduction has the following form:

> 10 per cent of the M's are P's
> S is a numerous set, taken at random, from among the M's
> Therefore, probably and approximately 10 per cent of the S's are P's.

Now,

> . . . the principle of statistical deduction is that these two proportions—namely, that of the P's among the M's, and that of the P's among the S's—are probably and approximately equal. If, then, this principle justifies our inferring the value of the second proportion from the known value of the first, it equally justifies our inferring the value of the first from that of the second, if the first is unknown but the second has been observed.[2]

An inference of the latter sort would be *inductive* and could be illustrated as follows:

> S is a numerous set, taken at random from among the M's
> 10 per cent of the S's are found to be P's
> Hence, probably and approximately 10 per cent of the M's are P's.

When, instead of a fractional value, such as 10 per cent, we have "all" or "none," the inference is an ordinary induction by simple enumeration, of the sort known as "imperfect."

In a manuscript on "The Varieties and Validity of Induction," which he wrote late in his career, Peirce distinguished three types of induction: crude, quantitative, and qualitative.[3] Qualitative induction is synonymous with hypothesis, or abduction, which we will analyze later. Quantitative induction is the same as the inversion of a statistical deduction; this type of induction "presumes that the value of the proportion, among the S's of the sample, of those that are P, probably approximates, within a certain limit of approximation, to the value of the real probability in question." [4] "Crude induction," the weakest kind, Peirce writes, is essentially Bacon's "*inductio quae procedit per enumerationem simplicem.*" However, the phrase does not exactly define crude induction "since in most cases no enumeration is attempted; and the enumeration, even if given, would not be the reasoner's chief reliance, which is rather the *absence* of instances to the contrary." [5]

Whatever the type of induction, however—whether it is an inverted statistical deduction, a quantitative, or crude induction, or whether the relation of S's and P's is fractional or universal—the logical structure, Peirce says, is the same: inferring from a sample to the whole lot sampled.[6] Induction of any type, then, will yield trustworthy results, Peirce writes, only if the sample used as a basis for the inference to the unobserved is *fairly chosen,* which means that it shall be "drawn at random and independently from the whole lot sampled. That is to say, the sample must be taken according to a precept or method which, being applied over and over again indefinitely, would in the long run result in the drawing of any one set of instances as often as any other set of the same number." [7] The sample, in other words, should be so taken that the different sorts of entities contained in the class sampled will be represented in the sample as they are in experiences of the content of that class in which "volition does not intermeddle at all." [8] And as a means of securing this, in cases where our will is likely to interfere in spite of ourselves, Peirce suggests the numbering of all the objects in the lot, and the drawing of numbers "by means of a roulette, or other such instrument. We may even go so far as to say that this method is the type of all random drawing." [9]

Another important requirement for trustworthy induction is that the character P, for which we propose to sample a class, should be designated in advance of the examination of the sample, for "suppose we were to draw our inferences without the predesignation of the character P; then we might in every case find some recondite character in which those instances would agree." And that to find such characters is simply a matter of sufficient ingenuity Peirce demonstrates by showing that the ages at death of the first five poets mentioned in a certain biographical dictionary all agree in several remarkable ways: for example that "the first digit raised to the power indicated by the second, and then divided by three, leaves a remainder of *one*" in each case, while yet "there is not the smallest reason to believe that the next poet's age would possess these characters." [10] Nevertheless, "if the number of instances be very great indeed, the failure to predesignate is not an important fault," for "the limitation of our imagination and experience amounts to a predesignation" far within the limits inside which the number of characters, or instances, would be great enough to increase the probable error appreciably.[11]

2. THE JUSTIFICATION OF INDUCTION

Peirce's justification of induction, which one writer has placed among the "practical vindications" of induction,[12] is usually prefaced by arguments against two traditional methods of justification, by the rule of succession and by the uniformity of nature doctrine.

The rule of succession, as a justification of induction, is linked with the classical theory of probability. Thomas Goudge, explicating Laplace's rule, writes:

If, for instance, we know that an event has occurred m times and failed to occur n times, under given circumstances, then the probability of its occurrence when these conditions are again fulfilled is $\frac{m+1}{m+n+2}$. Now if this rule were sound, it would obviously give a high probability to predictions based on cases where an event is known to have occurred a large number of times, and where no instance of its failure to occur is recorded. Thus (to use Quetelet's example) if I have observed the tide rise for 10,000 successive days, and know of no day that it did not rise, I am entitled to believe with a probability of $\frac{10{,}000+1}{10{,}000+2}$ that it will rise to-morrow. Since this fraction approaches very close to unity, the principle involved seems to offer a means of justifying induction by simple enumeration.[13]

Peirce generally opposes this type of justification of induction because it depends upon the classical theory of probability, which he believes contains at least two untenable assumptions, namely, that ignorance is denoted by the probability ½ and that, with respect to unknown probabilities, all ratios are equiprobable. (1) As we saw in the previous chapter, Peirce believes the view that ignorance is denoted by the probability ½, and therefore that in the absence of all knowledge the antecedent probability of an event should be assumed to be ½, really rests on a confusion between the subjective and the objective meanings of the word probability. The probability expressed under such circumstances by the fraction ½ is not the probability *of the event;* it is the probability *of our answering* yes rather than no if, remaining all the while in complete ignorance, we are asked to reply a number of times with yes or no to the question: Will the event occur? [14] (2) Of the assumption that all ratios are equiprobable, which in frequency terms means that antecedent to empirical evidence all frequencies are equiprobable, Peirce observes, "though it may be applied to any one unknown event, [it] cannot be applied to all unknown events without inconsistency." [15] Peirce, it should be pointed out, did not equate, as some classical theorists did, the equiprobability of ratios and the equiprobability of constitutions of the universe. "It is a very different thing to assume that all frequencies are equally probable from what it is to assume that all constitutions of the universe are equally probable"; [16] and, indeed, the latter assumption is much better than the former for, at least, it is not inconsistent. But certainly it would not solve the problem of induction either but vitiate the possibility of induction! On the assumption that any one constitution of the universe is as probable as any other, no inductive inference would be possible; for, on such a view, "the occurrences or non-occurrences of an event in the past in no way affect the probability of its occurrence in the future." [17]

Peirce characterized one version of J. S. Mill's uniformity of nature doctrine thus: "The whole force of Induction is the same as that of a syllogism of which the major premiss is the same for all inductions, being a certain 'Axiom of the uniformity of the course of nature'. . . . This was substantially Whately's theory of 1826." [18] Peirce rejects this view for numerous reasons. (1) If Mill means that nature is completely uniform, then apparently his assumption is factually false, since, "for every uniformity known, there would be no difficulty in pointing out thousands of non-uniformities." [19] (2) Mill's effort to justify induction deductively obliterates the distinction between induction and deduction, but clearly there is a difference between explicative (deductive) and ampliative (inductive) inference, etc.[20] (3) An induction, unlike a demonstration, rests not only on the facts observed, but also on the manner in which the facts have been collected.[21] (4) The uniformity of nature principle, which supposedly acts as a major premise, is so vague that the syllogism involves the fallacy of undistributed middle.[22]

Peirce's own justification of induction changed form during his philosophical career. At first, he also justified induction by assimilating it to an inverse form of deduction: induction, he thought, is justified if it is the inversion of a valid statistical deduction. "These two forms of inference, statistical deduction and induction," he writes, "plainly depend on the same principal of equality of ratios, so that their validity is the same." [23] Yet, even in this comparatively early Johns Hopkins paper, Peirce has the germ of his later method of justifying induction. The nature of the probability in statistical deduction and induction, he observed, is quite different. In the case of statistical deduction, probability means that, although any drawing may not conform to the predicted ratio, yet, "on continuing the drawing sufficiently, our prediction of the ratio will be vindicated at last." In the case of induction, on the other hand, probability means that, although the whole lot may not conform to the ratio of the sample, yet if we draw new samples the ratio in them will differ from that of the first sample and correct it till the ratio actually prevailing in the whole lot sampled is approximated, i.e., discovered at last.[24]

Peirce, in later writings, enlarged this "self-corrective" character of induction:

The true guarantee of the validity of induction is that it is a method of reaching conclusions which, if it be persisted in long enough, will assuredly correct any error concerning future experience into which it may temporarily lead us. This it will do not by virtue of any deductive necessity (since it never uses all the facts of experience, even of the past), but because it is manifestly adequate, with the aid of retroduction [hypothesis] and of deductions from retroductive suggestions, to discovering any *regularity* there may be among experiences, while *utter irregularity is not surpassed in regularity by any other relation of parts to whole*, and is thus readily discovered by induction to exist where it does exist, and the amount of departure therefrom to be mathematically determinable from observation where it is imperfect.[25]

[An] endless series must have some character; and it would be absurd to say that experience has a character which is never manifested. But there is no other way in which the character of that series can manifest itself than while the endless series is still incomplete. Therefore, if the character manifested by the series up to a certain point is not that character which the entire series possesses, still, as the series goes on, it must eventually tend, however irregularly, towards becoming so; and all the rest of the reasoner's life will be a continuation of this inferential process. This inference does not depend upon any assumption that the series will be endless, or that the future will be like the past, or that nature is uniform, nor upon any material assumption whatever.[26]

These passages, however, are not developed, and one must look at Peirce's famous article on "The Doctrine of Necessity Examined" to see the fullest development of his self-corrective justification of inductive inference. He there first of all emphasizes the point that the validity he claims for the inferences resulting from the sampling process is *provisional* and *experiential*. Considering as an example the sampling of a cargo of wheat in bulk, he tells us that, by referring to the inference as *experiential*, he means "that our conclusion makes no pretension to knowledge of wheat-in-itself," but only of such wheat as is the matter of possible experience. And, by speaking of the inference as *provisional*, he means "that we do not hold that we have reached any assigned degree of approximation as yet, but only that if our experience be indefinitely extended," and all the knowledge it contributes duly applied, our conclusion will correct itself, and will become in the long run indefinitely "close to the experience *to come*." And he concludes: "Now our inference, claiming to be no more than thus experiential and provisional, manifestly involves no postulate whatever." [27]

Unfortunately there are several crucial difficulties in Peirce's self-corrective defense of the validity of induction. At the outset of his discussion of the sampling of the wheat cargo, Peirce asks us to "suppose that by some machinery the whole cargo be stirred up with great thoroughness." [28] Now, that the stirring has been thus thorough is always more or less of a postulate, i.e., to use his own definition of a postulate, it is a proposition that we hope is true. It is only so far as it is true that the method of sampling that he indicates—namely, "that twenty-seven thimblefuls be taken equally from the forward, mid-ships, and aft parts, from the starboard, centre, and larboard parts, and from the top, half depth, and lower parts" of the hold—will constitute *random* sampling in the sense placed by him on that term, viz., sampling such that it "would in the long run result in the drawing of any one set of instances as often as any other set of the same number." [29] The requirement that the cargo be thoroughly stirred and then sampled as stated is, in its effect, equivalent to the device elsewhere suggested by him of numbering the members of a class to be sampled and drawing numbers by means of a roulette. Both are devices for virtually converting a class in which the antecedent probability of drawing any particular member *is not known to be the same* as that of drawing any other (no matter what the method used), into another class in which this antecedent probability *is known to be the same* when a certain method, viz., the roulette wheel, is used.

The sampling process, if, as required by Peirce, it is to be really at random, is thus not independent of an assumption concerning antecedent probability. And whether the assumption of antecedent probability that it does make is justified or not depends on a virtual conversion, by some such device as just described, of the class given into another; a conversion that is in very many cases impracticable or, at best, imperfect. It may be further noted in this connection that the assumption of equal antecedent probability in a random drawing of the various objects of a lot has exactly the same ultimate import as the more familiar assumption of the "uniformity of nature." Both, namely, amount to assuming that the character of the already observed is, under certain circumstances, more or less reliable evidence of the character of some realm of the as yet unobserved.

However, Peirce would doubtless say, the very purpose he had in mind when specifying that the result of ampliative inferences on the basis of sampling is only experiential and provisional, or self-corrective, was to obviate such criticisms as the above. This limitation of the scope claimed for the inferences does so in this manner: sampling enables us to make inferences as to future experience, but if, when that experience comes, it does not conform to the inference, it is at once seen to constitute part of the enlarged corrected data of sampling, since the inductive method "must avail itself of every sort of experience; and our inference, which was only provisional, corrects itself at last." [30] "At last," unfortunately, can only mean here: when the need for any inference has passed because there is nothing left to infer to. That is to say, the more nearly certain we become of the accuracy of our ratio, the less use we have for it, for the more nearly exhaustive is our examination of the class. For any practical purpose, and this is the crux of the whole matter, *there is a line to be drawn between the process of investigation and the application of its results.* What we seek is something that can be regarded as knowledge, i.e., as a starting point for deduction in terms of action, and not merely something to be regarded as provisional and as itself only one step in an indefinite process of inductive investigation. Perhaps such "knowledge" is never certainly to be had, but, if not, it is surely better to realize clearly that we merely "hope" we have some of it than to hedge the word knowledge about with qualifications that amount to a contradiction of any pragmatic sense of the term.

There is still another strand in Peirce's justification of induction—quite different, to be sure, from the self-corrective strand, yet closely related to it, too. Sometimes this second strand of Peirce's argument is called the predictionist justification. Inductive policy or procedure, this argument runs, has shown itself predictively reliable and so is self-justifying. In the case of synthetic inference, Peirce writes, however, we only know the *degree of trustworthiness* of our proceeding: "As all knowledge comes from synthetic inference, we must equally infer that all human certainty consists merely in our knowing that the processes by which our knowledge has been derived are such as must generally have led to true conclusions."[31] This type of argument amounts, of course, to an inductive justification of induction, and I am not aware that Peirce ever adequately met the criticism that this procedure is circular. Nevertheless, since it is quite clear that he did not think it circular, an interesting point follows about his self-corrective argument: it is not equivalent to contemporary "self-corrective" practical vindication of induction, since these arguments, unlike Peirce's, assume, along with Hume, that any inductive justification of induction is circular.

3. HYPOTHESIS

In the main part of his career, Peirce clearly distinguished two separate types of ampliative (synthetic) inference, namely, induction and hypothesis (which he also called abduction). In inductive inference, one infers the existence of phenomena *similar* to the phenomena that compose the evidential base of inference, while in hypothetical inference one assumes the existence of something *different* from what is observed—generally something not directly observable —which, if true, would account for what is observed.

Induction is where we generalize from a number of cases of which something is true, and infer that the same thing is true of a whole class. Or, where we find a certain thing to be true of a certain proportion of cases and infer that it is true of the same proportion of the whole class. . . . By induction [then], we conclude that facts, similar to observed facts, are true in cases not examined. By hypothesis, we conclude the existence of a fact quite different from anything observed, from which, according to known laws, something observed would necessarily result. . . . The great difference between induction and hypothesis is, that the former infers the existence of phenomena such as we have observed in cases which are similar, while hypothesis supposes something of a different kind from what we have directly observed, and frequently something which it would be impossible for us to observe directly.[32]

Peirce's examples of hypotheses that involve assumptions about unobservable entities range from the "in principle unobservable" entities of atomic theory to the "only practically or physically unobservable" entities that are assumed in any *historical* explanation. "Numberless documents and monuments refer to a conqueror called Napoleon Bonaparte. Though we have not seen the man, yet we cannot explain what we have seen, namely, all these documents and monuments, without supposing that he really existed. Hypothesis again."[33]

Peirce explicates the difference between induction and hypothesis in another fashion that is clearly related to the preceding *differentia*. He first sets up a deductive model:

Rule: All men are mortal.
Case: Enoch was a man.
Result: Enoch was mortal.

In these terms, the relation between induction and hypothesis (or abduction) may be characterized as follows: induction proceeds from Case and Result to Rule while hypothesis proceeds from Rule and Result to Case. The use of the inductive procedure leads to the discovery of *laws;* the use of the hypothetical procedure leads, on the other hand, to the discovery of *causes.*[34] In this case, however, for every inductive inference there is a corresponding hypothetical inference that does not involve any assumption whatever about unobservables.

In the later years of his philosophizing, Peirce relinquished his belief in the fundamental, even though not complete,[35] difference between induction and hypothesis and claimed instead an intimate union of the two elements in any ampliative inference. He does not elaborate greatly on this view, but there are several fragments whose import is this:[36] hypothesis *suggests* the theories that induction subsequently *verifies.* Hypothesis, then, is the only *source* of genuine synthetic knowledge, and induction is the procedure of *testing* it. This inductive procedure, of course, makes use of the deductive elaboration of the hypothesis; but even if all deduced consequences are found to occur the hypothesis is still only probably true.

> Induction is an Argument which sets out from a hypothesis, resulting from a previous Abduction, and from virtual predictions, drawn by Deduction, of the results of possible experiments, and having performed the experiments, concludes that the hypothesis is true in the measure in which those predictions are verified, this conclusion, however, being held subject to probable modification to suit future experiments.[37]

Throughout his essays, Peirce formulates criteria that a good hypothesis must meet, not particularly original, perhaps, except for the one that seems most dubious. (1) A sound hypothesis must give rise to numerous deductively elaborated consequences that can be checked by observation.[38] (2) One must select the set of predicted consequences at random; that is, one must not simply select a set that he knows antecedently will be verified.[39] (3) The testing procedure must be objective and unbiased; the negative instances must be honestly noted.[40] (4) The hypothesis, moreover, should be as "simple" as possible—Occam's razor, apparently. Surprisingly enough, however, Peirce came to interpret "simplicity" not in the sense of smallest number of independent variables but in the sense of psychological simplicity—what is more facile and natural to the imagination; what, in short, instinct suggests. Peirce came to this strange view because he was struck with the "fact" that in the relatively short time man has studied nature he has been able to discover so many true theories, a success that could not have occurred by "chance." How does a scientist ever come to discover a true theory, Peirce asks.

> You cannot say that it happened by chance, because the possible theories, if not strictly innumerable, at any rate exceed a trillion—or the third power of a million; and therefore the chances are too overwhelmingly against the single true theory in the twenty or thirty thousand years during which man has been a thinking animal, ever having come into any man's head. . . . If you carefully consider with an unbiased mind all the circumstances of the early history of science and all the other facts bearing on the question . . . I am quite sure you must be brought to acknowledge that man's mind has a natural adaptation to imagining correct theories of some kinds. . . . But if that be so, it must be good reasoning to say that a given hypothesis is good, as a hypothesis, because it is a natural one, or one readily embraced by the human mind.[41]

There are, certainly, a number of dubious assumptions here, but the use of the insufficiency of chance argument, which is the bulwark of the argument from design, is clear enough, although Peirce himself, in another context, succinctly rejects this type of argument: "Never mind how improbable these suggestions are; everything which happens is infinitely improbable."[42] Moreover, since false theories, for anything he shows to the contrary, would be just as impossible to account for by chance, they would suggest a natural adaptation for imagining incorrect theories, and hence "natural" is no longer a meaningful characteristic a good scientific hypothesis must have.

4. PROBABILITY

Peirce's contributions to probability theory, as Thomas Goudge shows, are extremely modern in appearance; and, indeed, following Venn, he was a forerunner in the technical formulation of the frequency theory, albeit, again, his contributions are sketchy and undeveloped.

A good deal of Peirce's writing in probability theory was negative in nature, showing up what he took to be inadequacies in the classical, subjective, or conceptualistic view. His objections, as we have seen in the uniformity of nature discussion, centered around the two assumptions that ignorance is denoted by the probability ½ and that, with respect to unknown probabilities, all ratios are equiprobable. The former, he thought, confuses the objective and subjective meanings of the word probability; and the latter, interpreted as "equiprobability of ratios," is self-contradictory, or interpreted as "equiprobability of constitutions of the universe" is destructive of the possibility of induction.

Peirce's characterization of probability emerges most clearly in "The Doctrine of Chances," "The Probability of Induction," "Varieties of Induction," and "Notes on the Doctrine of Chances."

In "The Doctrine of Chances" (1878), he writes that probability is a relative number,

> . . . namely, it is the ratio of the number of arguments of a certain genus which carry truth with them to the total number of arguments of that genus, and the rules for the calculation of probabilities are very easily derived from this consideration. . . . To find the probability that from a given class of premisses, A, a given class of conclusions, B, follows, it is simply necessary to ascertain what proportion of the times in which premisses of that class are true, the appropriate conclusions are also true. In other words, it is the number of cases of the occurrences of both the events A and B, divided by the total number of cases of the occurrence of the event A.[43]

But the ratio is only "ultimately" fixed.

> As we go on drawing inference after inference of the given kind, during the first ten or hundred cases the ratio of successes may be expected to show considerable fluctuations; but when we come into the thousands and millions, these fluctuations become less and less; and if we continue long enough, the ratio will approximate toward a fixed limit. We may, therefore, define the probability of a mode of argument as the proportion of cases in which it carries truth with it.[44]

Peirce reiterated this frequency viewpoint in his "The Probability of Induction," published in the same year, but added nothing new.[45]

In a later paper on "The Varieties of Induction" (1905), Peirce refined, mathematically, the concepts of "long run" and "limit":

> As it is, I will limit myself to a single needful explanation that, so far as I know, the reader could not find definitely stated in any of the books. It is that when we say that a certain ratio will have a certain value in "the long run," we refer to the *probability-limit* of an endless succession of fractional values; that is, to the only possible value from 0 to ∞, inclusive, about which the values of the endless succession will never cease to oscillate; so that, no matter what place in the succession you may choose, there will follow both values above the probability-limit and values below it; while if V be any *other* possible value from 0 to ∞, but *not* the probability-limit, there will be some place in the succession beyond which all the values of the succession will agree, either in all being greater than V, or else in all being less.[46]

In his "Notes on the Doctrine of Chances" (1910), which is, perhaps, his last word on this subject, Peirce stresses the nonempirical or, what is the same, the infinite nature of the "long run" in the frequency, or "materialistic," theory of probability. He admits that in his past work he had written as if the "long run" were an empirical series; he had, after all, defined probability as the ratio of the number of times an event occurred divided by the number of occasions it could have occurred. Now this definition, Peirce writes in his *Notes*, is clearly wrong, since probability and the long run refer to the future, and no one knows how many times a given die will be thrown in the future. Thus, if probability is the ratio of the occurrences of a specific event divided by the number of occurrences of its generic occasion, it must be the ratio that *would be* in the long run and has nothing to do with an empirical series. "This long run can be nothing but an endlessly long run; and even if it be correct to speak of an infinite 'number,' yet $\frac{\infty}{\infty}$ (infinity divided by infinity) has certainly, *in itself*, no definite value." [47]

Thomas Goudge believes this last statement exhibits Peirce's awareness of a fundamental difficulty with the frequency theory, namely, how an empirical determination of a ratio and an endless series can be compatible elements. Peirce may well have been aware of this difficulty, but he does not meet it, as Goudge suggests he does, with his concept of "would-be" and "habit."

Peirce equates a "would-be in the long run" with a "habit," so that to say that a die has a certain "would-be" amounts to saying that it has a property much like any *habit* that a man might have although much simpler;

> . . . and just as it would be necessary, in order to define a man's habit, to describe how it would lead him to behave and upon what sort of occasion—albeit this statement would by no means imply that the habit *consists* in that action—so to define the die's "would-be," it is necessary to say how it would lead the die to behave on an occasion that would bring out the full consequences of the "would-be." . . .[48]

But in order that the full effect of the die's habit can be manifested, Peirce says, it is necessary that the die should undergo an endless series of throws; [49] but the "full effect" is still empirically determined: "I really know no other way of defining a habit than by describing the kind of behavior in which the habit becomes actualized" [50]—so that the prima facie incompatibility of an empirically determined ratio and an endless series, so far as Peirce's concepts of "would-be" and "habit" help, remains unresolved.

Peirce correctly drew several corollaries of the frequency interpretation of probability. He observed that, on this view, it is meaningless to talk, except in an elliptical fashion, of the probability of an *individual* argument or event, for frequency judgments refer only to "leading principles" or "major premisses." [51] Since such frequency-probability judgments always involve, or are capable of being given, a numerical value,[52] it is also illegitimate to talk about the probability of scientific laws and hypotheses: "It is nonsense to talk of the probability of a law, as if we could pick universes out of a grab-bag and find in what proportion of them the law held good." [53] Peirce, himself, however, like anyone else, talks about the probable truth of a hypothesis and the relative probabilities of hypotheses, of the same hypothesis at different times, or of different hypotheses at the same time. As we have seen, he writes about induction and hypothesis:

> Induction is an Argument which sets out from a hypothesis, resulting from a previous Abduction, and from virtual predictions, drawn by Deduction, of the results of possible experiments, and having performed the experiments, concludes that the hypothesis is true in the measure in which those predictions are verified, this conclusion, however, being held subject to probable modification to suit future experiments.[54]

In this type of probability, where hypotheses "are true in the measure in which the predictions are verified," favorable evidence, as we have seen in the chapter on Jevons, increases the probability of the theory while unfavorable evidence decreases it; while in the frequency type of probability statement favorable evidence confirms it while unfavorable evidence decreases or increases it. Peirce, however, made no effort, or saw no need, to connect or relate his two notions of probability—showing that either one is reducible to the other, or that there are two distinct notions of probability—and thereby missed a central problem in the theory of probability.

Peirce, it should be noted, finally, formulates the frequency theory in terms of *arguments* rather than, in the more usual way, in terms of *events*. Thomas Goudge, in his excellent commentary on Peirce, compares Peirce with John Venn on this point. "It is clear that for both Venn and Peirce the fact of relative frequency lies at the heart of probability. Venn interprets it as the relative frequency of the occurrence of a given event within a series of events. Peirce takes it to be the relative frequency with which an argument yields true conclusions in the class of arguments to which it belongs." [55] However, as Goudge indicates, a translation from one terminology to the other can be readily effected, so apparently there is no fundamental difference between these two ways of formulating the matter, although Peirce thought his terminology avoided many of the pitfalls inherent in the "event" terminology.[56]

CHAPTER THIRTEEN

Chauncey Wright and the American Functionalists

CHAUNCEY WRIGHT's concepts of irregularity, accident, cosmic weather, and the ateleology and neutrality of science form an important part of his interesting philosophy of science and, as we shall see, plunge one into the midst of several problems that are receiving much attention in contemporary philosophy.

I. THE UNIVERSALITY OF CAUSALITY

Wright believed what Peirce later denied, that the universality of causality, the claim that every event has a cause, is a postulate or assumption of scientific inquiry. Wright, in Peirce's terms, was a hidebound necessitarian—not, it is true, in a state of case-hardened ignorance, however, because he did not have the advantage of Peirce's examination of necessitarianism! [1]

Wright's insistence on the universality of causality emerges most clearly in his discussion of the "palaetiological" sciences,[2] that is, sciences like geology that deal with concrete series of events rather than with isolated controlled systems. In his "Genesis of Species," a defense of Darwin against the Jesuit naturalist St. George Mivart, Wright said that natural selection, while it imitates mechanics in isolating and separating causes under controlled conditions—e.g., in experimentation with domesticated plants and animals—nevertheless should not be compared, as an explanation of the origin of species, with this science but rather with geology, meteorology, and political science, all of which exhibit *causal complexity* and *irregularity*.[3] The only genuine explanation in these sciences, Wright urged, is the deduction as far as possible of the concrete series from combinations of elementary laws discovered in control situations. "As far as possible" is an important consideration. In a concrete series of events causal chains are intermingled most intimately; and this complexity, unfathomed in precise detail, generally gives the appearance, but the appearance only, of irregularity in concrete events. Irregularity, in short, indicates not an abridgment of causality but only an abridgment of our knowledge of it.

An "uncontrolled" event, Wright believed,[4] is explained when it is derived from fixed principles or laws from the occasions that concrete causes present. Causes in concrete series of events are called *accidents* only when their occurrence, in turn, cannot be derived or predicted —which is frequently the case, again, as a result of the causal complexity in any concrete series. This unpredictability is Darwin's meaning of "accident," a notion that unfortunately, Wright said, is often erroneously interpreted as uncaused. "In referring any effect to 'accident' he only means that its causes are like particular phases of the weather, or like innumerable phenomena in the concrete course of nature generally, which are quite beyond the power of finite minds to anticipate or account for in general, though none the less really determinate or due to regular causes."[5] "Accident," then, is a characteristic not of events but of our knowledge of them; it does not mean that events are uncaused, but that we do not know the cause. That every event has a cause, Wright concludes, is the fundamental assumption of all scientific inquiry.

Mr. Mivart, like many another writer, seems to forget the age of the world in which he lives and for which he writes,—the age of "experimental philosophy," the very standpoint of which, its fundamental assumption, is the universality of physical causation. . . . The very hope of experimental philosophy, its expectation of constructing the sciences into a true philosophy of nature, is based on the induction, or, if you please, the *a priori* presumption, that physical causation is universal; that the constitution of nature is written in its actual manifestations, and needs only to be deciphered by experimental and inductive research; that it is not a latent invisible writing, to be brought out by the magic of mental anticipation or metaphysical meditation . . . and this is to presume that the order of nature is decipherable, or that causation is everywhere either manifest or hidden, but never absent.[6]

The accidental causes of science are only "accidents" relatively to the intelligence of a man . . . as if to the physical philosopher there could possibly be an absolute and distinct class, not included under the law of causation, "that every event must have a cause or determinate antecedents. . . ." [7]

It is somewhat difficult to determine precisely what epistemological status Wright is ascribing to "Every event has a cause." When he writes that it could be called an a priori assumption of science, he is not claiming that the statement is synthetic and necessarily true, for he accepted the Humean meaning criterion. Apparently he would not admit that the notions of truth and falsity can be predicated legitimately of this sentence at all. Wright's claim that universal causality is "presupposed" in scientific inquiry apparently amounts to claiming, not that it is a true statement about the universe, but that it is a rule—and so not true or false—that is necessary for inductive inference just as certain rules are required to make deductive inferences possible. "Put thus, however," G. J. Warnock writes, "this contention looks obviously wrong." [8]

It is probably true that most people who use inductive arguments do assume, or perhaps half-unconsciously take for granted, that if they try hard enough they will succeed in their quest; but it is not by any means *necessary* that they should assume this. One might well continue to do the best that one could, with failure accepted as a constant lurking possibility; and certainly one might accept without any disquiet the idea that inquirers in other fields might always fail. H. W. B. Joseph's assertion [*Introduction to Logic*, p. 420] that to accept this idea is to "despair of reason and thought" is dramatic, but an exaggeration. Failure and despair in some cases are compatible with optimism and success in others.[9]

Wright, nevertheless, while he did not adequately characterize the epistemological status of "Every event has a cause," met several of the traditional criticisms of necessitarianism.

Peirce objected to the doctrine of necessity thus: "When I ask the necessitarian how he would explain the . . . irregularity of the universe, he replies to me out of the treasury of his wisdom that irregularity is something which from the nature of things we must not seek to explain." [10] On the contrary, for Wright, since irregularity is a function of causal complexity, it is a remnant that always challenges further explanation; and even in the cases where an explanation is impossible one can sometimes give what appears to be a reasonable explanatory sketch.

Peirce, again, objected to "necessitarianism" because, he said, it requires that the novelty and complication of the universe be now no greater than at any previous time; everything to come is built in, so to speak, from the beginning. In Wright's cosmology, however, there was no beginning; he believed, with Aristotle, he said,[11] that the universe is uncreated and eternal—and this, again, reflects his nondevelopmental viewpoint. However, the problem still remains of whether at any given time the universe contains more variety and novelty than at any previous time. Wright did not think that there is novelty in Peirce's sense of the word, for this sense requires that a "novel" event have no antecedent at all, and Wright never admitted an abridgment of causality. He specified, however, several other senses in which it is perfectly correct for a determinist to talk about novelty.[12] His distinctions are, variously, ontological, epistemological, and scientific in nature. Ontologically an event is novel or genuinely new simply when it first occurs providing that it is discontinuous with previous events (i.e., with no continuity in kind) and is involved only potentially in antecedent events. Further, he saw that, epistemologically and psychologically, novelty or newness, or its absence, is not an absolute property of an event but relative to a theory or composition rule.

The word "evolution" conveys a false impression to the imagination, not really intended in the scientific use of it. It misleads by suggesting a continuity in the *kinds* of powers and functions in living beings, that is, by suggesting transition by insensible steps from one *kind* to another, as well as in the *degrees* of their importance and exercise at different stages of development. The truth is, on the contrary, that according to the theory of evolution, new uses of old powers arise discontinuously both in the bodily and mental natures of the animal, and in its individual developments, as well as in the development of its race, although, at their rise, these uses are small and of the smallest importance to life. . . . Their services or functions in life, though realized only incidentally at first, and in the feeblest degree, are just as distinct as they afterwards come to appear in their fullest development. The new uses are related to older powers only as *accidents*, so far as the special services of the older powers are concerned. . . .[13]

The appearance of a really new power in *nature* (using this word in the wide meaning attached to it in science), the power of flight in the first birds, for example, is only involved potentially in previous phenomena. In the same way, no act of self-consciousness, however elementary, may have been realized before man's first self-conscious act in the animal world; yet the act may have been involved potentially in pre-existing powers or causes. The derivation of this power, supposing it to have been observed by a finite angelic (not animal) intelligence, could not have been foreseen to be involved in the mental causes, on the conjunction of which it might, nevertheless, have been seen to depend. The angelic observation would have been a purely empirical one.[14]

Experimental science, as in chemistry, is full of examples of the discovery of new properties or new powers, which, so far as the conditions of their appearance were previously known, did not follow from antecedent conditions, except in an incidental manner—that is, in a manner *not then foreseen* to be involved in them; and these effects became afterwards predictable from what had become known to be their antecedent conditions only by the empirical laws or rules which inductive experimentation had established. Nevertheless, the phenomena of the physical or chemical laboratory, however new or unprecedented, are very far from having the character of miracles, in the sense of supernatural events. . . . Scientific research implies the *potential* existence of the natures, classes, or kinds of effects which experiment brings to light through instances, and for which it also determines, in accordance with inductive methods, the previously unknown conditions of their appearance. This research implies the *latent* kinds or natures which mystical research contemplates (erroneously, in some, at least, of its meditations) under the name of "the supernatural." [15]

Wright, it will be noted, in talking about novelty used the notion of potentiality twice—"really new powers" are involved only potentially in previous phenomena; and scientific research implies the potential existence of classes which experiment brings to light through instances. Statements about potentiality and counterfacts are isomorphic, so it is not surprising to find Wright explicating the concept "natural or caused event" in terms of counterfactual inference.

Nature means more than the continuance or actual repetition of the properties and productions involved in the course of ordinary events, or more than the *inheritance* and reappearance of that which appears in consequence of powers which have made it appear before. It means, in general, those kinds of effects which, though they may have appeared but once in the whole history of the world, yet appear dependent on conjunctions of causes which *would always* be followed by them. . . . Certain "physical constants," so called, were so determined, and are applied in scientific inference with the same unhesitating confidence as that inspired by the familiarly exemplified and more elementary "laws of nature," or even by axioms.[16]

One might criticize Wright's use of counterfactual inference as an irreducible notion (as one commentator has criticized Peirce),[17] claiming that it is incompatible with his Humean, pragmatic, or positivistic views on meaning; but the justice in doing so, I suspect, is extremely doubtful. The whole procedure smacks of projecting contemporary issues and frameworks back into a historical context that was not made to accommodate them.

The contemporary problem of counterfactual inference stems from the alleged impossibility for a contemporary Humean of explicating the concept "causal law," which supports counterfactual inference, within the conceptual framework of *Principia mathematica;* counterfactual inference, the objection goes, is rendered either self-contradictory or vacuous.[18] The Humeans have replied, however, that the difficulty only occurs when one tries to translate "law" by the notion of material implication alone; which, they say, they do not do.[19] Again, others have tried to distinguish between causal laws and accidental correlations, within a Humean framework, with such concepts as infinite scope, purely qualitative predicate, entrenched predicate, projectibility index, theoretic concepts, etc.[20] The connection between counterfactual inference and lawlikeness is peculiarly intimate certainly, but their precise relationship is not agreed upon; some writers believe that the criterion of lawfulness would suffice as an account of counterfactual inference; others, that it is only a part of the latter notion—a step in its analysis but not a complete analysis.[21]

All of these positions must be evaluated and the relationship between lawfulness and counterfactuals clarified before one can justly conclude that counterfactual inference and potentiality are incompatible with a positivistic, pragmatic, or Humean position in philosophy. Consequently, since the issue is still a going concern in contemporary philosophy and perhaps mooted temporarily, it seems impetuous to conclude that Wright's or Peirce's meaning criteria and views of lawfulness are incompatible; and it seems particularly pointless when neither they, nor any one else at the time, thought it was a problem. One might say that, in spite of their logical genius, the problem did not occur to them, not because they did not sufficiently analyze the notion of counterfactual inference,[22] but because they were not philosophizing with *Principia mathematica* as an epistemological model. And, if I may indulge in counterfactual inference instead of talking about it, I suspect that Wright, if he were philosophizing now, would see clearly that the Humean's inability to analyze the notion of lawfulness—granting, as he might or might not, that a Humean is unable to do so—does not by indirect proof establish the existence of ontological ties and potencies, as some writers imagine it does. The difficulties with the view of ontological connections still remain and must be resolved in their own right.

2. THE NEUTRALITY OF SCIENCE

Philip P. Wiener, in a recent article and book,[23] has cogently claimed that Wright's *central* contribution to philosophy is showing the metaphysical neutrality of scientific investigation. By "scientific neutrality" Wright meant that science is uncommitted to any particular metaphysical or theological interpretation of its findings and free from all forms of control imposed by metaphysical and theological authorities. After all, orthodoxy shows great resource in reconciling *any* scientific results with its own cherished convictions. According to Wright, since results of science can be accounted for as true by alternate ontologies, the establishment of scientific concepts and laws is independent of any particular ontology. These thoughts were communicated to Francis Abbot.

> Secondly, you surprise me by asking if Idealism is not "the very negation of objective science?" By objective science, I understand the science of the objects of knowledge, as contradistinguished from the processes and faculties of knowing. Does Idealism deny that there are such objects? Is not its doctrine rather a definition of the nature of these objects than a denial of their existence? There is nothing in positive science, or the study of phenomena and their laws, which Idealism conflicts with. (See Berkeley.) Astronomy is just as real a science, as true an account of phenomena and their laws, if phenomena are only mental states, as on the other theory.[24]

However, Wright did not follow the line of most modern empiricists who hold that, since scientific concepts and laws are true independently of any ontology, then this entails that no scientific fact in turn can establish any particular ontological contention. This idea is often expressed by saying that scientific assertions occur already in an epistemologically constituted universe. In Wright's case we must not lose sight of his allegiance to nineteenth-century positivism where "positivism" refers not to a specific doctrine but to the general elimination of metaphysics. Although differing in many details

Wright agreed with Mill and Comte on the meaninglessness of metaphysics and the way its concepts can be eliminated. This method of eliminating metaphysics is to give a genetic account of the occurrence of the allegedly irreducible ontological concept. On the question of what empirical laws eliminate metaphysics there is a difference among the nineteenth-century positivists. Comte's genetic method uses historio-comparative laws while Mill uses the concepts and laws of associationistic psychology. Wright's particular kind of genetic elimination differs from both Mill's and Comte's in that his causal analyses are usually in the framework of Darwin's biological evolution. For example, according to Wright, metaphysical devotion and submission in the face of mystery and reverence for the unknown are conditionally useful. They were no doubt important to the savage just as they are for children in fixing attention on serious problems and earnest considerations. Unfortunately, Wright continues, "the metaphysical modes of thought and feeling retain these early habits in relations in which they have ceased to be serviceable to the race, or to the useful development of the individual, especially when in the mystic's regard obedience has acquired an intrinsic worth, and submission has become a beatitude." [25]

One writer doubts that there is any unique meaning in the notion of "genetic method" in philosophy.[26] In my account I have not tried to show that there is a common core of meaning in the genetic methods but have simply claimed that the several genetic methods were used by nineteenth-century positivists for the same negative purpose, namely that of eliminating metaphysics.

Another feature of the genetic approach is its use to *establish* certain philosophical positions as well as disestablish others. Wright seems to have done the former as well as the latter. For example, in his essay on the evolution of self-consciousness it seems clear that he believes that his genetic account of the origin of self-consciousness directly establishes a neutral view between realism and idealism, a view interestingly akin to James's doctrine of "pure experience." [27] In this way Wright was again true to his contemporary positivism.

The upshot of this discussion is that Wright seems to have claimed only half of what the modern proponent of the metaphysical neutrality of science is claiming, i.e., that ontologies do not affect the truth of scientific statements, but, as far as the counterrelation was concerned, he was true to his nineteenth-century empiricism.

3. SCIENCE AND TELEOLOGY

Wright rejected teleology in science in the obvious sense of planned adjustments, but he also rejected as teleological the ascription to the natural world of categories that correctly characterize only conscious behavior, even when there is no accompanying assumption about a world mind whose existence and directive nature these categories reflect. Wright felt that teleology was a subtle poison, and all his criticisms of Spencer's philosophy and the nebular hypothesis center around the less obvious forms. For Wright, any evolution with ultimate progress or even directionality is a perversion of biological evolution—which he preferred to call simply descent with modification—into evolutionism, and so inconsistent with the principle of countermovements, the lack of discernible tendencies on the whole, which he thought was the drift of the logic of science, and, consequently, also inconsistent with his cosmogonic views, to which he applied the apt metaphor "cosmical weather."

Cosmogonic speculation about the production of systems of worlds, Wright thought,[28] belongs to the palaetiological category of science. The cosmogonist uses the laws of physics, particularly the principles of gravity and thermodynamics, discovered in controlled situations, to explain the physical history of the system of worlds where there has been an uncontrolled, complex interpenetration of the principles at work. The result of this causal complexity again is apparent irregularity. "The constitution of the solar system is not archetypal, as the ancients supposed, but the same corrupt mixture of law and apparent accident that the phenomena of the earth's surface exhibit. . . ." [29] This irregularity rather than regularity, Wright urged against Laplace, is the proper evidence that the solar system is a product of physical or natural causes and not the result of a creative fiat. Ordinary weather phenomena exhibit the same logical features of causal complexity and apparent irregularity, and the effort to predict them runs into the same difficulties as the cosmogonist's effort to explain the production of systems of worlds—all of which provided Wright with his metaphor "cosmical weather."

But the metaphor meant more than this for Wright; for he believed that the production of systems of worlds, like ordinary weather, shows on the whole no development or any discernible tendency whatever but is a doing and undoing without end—a kind of weather, cosmical weather. Wright based his ateleological view on what he called the principle of countermovements, "a principle in accordance with which there is no action in nature to which there is not some counter-action," which he contended is a likely generalization from the laws and facts of science.[30] Concrete courses of events in nature and interstellar space do not exhibit the dramatic unities, do not have a beginning, middle, and end; and the exhibition of these unities, and the consequent ignoring of countermovements, in the nebular hypothesis, Wright felt, renders this cosmological view nugatory.[31]

Wright worked out a technical and elaborate hypothesis about the nature of cosmical weather,[32] a system that exhibited the principle of countermovements and avoided what Wright believed were the teleological elements of the nebular hypothesis. Quite briefly, Wright, impressed with the conservation of energy principle, accounted for the origin of the sun's heat and the positions and movements of planets by the first law of thermodynamics and the conservation of angular momentum. The spiral fall of meteors into the sun, he thought, is the cause of its heat. The sun does not rapidly increase in size, he suggested, because the heat of the sun is reconverted into mechanical energy. Part of the heat is consumed in vaporizing the meteors and parts of the mass of the sun, while the rest is expended in further heating, expanding, and thus lifting the gaseous material to the heights from whence it spiraled into the sun. There it cools and condenses, and the cycle begins again.

Wright apparently thought that mechanical energy and heat energy are not only convertible but reversible. However, he was not unaware of the second law of thermodynamics, at least not when he wrote "The Philosophy of Herbert Spencer" a year later.[33] In a section of this article devoted to the nebular hypothesis, Wright says that the most obvious objection to his hypothesis is Thomson's theory that there is a universal tendency in nature to the dissipation of mechanical energy,

> . . . a theory well founded, nay, demonstrated, if we only follow this energy as far as the present limits of science extend. But to a true Aristotelian this theory, so far from suggesting a dramatic *dénouement*, such as the ultimate death of nature, only

propounds new problems. What becomes of the sun's dynamic energy, and whence do the bodies come which support this wasting power? [34]

4. THE LOGIC OF PSYCHOLOGY

Wright, it is true, did not discuss the subject matter of psychology from a methodological point of view, perhaps because he was involved in developing the body of knowledge itself. On the other hand, his discussions of physics and chemistry, areas in which the subject matter was well developed, took a methodological turn.[35] Regardless of Wright's own practice, however, the methodological implications of his contribution in actual theory to "empirical" psychology remain, and drawing these implications is one means by which I shall show the significance of his work relative to the development of psychological concepts and laws. I shall consider in what specific ways his position is similar to, or different from, the later American functionalists' views on these matters and on the problem of causal interaction between physical and phenomenal events. Any similarity will perhaps gain in significance if, as it is further claimed, there is an actual historical influence via James of Wright on Dewey, the most noted functionalist.

The basis of our discussion of Wright's psychology is his essay, "The Evolution of Self-Consciousness," which one commentator has characterized as having given a fresh impetus to empirical psychology in America.[36] In this essay Wright considered the problem of whether self-consciousness has evolved from other mental powers or, as the theologians claim, is irreducible. He thought that self-consciousness was a function of natural causes, emerging from the extension of such already existing mental powers as memory and attention, powers common in different degrees both to man and to the lower animals, but that it was discontinuous with these causes—that is, exhibited important new characteristics. He called his scientific enterprise "psychozoology."

Wright first distinguished scientific (reflective) thinking from enthymematic inference. The former, which is peculiar to the minds of men and distinguishes them from the minds of other animals, brings particular facts under explicit general principles or major premises. The latter goes from minor premises to conclusions, skipping major premises. In such cases the data of experience, which if consciously formulated would be the major premises, are *causally* effective in suggesting, more or less clearly, conclusions from minor premises. Enthymematic reasonings are exhibited in inference from signs and likelihoods as in prognostication of the weather and in orientations of many animals. In enthymematic inference signs are harbingers of events without recognition of the relation between the sign and the thing signified; in other words, the semantical capacity of the sign is unrecognized. In scientific inference, however, signs themselves are objects of reflective attention, and a sign "is recognized in its general relations to what it signifies, and to what it has signified in the past, and will signify in the future." [37]

Internal images as well as outward perceptions, Wright thought, are operative as signs in inference, and the *recognition* of them is the crucial step in achieving scientific knowledge. Internal images, he says, are "representative imaginations" that represent all the particular objects or relations of a kind, like the visual imaginations called up by such general names of objects as "dog," "tree," etc. These images are vague and feeble in intensity but effective as signs or directive elements in thought. The image of "men" as a sign of mortality leads one from the sign of this man's human nature to the expectation that he will die; but in enthymematic inference the internal image "men," because of its weak nature, falls out of consciousness, and the present sign leads on directly to the anticipation of mortality. The internal sign is lost sight of in the onrush of attention to the thing signified. However, with an extension of the range of memory power together with a corresponding increase in the vividness of its impressions (variations useful in other directions and thus likely to be secured by natural selection), a person is able to fix attention on both internal and present signs, and so become aware not only of the functioning of internal signs as major premises, but of a simultaneous internal and external suggestion of the same thing; for example, the realization that the internal image "men" and the perception of "this man" both signify "mortality." "And the contrast of thoughts [memory images] and things [present perceptions], at least in their power of suggesting that of which they may be coincident signs, could, for the first time be perceptible. This would plant the germ of the distinctively human form of self-consciousness." [38] This germ of self-consciousness depends for its extension on the use of language: "it must still be largely aided by the voluntary character of outward signs,—vocal, gestural, and graphic,—by which all signs are brought under the control of the will. . . ." [39]

Wright educed a number of philosophical implications from this account of the origin of self-consciousness. One is that the distinction of subject and object is a classification through observation and analysis and not, therefore, as the metaphysicians believe, an intuitive distinction. While the classification may be intuitive in the sense of unlearned or instinctive, this meaning, he says, is not the metaphysician's sense of the word. The metaphysician's doctrine that the distinction between subject and object is intuitive "implies that the cognition is absolute; independent not only of the individual's experiences, but of all possible previous experience, and has a certainty, reality, and cogency that no amount of experience could give to an empirical classification." [40] Wright claims, however, that only lexical or logical statements are necessary or certain and that this necessity stems from their nature as identity statements, or tautologies. He also concludes that phenomena before being empirically classified into subject and object do not belong to either the mental or physical worlds; they are neutral phenomena. And the categories of subjective and objective, after they arise, are functional, not substantive, distinctions.

Wright's position is essentially a neutral monism and comes to light particularly clearly in his criticism of natural realism and idealism. Natural realism "holds that both the subject and object are absolutely, immediately, and equally known through their essential attributes in perception." [41] Wright, however, objected to an unqualified "immediately known." An unattributed phenomenon, he says, if not referred to its cause or classified as sensation or emotion, belongs to neither world exclusively. While Wright recognizes that there may be no such unattributed phenomena in present experience, he claims that the classification into subject and object is not independent of all experience. It is, in part at least, instinctive and probably naturally selected from our progenitors. If the natural realist does not make such a concession to empiricism and fallibilism but remains an absolute intuitionist, then he renders the facts of illusion inexplicable.

Idealism unlike realism, Wright says, claims that the conscious subject is immediately known, and its phenomena are known intuitively to belong to it, whereas objects are known only mediately by their effects on us. He thought that idealism confuses physiological or genetic subjectivity with phenomenal subjectivity. And, again, he asserts that originally, in the experience of our progenitors, phenomena were unclassified or unattributed. About subjective phenomena he writes, "Instead of being, as the theories of idealism hold, first known as a phenomenon of the subject *ego* . . . its first unattributed condition would be, by our view, one of neutrality between the two worlds." [42]

Wright's concern with the problems of evolution and his use of the notions of chance variations and natural selection are indicative of his general Darwinian orientation. As far as the methodology of psychology is concerned, however, Wright's following Darwin in his general areas of investigation is of greater significance. In his work in biology Darwin observed that there are in organisms minute variations that result in different reactions, some of which meet the contingencies of the environmental conditions more adequately than others. He also found that the reactions adequate to the environment persisted or were "naturally selected." Darwin's attention, then, was directed toward two general areas—the reactions of the organism and the environs that elicit the reactions; in our terms, two "variables," "stimulus" and "response," respectively. Wright was similarly oriented in his efforts to account for the occurrence of a particular naturally selected reaction, self-consciousness. As we have seen, he was interested in tracing out genetically the development of certain mental powers, memory and imagination, and in this undertaking he dealt both with signs (stimuli)—either external stimulation, including verbal stimulation, or internal images that have their origin in external stimulation—and with the reactions (responses) of the organism to such signs. It should be noted that in talking about Darwin I have used "stimulus" to refer to the physical environment, but the term as used throughout this section, as in the case of Wright, is not restricted to this meaning. It may refer to physical environment and/or phenomenal experience. The stimulus, then, is characterized simply as that to which the organism is responding.

Wright's stimulus-response orientation becomes all the more significant if we note that at the same time he was developing his psychozoological principles the associationists (J. S. Mill and Alexander Bain) and the Wundtians were engaged in building their psychologies in terms of a single variable, mental contents (sensations, feelings, ideas, etc.). In contrast, the American psychologists who came after Wright—William James, James Mark Baldwin, James McKeen Cattell, and the functionalists—showed the stimulus-response orientation. The functionalists, however, as we shall see, differed from Wright in giving a teleological interpretation to the relation between the variables in order to avoid a dualistic metaphysics of psychology. Wright also avoided dualism but without characterizing the stimulus-response relationship teleologically.

Let us turn from the matter of variables to the coordinate matter of lawfulness. In the matter of lawfulness Wright again did not indulge in methodological analysis of what he was doing although he did explicitly characterize the distinctively human form of knowledge as scientific and, in the description of what constitutes scientific knowledge, showed it to be, for one thing, a seeking after what we would now call process laws.[43] A process law consists of a general statement of regularity between variables as a function of time. Given such a law and the statement of present conditions, the prediction of subsequent conditions can be made. (In contrast, the Wundtians or the structuralists, as they came to be known in this country, sought syndromatic laws, which have the form: if x is present, then other specifiable sensory elements y and z coexist with x.) There are process laws of varying degrees of generality, from the simple *if A then B* to a complicated mathematical formula. It was the *if A then B* sort of thing that Wright was after when he tried to find the conditions under which certain responses will occur and so to account for the evolution of self-consciousness.

The British associationists sought for historical laws that are a form of process law. The structure of the process law as it occurs in associationistic psychology would be, for example: if idea A has been followed n times by idea B (contiguity) under conditions C (such "secondary laws" as set, interest, etc.), and if A occurs now under conditions C, then B will occur again. The conditions would be established historically; that is, the information concerning the past experience of the organism plays a part in the description of present conditions. A law with this historical feature is still in principle a process law if past experience is assumed to be a matter of present conditions by virtue of the presence in the organism of the physiological traces of the past experience.

Although Wright was interested in the same kind of laws as the associationists, he departed from them in the matter of variables. The associationists dealt only with variables of one kind, namely the "impressions" or "ideas" that become united into complex perception according to such laws as contiguity and similarity. Wright's important function, then, was in applying Darwin's stimulus-response conception of variables to psychology and in doing this also preserving the orientation toward process laws exhibited by the British associationists.

So far the points in Wright's development of psychozoological principles to which I have attributed methodological significance can be briefly summarized. First, Wright took over the Darwinian orientation to stimulus-response variables in psychological investigation at a time when not only the traditional associationistic psychologists, but also the new experimental psychologists in Germany, the Wundtians, were preoccupied with a single class of variables. In addition, Wright also worked in terms of the process notion of lawfulness in which the British associationists were interested.

Having considered Wright's psychozoology primarily in relation to the ideas of his contemporaries, let us turn to a more detailed comparison of Wright with the later American psychologists who became known as the functionalists. Of primary importance here is the matter of variables.

The functionalists were, like Wright, oriented toward two variables, stimulus and response; but, unlike Wright, they conceived the variables as standing in a teleological or mutually dependent relation to one another. For example, James Rowland Angell in his systematic paper on functionalism [44] speaks of the "accommodatory service" that the response bears to the stimulating conditions. And Dewey, in an earlier paper on functionalism,[45] had developed this position in his criticism of the reflex-arc concept. He formulated the teleological relation between stimulus and response as a kind of "coordination." Functionally the stimulus was conceived to be that part of a coordination or adaptation which characterizes the problematic situation that the response (as functionally distinguished) is designed to resolve successfully. Furthermore, the stimulus is not mere sensation but is constituted by various orientation movements. There is no datum per se for Dewey; but a datum is always to be discovered and finally understood as requiring, being *for*, some response. In short, stimulus and response are meaningfully related; one does not exist without the other.

What, if any, are the implications of this teleological view of variables for lawfulness? Dewey's and Angell's particular interpretation of stimulus-response variables is such that a response could not be predicted from given physical conditions as seen through the eyes of the experimenter. But rather one would have to know how a stimulus is "seen" or "understood"—what it "means" to the responding organism—before one could know what the stimulus actually is. In other words, a response would have to occur before the stimulus could be constituted as meaningful. Such a requirement does not mean that the discovery of process laws is not possible, but it does mean that a functionalist has a basis for prediction different from the physical environment as seen through the eyes of the experimenter. It means that in the actual scientific situation the conditions for prediction would be found in the perceptual response of the subject (that which determines the stimulus), and this perceptual response, in turn, would constitute the conditions from which a subsequent response could be predicted. There must be prediction of one response of the subject from another response. This sort of procedure would be required of the functionalists if they were in actual practice to be consistent with their theory.

However Dewey, in particular, was not interested primarily in the nature of psychological laws but wanted to provide a teleological interpretation of the variables that would avoid a metaphysical dualism with its attendant difficulties for psychology of how such different substances as mind and matter could ever stand together in a causal relationship.[46] We shall discover that Wright also avoided dualism but at the same time avoided a teleological view of variables.

In his reflex-arc paper Dewey attacked the notions of a mental stimulus (sensation) and a physical response as the reintroduction of dualism into philosophy and psychology. Dewey claimed that the stimulus-response relation, a coordination, is a single process that can be described in either of two ways, physically or phenomenally. The series of events from stimulus to response can be described in terms of physical energies impinging upon receptors and continuing in the physical activity of motor nerves and muscles. As Dewey says, it is one uninterrupted continuous redistribution of mass and motion. Or the same series of events can be described purely from the psychical side: "It is now all sensation, all sensory quale; the motion, as psychically described, is just as much sensation as is sound or light or burn." [47] The avoidance of the mental-physical dichotomy is the main point of Dewey's effort to characterize the stimulus-response variables as a coordinate unity.

Wright also attempted to eliminate dualism and the attendant problem of how the two realms affect each other, but without committing himself to a teleological view of variables. His method was to show that the distinction between thought and things, the mental and the physical, is not an ontological distinction, i.e., a distinction between ultimately real constituents of the world, but one that is superimposed on the world by experience, by learning. And Wright believed that he eliminated the ontological dualism by giving a genetic account of the occurrence of dualism within the framework of phenomenal experience. According to Wright, the distinction between thought and things results only when representative images are attended to and intensified so that outward signs may be recognized as substitutes for them, thereby enabling one to be conscious of the simultaneous internal and external suggestion of the things signified. In prereflective sign reasoning, instead of anything's being "first known as a phenomenon of the subject *ego*, or as an effect upon us of an hypothetical outward world, its first unattributed condition would be, by our view, one of neutrality between the two worlds." [48] Wright's genetic account throws doubt on such metaphysical dualisms by showing that dualism is a matter of learning and thus a distinction only within the phenomenal world: "The distinction of subject and object becomes . . . a classification through observation and analysis, instead of the intuitive distinction it is supposed to be by most metaphysicians." [49]

Finally, let us turn from methodological considerations to the thread of historical continuity between Wright and Dewey via James as the connecting link. Ralph Barton Perry and others have already pointed out a historical connection between Wright and James in their psychological doctrines. Perry mentions, for example, that in

James's early essay on "Brute and Human Intellect" (1878) the author draws on Wright's psychozoological distinction between sign responses and self-consciousness to distinguish between animal and human intelligence.[50] There are more similarities between Wright's genetic account of reflection and James's scientific discussion of reasoning in his *Principles of Psychology*. Like Wright, James discusses sign reasoning in which the sign is lost sight of in the onrush of attention to the thing signified. Wright had written that association by contiguity and association by similarity are characteristic, respectively, of the child's mind and of inventive thinking.[51] James also claims that association by similarity rather than contiguity is characteristic of higher mental processes. Both Wright and James in this matter had followed the younger Mill in reintroducing the importance of similarity as a basic law of learning after it had been banished as a basic law by James Mill.[52] Concerning the evolution of self-consciousness James drew explicitly on Wright's work, as he indicated in a footnote.[53] But it is to be noted that Wright influenced James in these scientific matters and not in his over-all philosophy of chance and will to believe. And Dewey in his article "The Development of American Pragmatism" asserts that the instrumentalists found their genetic approach, their functional ideas, in the *Principles* rather than in James's more philosophical work.

> It is particularly interesting to note that in the "Studies in Logical Theory" (1903), which was their first declaration, the instrumentalists recognized how much they owed to William James for having forged the instruments which they used. . . . But it is curious to note that the "instruments" to which allusion is made, are not the considerations which were of the greatest service to James. They precede his pragmatism and it is in one of the aspects of his "Principles of Psychology" that one must look for them.[54]

The force of the influence, according to Dewey, is contained in James's chapters on attention, discrimination and comparison, conception, and reasoning.

Whatever influence, then, Wright's genetic account of reasoning had on James himself increases its significance in its secondary and, of course, subsidiary influence, via James, on Dewey's functional psychology and thus on his instrumental logic.

Notes

CHAPTER ONE

Natural Science in the Renaissance

1. Cf. Pierre Duhem, *Les précurseurs parisiens de Galilée*, 3rd ser., *Études sur Léonard de Vinci* (Paris: F. de Nobèle, 1955; originally published by A. Hermann, 1906–13), p. xiii.

Whenever we venture into the history of science for background material, particularly when we talk about the antecedents and early phases of modern science, we are greatly indebted to the monumental work of Pierre Duhem and the more recent researches of Anneliese Meier, Marshall Clagett, and Ernest A. Moody inspired by it. Books on the history of science are legion. The reader may find some of the following ones valuable if he wishes to pursue any aspect of the history of science mentioned in our text. H. Butterfield, *The Origins of Modern Science 1300–1800* (London: G. Bell and Sons, 1950); A. C. Crombie, *Augustine to Galileo* (London: Falcon Press, 1952), *Robert Grosseteste and the Origins of Experimental Science 1100–1700* (Oxford: Clarendon Press, 1953); William Dampier, *A History of Science* (Cambridge, Eng.: Cambridge University Press, 1936); A. R. Hall, *The Scientific Revolution 1500–1800* (New York: Longmans, Green and Co., 1954); Charles H. Haskins, *Studies in the History of Medieval Science* (Cambridge, Mass.: Harvard University Press, 1927); S. F. Mason, *Main Currents of Scientific Thought* (New York: Henry Schuman, 1953); P. Brunet and A. Mieli, *Histoire des sciences: Antiquité* (Paris: Payot, 1935); H. T. Pledge, *Science since 1500* (New York: Philosophical Library, 1947); George Sarton, *A History of Science* (Cambridge, Mass.: Harvard University Press, 1952); Lynn Thorndike, *A History of Magic and Experimental Science* (6 vols.; New York: Macmillan Co., 1929–41); F. W. Westaway, *The Endless Quest* (London: Blackie and Son, 1934); A. Wolf, *A History of Science, Technology, and Philosophy in the Sixteenth and Seventeenth Centuries* (new ed.; London: George Allen and Unwin,

1950) and *A History of Science, Technology, and Philosophy in the Eighteenth Century* (New York: Macmillan Press, 1939).

2. Duhem, *Les précurseurs* . . . , p. v.

3. *Ibid.*, p. 181.

4. John Herman Randall, Jr., *The Making of the Modern Mind* (rev. ed.; Boston: Houghton Mifflin Co., 1940), p. 212.

5. Duhem, *Les précurseurs* . . . , pp. 123–24.

6. Francesco Petrarca, *Opera* (Basel: Henrichus Petri, 1554), line 1038.

7. Erasmus, *The Praise of Folly*, trans. H. H. Hudson (Princeton, N.J.: Princeton University Press, 1941), pp. 54–55.

8. *Ibid.*, pp. 76–77.

9. *The Works of François Rabelais*, trans. Urquhart and LeMotteux, ed. A. J. Nock and C. R. Wilson (2 vols.; New York: Harcourt, Brace and Co., 1931), I, 428.

10. Ernst Cassirer, *Das Erkenntnisproblem* (2 vols.; Berlin: Verlag Bruno Cassirer, 1922), I, 85. For selected writings from Ficino *et al.*, see *The Renaissance Philosophy of Man*, ed. Ernst Cassirer, Paul Oskar Kristeller, and J. H. Randall, Jr. (Chicago: University of Chicago Press, 1948).

11. Giovanni Francesco Pico della Mirandola, *Examen vanitatis doctrinae gentium, et veritatis Christianae disciplinae* (Mirandola, 1520), Book II, chap. xxiv.

12. *Ibid.*, Book IV, chap. xii.

13. *Ibid.*, Book V, chap. ii.

14. Juan Luis Vives, *De causis corruptarum artium*, Book III, line 377, in *Opera* (2 vols.; Basel, 1555), Vol. I. All references to Vives are to this edition.

15. *Ibid.*

16. Vives, *De censura veris*, Book II, line 605.

17. Vives, *De causis* . . . , Book III, line 377. Cf. also T. G. A. Kater, *Johann Ludwig Vives und seine Stellung zu Aristoteles* (Erlangen: E. T. Jacob, 1908).

18. Vives, *De causis* . . . , Book V, line 413.

19. Duhem, *Les précurseurs* . . . , pp. 180–81.

20. *Ibid.*, p. xii.

21. Randall, *The Making of the Modern Mind*, p. 216.

22. *Ibid.*, pp. 216–17.

23. *Ibid.*, p. 217.

24. Duhem, *Les précurseurs* . . . , p. vi.

25. *Ibid.*, pp. xii–xiii.

26. Edmondo Solmi, *Nuovi studi sulla filosofia naturale di Leonardo da Vinci* (Mantua: G. Mondovi, 1905), p. 20.

27. *The Literary Works of Leonardo da Vinci Compiled and Edited from the Original Manuscripts*, ed. Jean Paul Richter and Irma A. Richter (2 vols.; 2nd ed. enl. and rev.; London: Oxford University Press, 1939), I, 371–72.

28. *Ibid.*, I, 372.

29. *Ibid.*, II, 241.

30. *Ibid.*, I, 116.

31. *Ibid.*

32. *Ibid.*, II, 239.

33. *Ibid.*, II, 240.

34. *Ibid.*, I, 119.

35. *Ibid.*, II, 240.

36. *Ibid.*, I, 211.

37. *Ibid.*, I, 239. Cf. "questa ragione si vede manifestamente conferma dalla esperienzia." MSS de l'Institut, MS A, fol. 46 recto, quoted by Solmi, *Nuovi studi* . . . , p. 53. Also "Qui l'esperienzia ne mostra," quoted by Solmi, *ibid.*, p. 54, n. 3.

38. MSS de l'Institut, MS A, fol. 31 recto, quoted by Solmi, *Nuovi studi* . . . , pp. 53–54.

39. Richter, *Literary Works of Leonardo da Vinci*, I, 305. Solmi, *Nuovi studi* . . . , sees in this passage an anticipation of Descartes' second rule of method (that of "division"). Solmi also quotes Leonardo's "la cosa piu facile sia scala e guida alla men facile" (MSS de l'Institut, MS E, fol. 54 recto) as a parallel to Descartes' third rule of method (that of "order"). It should be remembered, however, that these and other precepts of Leonardo are given by him simply for the specific guidance of students of painting, and not with reference to scientific investigation in general.

40. Richter, *Literary Works of Leonardo da Vinci*, II, 250–51. Cf. *The Notebooks of Leonardo da Vinci*, ed. and trans. Edward MacCurdy (2 vols.; New York: Reynal and Hitchcock, 1938), I, 88–89.

41. Cf. Solmi, *Nuovi studi* . . . , pp. 29 ff.

42. *Ibid.*, pp. 39 ff.

43. Richter, *Literary Works of Leonardo da Vinci*, I, 367.

44. *Ibid.*, I, 371.

45. *Ibid.*

46. *Ibid.*, II, 85.

47. *Ibid.*, II, 134.

48. *Ibid.*, I, 124.

49. Cf. Solmi, *Nuovi studi* . . . , pp. 42 ff.

50. Leonardo da Vinci, *Frammenti letterari e filosofici*, ed. Edmondo Solmi (Florence: G. Barbèra, 1925), p. 88.

51. *Ibid.*, p. 89.

52. Cf. "Pruova a fare uscire l'acqua da diverse qualità di spiracoli, lunghi e corti, smussi di fuori e dentro, tardi e quadri, sottili e grossi, e farla battere in diverse oppositioni, che cosi avrai infinite esperienzie da notare, e farne regola" (MSS de l'Institut, MS A, fol. 58 verso, quoted by Solmi, *Nuovi studi* . . . , p. 54, n. 4).

53. Cf. Solmi, *Nuovi studi* . . . , p. 57.

54. *Ibid.*, pp. 56–57.

55. *Ibid.*, pp. 58 ff.

56. Cf. "Io stimo che sia per cagion d'altri piccoli sonagli, che sian congiunti ad esso sonaglio maggiore . . . ma non ho ancora investigata la causa" (MSS de l'Institut, MS F, fol. 28 recto, quoted by Solmi, *Nuovi studi* . . . , pp. 49–50).

57. Cf. the example quoted by Solmi, *Nuovi studi* . . . , p. 50.

58. Da Vinci, *Frammenti* . . . , p. 90.

59. Cf. "questa isperienzia è nata dalla ragione" (MSS de l'Institut, MS F, fol. 29 recto, quoted by Solmi, *Nuovi studi* . . . , p. 54).

60. Da Vinci, *Frammenti* . . . , p. 96.

61. *Ibid.*, p. 94.

62. *Ibid.*, p. 84.

63. Richter, *Literary Works of Leonardo da Vinci*, I, 117.

64. Da Vinci, *Frammenti* . . . , p. 83.

65. *Ibid.*, p. 86.

66. *Ibid.*, pp. 85–86.

67. *Ibid.*, p. 95.

68. *Ibid.*, p. 86.

69. Richter, *Literary Works of Leonardo da Vinci*, I, 112.

70. MacCurdy, *Notebooks of Leonardo da Vinci*, I, 88.

71. Da Vinci, *Frammenti* . . . , p. 95.

72. *Ibid.*, p. 88. Cf. *ibid.*, p. 87: "Ricordati quando commenti l'acqua d'allegar prima la sperienza e poi la ragione."

73. MSS de l'Institut, MS A, fol. 33 recto, quoted by Solmi, *Nuovi studi* . . . , p. 54.

74. *Ibid.*, fol. 57 recto, quoted by Solmi, *Nuovi studi* . . . , p. 54. Cf. Cassirer, *Das Erkenntnisproblem*, I, 324–25: "So gründet Leonardo seinen Idealbegriff der Wahrheit und der Vernunft in dem fruchtbaren 'Bathos der Erfahrung,' während umgekehrt der Erfahrungsbegriff selbst seinen Wert aus dem notwendigen Zusammenhang erhält, in dem er mit der Mathematik steht. . . . Die Erfahrung selbst ist nichts anderes als die äussere Erscheinungsform der Vernunftbeziehungen und Vernunftgesetze."

75. Il codice atlantico, fol. 203 recto, cited by Solmi, *Nuovi studi* . . . , pp. 66–67.

76. Da Vinci, *Frammenti* . . . , p. 87.

77. *Ibid.*

78. Richter, *Literary Works of Leonardo da Vinci*, I, 129–30.

79. *Ibid.*, p. 130.

80. Da Vinci, *Frammenti* . . . , p. 113.

81. *Ibid.*, p. 112.

82. MacCurdy, *Notebooks of Leonardo da Vinci*, I, 69.

83. Il codice atlantico, fol. 169 verso, quoted by Solmi, *Nuovi studi* . . . , p. 64.

84. Da Vinci, *Frammenti* . . . , p. 85.

85. MacCurdy, *Notebooks of Leonardo da Vinci*, I, 75.

86. *Ibid.*, p. 79.

87. *Ibid.*, p. 82.

88. Richter, *Literary Works of Leonardo da Vinci*, II, 100.

89. *Ibid.*, I, 119–20.

90. Da Vinci, *Frammenti* . . . , pp. 97–98.

91. MacCurdy, *Notebooks of Leonardo da Vinci*, I, 77.

92. Da Vinci, *Frammenti* . . . , p. 119.

93. *Ibid.*, p. 120.

94. MacCurdy, *Notebooks of Leonardo da Vinci*, I, 74.

CHAPTER TWO

Theory of Hypothesis among Renaissance Astronomers

1. Cited by Pierre Duhem, *Le système du monde: Histoire des doctrines cosmologiques de Platon à Copernic* (6 vols.; Paris: A. Hermann et fils, 1913–54), III, 451–52.

2. Cf. Pierre Duhem, "ΣΩΖΕΙΝ ΤΑ ΦΑΙΝΟΜΕΝΑ, Essai sur la notion de théorie physique de Platon à Galilée," *Annales de Philosophie Chrétienne*, Ser. 4, VI (1908), 360–61.

3. Nicolaus Copernicus, *De revolutionibus orbium caelestium libri sex*, including also Joachim Rheticus, *De libris revolutionum narratio prima* (Thorn: Society of Copernicus, 1873), p. 6: "Inde igitur occasionem nactus, coepi et ego de terrae mobilitate cogitare. Et quamvis absurda opinio videbatur, tamen quia sciebam aliis ante me hanc concessam libertatem, ut quoslibet fingerent cirulos ad demonstrandum phaenomena astrorum, existimavi mihi quoque facile permitti, ut experirer, an posito terrae aliquo motu

firmiores demonstrationes, quam illorum essent, inveniri in revolutione orbium caelestium possent."

4. *Commentariolus de hypothesibus motuum caelestium a se constitutis,* cited by Pierre Duhem, *La théorie physique* (Paris: Marcel Rivière & Cie.; 1914), p. 57. Cf. Philip P. Wiener's timely translation, *The Aim and Structure of Physical Theory* (Princeton, N.J.: Princeton University Press, 1954).

5. *De revolutionibus* . . . , p. 5.

6. *Ibid.*, p. 6.

7. *Ibid.*, p. 30.

8. Rheticus, *De libris revolutionum narratio prima,* p. 464. Cf. Edward Rosen, *Three Copernican Treatises* (New York: Columbia Universtiy Press, 1939).

9. Cf. Duhem, "Essai . . . ," p. 377.

10. This letter was first published in *Epistolae ad naturam ordinatarum plenius intelligendarum pertinentes* (1615).

11. Cited by L. Prowe, *Ueber die Abhängigkeit des Copernicus von den Gedanken griechischen Philosophen und Astronomen* (Thorn, 1865), pp. 38–39 n.

12. Cf. Duhem, "Essai . . . ," pp. 485–86.

13. Cf. Letter of Osiander to Copernicus (April 20, 1541) cited in Johann Kepler, *Apologia Tychonis contra Ursum,* contained in *Opera omnia,* ed. Ch. Frisch (8 vols.; Frankfurt and Erlangen: Heyder and Zimmer, 1858–71), I, 246.

14. *De revolutionibus* . . . , pp. 1–2.

15. Cf. Duhem, "Essai . . . ," pp. 486–511. He cites as adherents of this view: in the Netherlands, Gemma Frisius (1541); in Germany, Erasmus Reinhold (1542), Gaspard Peucer (1551), Oswald Schreckenfuchs (1556), Christian Wursteisen (1568); in Italy, Alessandro Piccolomini (1563), Andrea Cesalpinus (1569), Francesco Giuntini (1577–78), Giovanni Battista Benedetti (1585); at Paris, Ariel Bicard (1549), Pierre de la Ramée (1562), who, although not fully accepting the Copernican system, regards it as a possible and discussible supposition, and Jean Hannequin (1586), who writes that "the most foolish of men are those according to whom none but fools can doubt that the earth is at rest." For the attitude of the masters of the University of Paris toward the Copernican system see Duhem, *Les précurseurs parisiens de Galilée,* 3rd ser., *Études sur Léonard de Vinci* (Paris: F. de Nobèle, 1955; originally published by A. Hermann, 1906–13), pp. 247–254.

16. Cited by Dorothy Stimson, *The Gradual Acceptance of the Coperni-*

can Theory of the Universe (New York: Baker and Taylor Co., 1917), from Luther, *Tischreden,* IV, 575.

17. *Epistola B. Mithobio,* October 16, 1541, in Philipp Melanchthon, *Opera quae supersunt omnia,* ed. Karl Gottlieb Bretschneider. In *Corpus reformatorum* (Halle: C. A. Schwetschke and Son, 1834–60), IV, 679: "Vidi dialogum, et fui dissuasor editionis. Fabula per sese paulatim consilescet. . . . Profecto sapientes gubernatores deberent ingeniorum petulantiam cohercere."

18. *Initia doctrinae physicae,* in *Opera . . .*, XIII, 232.

19. *Ibid.,* p. 244.

20. *Ibid.*

21. *Ibid.,* p. 216: "Sed hic aliqui vel amore novitatis, vel ut ostentarent ingenia, disputarunt moveri terram. . . . Nec recens hi ludi conficti sunt." *Ibid.,* p. 292: "Sumus autem secuti in describendis illis Ptolemaei hypotheses, quae tot seculorum testimonio comprobatae, non temere convelli debent." *Ibid.,* p. 216: "Etsi autem artifices acuti multa exercendorum ingeniorum causa quaerunt, tamen adseverare palam absurdas sententias, non est honestum, et nocet exemplo. Bonae mentis est veritatem a Deo monstratam reverenter amplecti, et in ea acquiescere, et Deo gratias agere. . . . Quanquam autem rident aliqui physicum testimonia divina citantem, tamen nos honestum esse censemus philosophiam conferre ad coelastia dicta et in tanta caligine humanae mentis autoritatem divinam consulere, ubicumque possumus."

22. *Prutenicae tabulae coelestium motuum.*

23. *Logistice scrupulorum astronomicorum.* Praecepta calculi motuum coelestium, p. xxi, cited by Duhem, "Essai . . . ," p. 491. But at the same time Gaspard Peucer, another member of the Wittenberg school, although allowing the employment of astronomical *tables* calculated according to the Copernican system, was so far influenced by Melanchthon's example as to forbid to the schools any direct use of the *hypotheses* of that system, on the ground of their absurdity and opposition to the truth.

24. Christopher Clavius, *In Sphaeram Ioannis de Sacro Bosco commentarius* (Lyon: Gabiano, 1602).

25. *Ibid.,* p. 518.

26. *Ibid.,* p. 520.

27. *Ibid.,* p. 519.

28. *Ibid.,* p. 502. This is argued at length.

29. *Ibid.,* p. 518.

30. *Ibid.*

31. *Ibid.*

32. *Ibid.*, pp. 518–19.
33. *Ibid.*
34. *Ibid.*, p. 502.
35. *Ibid.*, p. 517.
36. *Ibid.*, pp. 519–20.
37. *Ibid.*, p. 520.
38. *Ibid.*
39. *Ibid.*, p. 517.
40. *Ibid.*, p. 522.
41. *Ibid.*, p. 520.

42. It is for this reason that he rejects the Copernican system. Cf. *Astronomiae instauratae progymnasmata* in Tycho (Tyge) Brahe's *Opera omnia*, ed. I. L. E. Dreyer (Copenhagen: Libraria Gyldendaliana, 1913), II, 14: "Rationes cur haec summi illius Copernici, utut admodum ingeniose, et concinne excogitata circuituum in Mundanis corporibus apparentium dispositio, re ipsa veritati non correspondeat, alias sufficienter ostensuri."

43. The reference seems to be to the "fictional" theory of hypotheses.

44. Tycho Brahe, *De mundi aetheri recentioribus phaenomenis liber secundus*, in *Opera omnia*, IV, Fasc. 1, 156: "When I had realized that Ptolemy's ancient arrangement of the celestial orbits did not sufficiently agree with the phenomena, and that the assumption of so many and such epicycles, although they explained the relation of the planets to the sun and their backward movements and stoppings, together with some part of their apparent lack of uniformity of motion, was superfluous, and, what is more, that these hypotheses sinned against the very first principles of science, inasmuch as they inconsistently admit the possibility of a uniform circular motion not about its own center, as should be the case, but about the center of another and eccentric sphere . . . and when also I had considered the recent innovation in these matters introduced by that great genius Copernicus in accordance with the opinion of Aristarchus of Samos, how although it cut away the superfluities of the Ptolemaic system, and removed the disagreements with the phenomena, in fact very cleverly provided against them, and in no way transgressed the principles of mathematics, nevertheless, by making that heavy, sluggish body of the earth, so inapt for movement, to be agitated in a course of motion no less abondoned [? *haud dissolutiore tenore motus*] than those aetherial luminaries (and what is more with a triple motion), not only contradicted the principles of physics, but also the authority of Sacred Scripture which so often affirms the stability of the earth . . . when, I say, I had observed that both these hypotheses thus admitted no light absurdities, I began to consider more deeply whether it were not possible to invent another system of hypotheses that should stand at all points in accord

with the principles both of mathematics and of physics, and should not make use of subterfuges in order to escape the censures of theology [*neque etiam theologicas censuras subterfugeret*], and at the same time should fully satisfy the celestial appearances."

45. Kepler, *Opera omnia*, I, 44.

46. *Ibid.*, p. 159.

47. *Tractatus in arithmeticam logisticam.* Introductio in geometriam. Explicatio brevis ac perspicua doctrinae sphericae in quattuor libris distributa. Cited by Duhem, "Essai . . . ," pp. 568–69.

48. P. Ramus (Pierre de la Ramée), *Scholarum mathematicarum libri unus et triginti* (Frankfurt, 1599), Book II. Cf. Francesco Maurolico, *De sphaera* in *Opuscula mathematica* (1575): "Toleratur et Nicolaus Copernicus, qui scutica potius aut flagello, quam reprehensione dignus est."

49. *The Works of Francis Bacon*, ed. James Spedding, Robert Leslie Ellis, and Douglas Denon Heath (14 vols.; London: Longman and Co., 1857–59), IV, 373.

50. Cf. section 4, this chapter.

51. Cf. *La cena de le ceneri* (1584) in *Le opere italiane di Giordano Bruno*, reprinted by Paola de Lagarde (2 vols.; Gottinga, 1888), I, 151: ". . . certa Epistola superliminare attaccata non sò da chi asino ignorante, et presuntuoso, il quale (come uolesse iscusando faurir l'authore, o' pur a' fine che ancho in questo libro gl' altri asini trouando anchora le sue lattuche, et fruticelli: hauessero occasione di non partirsene à fatto deggiuni) in questo modo le auuertisce auanti che cominciano ad leggere il libro, et considerar le sue sentenze."

52. *Ibid.*, p. 151: "Il Nolano disse che se Copernico per questa causa sola [sc. per la commodità de le supputationi] disse la terra mouersi, et non anchora per quell' altra: lui ne intese poco, et non assai. Ma è certo che il Copernico la intese come la disse, et con tutto suo sforzo la prouò." *Ibid.*, p. 152: "non solo fà ufficio di mathematico che suppone: ma ancho de physico che dimostra il moto de la terra."

53. Kepler, *Opera omnia*, I, 245: "Atqui o Osiander, quid eo te desperationis adegit, ut ex astronomia de vera mundi facie nihil certi colligi posse diceres? An, qui peritissimus eras harum rerum, etiamnum dubitabas de rebus evidentissimis? Incerta tibi etiamnum erat proportio Solis ad Terram prodita ab astronomis et sexcenta hujusmodi? Si haec ars causas motuum simpliciter ignorat, propterea quia nihil credis, nisi quod vides, quid medicinae fiet, in qua quis unquam medicus causam morbi vidit intus latentem aliter, nisi ex signis et symptomatis extra corpus in sensus incurrentibus, perinde ut astronomus ex stellarum situ aspectabili de forma motus eorum ratiocinatur."

54. Kepler, *Prodromus dissertationum cosmographicarum, continens mysterium cosmographicum*, in *Opera omnia*, I, 113.

55. *Ibid.*: "Neque dubito affirmare, quicquid a posteriori Copernicus collegit et visu demonstravit, mediantibus geometricis axiomatis, id omne vel ipso Aristotele teste, si viveret . . . a priori nullis ambagibus demonstrari posse."

56. *Ibid.*, p. 106: "Jamque in eo eram, ut eidem etiam Telluri motum Solarem, ut Copernicus mathematicis, sic ego physicis, seu mavis, metaphysicis rationibus ascriberem."

57. Kepler, *Opera omnia*, I, 46–47: ". . . harmonia et dispositionum analogia ex posteriori, ubi motus et motuum occasiones ad amussim constiterint, non autem ex priori, uti et tu et Maestlinus voluistis, petenda erit; quin et hic difficillime invenienda."

58. *Ibid.*, VI, 119: "Est pars physices, quia inquirit causas rerum et eventuum naturalium et quia inter ejus subjecta sunt motus corporum coelestium, et quia unus finis ejus est, conformationem aedificii mundani partiumque ejus indagare."

59. *Ibid.*, p. 121: ". . . oportet ut etiam causas reddere possis probabiles hypothesium tuarum, quas pro veris apparentiarum causas venditas, et sic astronomiae tuae principia prius in altiori scientia, puta physica vel metaphysica, stabilias. . . ."

60. *Ibid.*: ". . . non interclusus tamen nec ab iis argumentis geometricis, physicis vel metaphysicis, quae tibi suppeditantur ab ipsa diexodo disciplinae propriae super rebus, ad altiores illas disciplinas pertinentibus, dummodo nullam principii petitionem admisceas."

61. Ernst Cassirer, *Das Erkenntnisproblem* (3 vols.; Berlin: Verlag Bruno Cassirer, 1922), I, 342–43.

62. Nicolai Raimari Ursi *Tractatus astronomicus de hypothesibus astronomicis* (1597).

63. First published by Frisch in Kepler, *Opera omnia*, Vol. I.

64. Kepler, *Opera omnia*, I, 241–42: "Jam Ursi absurdas sententias paulisper expendamus. Primo hypotheses astronomicas, ait, esse effictam delineationem imaginariae non verae et genuinae formae systematis mundani. Quibus verbis aperte negat hypothesin esse, quae non sit falsa. Confirmat hanc suam sententiam paulo post, ubi ait, nihil aliud esse hypotheses, quam commenta quaedam. Et inferius ait, non fore hypotheses si verae sint. Item: proprium esse hypothesium, ex falso verum sciscitari." *Ibid.*, p. 238: "In descriptione hypothesium sic loquitur Ursus, quasi non ad aliud comparatae essent hypotheses, quam ad ludos hominibus faciendos."

65. Carl von Prantl, "Galilei und Kepler als Logiker," *Sitzungsberichte*

der philosophisch-philologischen und historischen Classe der k.b. Akademie der Wissenschaften zu München, II, No. 4 (1875), 405.

66. *Apologia Tychonis . . .*, in *Opera omnia,* I, 239: "Hypothesin autem in genere dicimus, quidquid ad quamcunque demonstrationem pro certo et demonstrato affertur."

67. *Ibid.,* p. 238.

68. *Ibid.:* ". . . certa et apud omnes homines confessa . . . quae haberent auctoritatem apud omnes."

69. *Ibid.:* ". . . quamvis non ab omnibus crederentur, ipsis tamen auctoribus satis erant nota, ut non dubitarent ea alia demonstratione certa et vera ostendere, id autem ut in sibi proposita demonstratione . . . a discente sibi concedi a principio postularent."

70. *Ibid.*

71. *Ibid.,* p. 239: "Ita in astronomicis, quoties ex iis, quae in coelo diligenter et apte attenti vidimus, aliquid numerorum et figurarum ope de stella, quam observaveramus, demonstramus: tunc ea, quam dixi observatio, in instituta demonstratione fit hypothesis, super quam praecipue demonstratio exstruitur."

72. *Ibid.:* "In specie tamen, cum numero multitudinis astronomicas dicimus hypotheses, facimus id more scholarum hujus seculi, designantes summam quandam conceptionum celebris alicujus artificis, ex quibus totam ille rationem motuum coelestium demonstrat: qua in summa insunt cum physicae tum geometricae propositiones omnes, quotquot ad totum opus astronomo illi propositum asciscuntur: sive illas aliunde ad suum commodum transtulerit artifex, sive antea ex observationibus demonstraverit, et jam via reciproca secum demonstrata a discente sibi ceu hypotheses concedi postulet, ex quibus illi et eos (quibus initio secum hypothesium loco usus erat) observatos stellarum situs et quos porro similiter apparituros sperat, necessitate syllogistica demonstraturum polliceatur."

73. *Ibid.,* p. 246: "At vero Osiander Copernicum vel potius ejus lectores aequivocatione ludificatus est, ea, quae in geometricis hypothesibus vera sunt, ad astronomicas transferens, cum diversissima sit utrarumque ratio."

74. *Ibid.,* pp. 246–47.

75. *Ibid.,* p. 112.

76. *Ibid.,* p. 240.

77. *Ibid.*

78. *Ibid.*

79. *Ibid.,* p. 113.

80. *Ibid.,* p. 240.

81. *Ibid.*

82. *Ibid.*, p. 241.

83. *Ibid.*, p. 242.

84. Kepler, *Opera omnia*, VI, 121: "Non enim mera debet esse licentia astronomis, fingendi quidlibet sine ratione. . . ."

85. *Ibid.*, I, 242.

86. *Ibid.*, p. 118.

87. *Ibid.*, p. 239.

88. Ramus, *Scholarum mathematicarum*, Book II, p. 50, cited by Prantl, "Galilei und Kepler als Logiker," p. 403.

89. Kepler, *Opera omnia*, III, 136.

90. Galileo Galilei, *Trattato della sfera ovvero cosmografia*, in *Le opere di Galileo Galilei* (Edizione Nazionale; 20 vols.; Florence: G. Barbèra, 1890), II, 211–12.

91. *Istoria e dimostrazioni intorno alle macchie solari*, in *Le opere*, V, 102. Cf. also the reference (*Errori . . . Giorgio Coresio*, IV, 248) to "il libro della natura, dove le cose sono scritte in un modo solo. . . ." It is interesting to compare the above passage with the very similar expressions of Galileo's English contemporary, William Gilbert (of Colchester), in *On the Loadstone and Magnetic Bodies, and on the Great Magnet the Earth*, trans. P. Fleury Mottelay (New York: John Wiley and Sons, 1893), pp. 321–27: "This *primum mobile* presents no visible body, is in no wise recognizable; it is a fiction believed in by some philosophers, and accepted by weaklings. . . . But now this inadmissible *primum mobile*, this fiction, this something not comprehensible by any reasoning and evidenced by no visible star, but purely a product of imagination and mathematical hypothesis, accepted and believed by philosophers, and reared into the heavens. . . . From these arguments, therefore, we infer, not with mere probability, but with certainty, the diurnal rotations of the earth; for nature ever acts with fewer than with many means; and because it is more accordant to reason that the one small body, the earth, should make a daily revolution than that the whole universe should be whirled around it."

92. Roberto Bellarmino a Paolo Antonio Foscarini, Roma, 12 aprile 1615, in Galileo Galilei, *Le opere*, XII, 171–72.

93. *Considerazioni circa l'opinione copernicana*, in *Le opere*, V, 351 ff.: ". . . vedrò di rimuovere due concetti . . . i quali concetti, se io non erro, sono diversi dal vero."

94. *Ibid.*

95. *Ibid.* Cf. Domenico Berti, *Copernico e le vicende del sistema copernicano in Italia nella secondo metà del secolo XVI e nella prima del secolo XVII* (Rome, 1876), pp. 132–33.

96. Cf. *Le opere*, VII, 296; V, 356–57. Here the point is that there

is no middle ground—either the earth moves and the Copernican system is right, or it stands still and that of Ptolemy is vindicated (Duhem, "Essai . . . ," pp. 583–84).

97. *Le opere,* V. 369. Cf. the very similar statement of G. W. Leibniz, *Die philosophischen Schriften,* ed. C. J. Gerhardt (7 vols., Berlin, 1880), IV, 356: "Nihil autem aliud de rebus sensibilibus aut scire possumus aut desiderare debemus, quam ut tam inter se quam cum indubitatis rationibus consentiant. . . . Alia in illis veritas aut realitas frustra expetitur, quam quae hoc praestat, nec aliud vel postulare debent Sceptici vel dogmatici polliceri."

98. Augustini Oregii S. R. E. Cardinalis, Archiepiscopi Beneventani, *Ad suos in universas theologiae partes tractatus* (1637), p. 119; *De Deo uno* (1629), p. 194. Cf. Berti, *Copernico e le vicende del sistema copernicano . . .* , pp. 138–39; Duhem, "Essai . . . ," pp. 585–86.

99. *Le opere,* VII, 488–89. *Mathematical Collections and Translations,* trans. Thomas Salusbury (2 vols.; London, 1661), I, 423–24.

CHAPTER THREE

Francis Bacon's Philosophy of Science

1. Prooemium to *The Great Instauration.* The edition of James Spedding, Robert Leslie Ellis, and Douglas Denon Heath, *The Works of Francis Bacon* (14 vols.; London: Longman and Co., 1857–59), is the one referred to throughout this chapter.

2. See Robert Leslie Ellis' Preface to the *Novum organum.*

3. *Ibid.*

4. *Novum organum,* I, 40.

5. *Ibid.,* I, 3.

6. *Ibid.,* I, 124. In II, 4, he says: "What in operation is most useful, that in knowledge is most true." See also *ibid.,* I, 81, and "Plan of the Instauration," Argument of Part III. The epistemological import of these statements of Bacon, it should be noted, is a pragmatism of a different sort from that of William James. Bacon virtually asserts: *Unless* an idea "works," it does not constitute real knowledge; while James's position is rather: *If* an idea "works," it constitutes knowledge, truth. In *Valerius terminus,* chap. xii, Bacon says that you may not always conclude "that the axiom which discovereth new instances is true, but contrariwise you may safely conclude that if it discover not any new instance it is in vain and untrue." This statement may be compared with James's assertion that there is no difference anywhere that does not make a difference somewhere. Baron Liebig, in his well-known arraignment of Bacon's views (English translation in *McMillan's Magazine* [July and August, 1863]), obviously misconceives the nature of Bacon's

pragmatism no less grossly than that of other parts of his doctrine. Concerning Liebig's unfairness to Bacon, see for instance Kuno Fischer's *Bacon*, English translation by Oxenford (London, 1857).

7. *Novum organum*, I, 9; also I, 67, 126.

8. *Ibid.*, I, 13.

9. *Ibid.*, I, 105. Compare the similar statement in the "Plan of the Instauration," Argument of Part II.

10. I, 19. Whewell lays great stress on the merit of Bacon's insistence on the necessity of a graduated and successive induction (*Philosophy of Discovery* [London: John W. Parker and Son, 1860], chap. xv, sec. 6). The soundness of this precept, considered in the light of the subsequent history of the physical sciences, is, Whewell remarks, most clearly apparent when the successive generalizations of such sciences are exhibited in the form of tables as given by him in the *Novum organon renovatum*, Book II, chap. ix, which he says constitute the only adequate illustrations, up to his day, of what Bacon anticipated. Cf. Sir J. Herschel, *Discourse on the Study of Natural Philosophy* (London: Longman, Brown, Green, and Longmans, 1830), sec. 96. See, however, in the article on Bacon in the *Encyclopaedia Britannica*, 9th ed., Adamson's assertion that "the inductive formation of axioms by a gradually ascending scale is a route which no science has ever followed, and by which no science could ever make progress."

11. *Novum organum*, I, 105.

12. *Ibid.*, I, 106. See also *Valerius terminus*, chap. xi, where, after stating his sixth "direction" for whiteness, Bacon adduces as evidence of its truth the fact that out of it "are satisfied a multitude of effects and observations" from which it was not itself derived. Also *ibid.*, chap. xii: "The only trial of the truth of knowledge to be accepted is where particulars induce an axiom of observation, which axiom found out discovereth and designeth new particulars . . . if it discover not any new instance it is in vain and untrue." Cf. also *Novum organum*, I, 25 and 82: "The true method of experience . . . commencing as it does with experience duly ordered and digested, not bungling or erratic, and from it educing axioms, and from established axioms again new experiments." Also *ibid.*, I, 103, 117, and II, 20; and *New Atlantis*, the end.

13. These conditions, he says, must moreover be so explicitly described that others can repeat the experiments and check their results: "In any new and more subtle experiment the manner in which the experiment was conducted should be added, that men may be free to judge for themselves whether the information obtained from that experiment be trustworthy or fallacious" (*Parasceve*, aphorism 9). See also "Plan of the Instauration," Argument of Part III.

14. "Plan of the Instauration," Argument of Part III.

15. *Ibid.* Cf. *Valerius terminus,* chap. xv; *Parasceve,* aphorism 5; and *Novum organum,* I, 98. For Bacon's description of "Experientia Literata"—i.e., of the various methods of experimenting—see *De augmentis scientiarum,* Book V, chap. ii.

16. *Novum organum,* I, 99; see also p. 73.

17. *Ibid.,* I, 99.

18. The work of Gilbert on the magnet, which in fact was a more concrete contribution to science than resulted from any of Bacon's own investigations, seems to have been frequently present to Bacon's mind. He comments upon it repeatedly, and in the main adversely so far as concerns the method of it, although he has difficulty in making out much of a case against it on the basis of his own fundamental principles. Gilbert had been appointed royal physician by Queen Elizabeth, who had moreover encouraged his scientific researches by settling a pension upon him, while Bacon had desired similar help in vain. But there appears no need to ascribe Bacon's unfairness to Gilbert either to personal or to scientific jealousy. It is fully explicable by the fact that Bacon, who did not possess on Gilbert's work the perspective that is ours today, was both too much engrossed in his own method and too unclear concerning the difficulties in the way of applying it in the concrete properly to evaluate researches that appeared to make no use of it, or of some of the features of it that were then most prominent in his mind.

19. *Novum organum,* I, 88.

20. *Ibid.,* II, 10.

21. For the meaning of these terms, see section 4 of this chapter.

22. *Novum organum,* II, 11.

23. *Ibid.,* II, 12.

24. *Ibid.,* II, 13.

25. *Ibid.,* II, 15.

26. *Ibid.,* II, 16.

27. *Ibid.,* II, 19.

28. *Ibid.,* II, 20. Ellis asserts that this first vintage is offered by Bacon *only* on the basis of the three tables of presentation, and independently of the process of exclusion: "In this Vindemiatio," he says, "we accordingly find no reference to the method of exclusion; it rests immediately on the three tables of Comparentia; and though of course it does not contradict the results of the Exclusiva, yet on the other hand it is not derived from them" (General Preface to the *Philosophical Works,* sec. 9). It is difficult, however, to see how this can be maintained. Not only does Bacon introduce his first vintage *after* a beginning of the process of exclusion has

been carried out, but he also speaks of the first vintage as to be attempted "after the three Tables of First Presentation . . . have been made *and weighed*" (*Novum organum*, II, 20; italics ours). This weighing can hardly be anything but the process of exclusion of which he has himself given an example after setting forth his tables of presentation. Decisive, however, seems Bacon's own statement at the beginning of aphorism 21: "The Tables of First Presentation and the Rejection or process of Exclusion being completed, and also *the First Vintage being made thereupon*, we are to proceed to the other helps of the understanding . . ." (italics ours). It is true that Bacon does not *complete* the process of exclusion, but merely exemplifies it. This is no less true, however, of the tables of presentation, which he declares "are meant only for examples"; and, if all of these things are admitted as fair examples, there is no reason then to deny the same status to Bacon's first vintage. Doubtless, Ellis is right in saying that the first vintage "is not to be taken as the type of the final conclusion of any investigation which he [Bacon] would recognize as just and legitimate" (General Preface to the *Philosophical Works*, sec. 9), but this is owing to the fact that the process of exclusion cannot be made accurate until we possess sound notions of simple natures (*Novum organum*, II, 19), *not* owing to its not having been made use of in arriving at the example of the first vintage. A *final* conclusion would be attainable only after our notions had been corrected, not only by the use of the prerogative instances (to one of which at least some appeal is made in reaching the first vintage), but presumably also by the use of the "other helps of the understanding," which Bacon promises but does not give.

29. *Novum organum*, II, 52.

30. Cf. Herschel, *Discourse on the Study of Natural Philosophy*, sec. 191. There is little reason to think that Bacon supposed it to be necessary to classify one's observations under one or another of his headings before one could use them. Rather, what he attempts is a classification of the facts of which we do make use when they present themselves to our observation.

31. In spite of his recognition of the desirability of quantitative determinations, to which the above testify, it must be acknowledged that Bacon wholly failed to realize the importance of mathematics in scientific investigations, and this failure constituted one of the gravest *lacunae* in his doctrine.

32. *Novum organum*, II, 9; also *De augmentis scientiarum*, Book III, chap. iv.

33. *Novum organum*, II, 4 and 5.

34. *De augmentis scientiarum*, Book III, chap. iv.

35. *Ibid.*

36. *Ibid.*

37. Professor John Nichol ("Bacon," in *Philosophical Classics for English Readers,* ed. W. Knight [Edinburgh: W. Blackwood and Sons, 1888–89], II, 188 n.) says: "We can attach no meaning to the form of a 'simple nature,' which is by definition irresolvable." It is irresolvable, however, only if the relation of form to simple nature is conceptual, but not if it is causal as appears to be Bacon's meaning (see below). Professor Nichol strangely appears to think that for Bacon the discovery of forms and the analysis of complex natures into simple were the same thing. Thus he writes: "Defeated in his search for simple natures, he [Bacon] was thrown back . . ." (p. 193).

38. Cf. Ellis' General Preface to the *Philosophical Works,* sec. 8.

39. The similarity remains in spite of the fact that Bacon, in more than one place, expresses his divergence from the views of Anaxagoras (e.g., *De augmentis scientiarum,* Book II, chap. xiii, first example), chiefly, however, with regard to the infinite number of the *Homoeomeriae.* As will appear from the sequel, he also differs from him in regarding as objectively existing only the primary qualities. But, as Berkeley pointed out, the distinction between the primary and the secondary qualities is difficult to maintain in the world as it is known.

40. Also *Novum organum,* I, 105: "And this induction must be used not only to discover axioms, but also in the formation of notions"; and p. 69: "Notions are ill-drawn from the impressions of the senses, and are indefinite and confused, whereas they should be definite and distinctly bounded." See also p. 60.

41. Or it might perhaps be said that the process he gives for the induction of axioms (by tables and exclusions) is rather a process for the induction of notions (by abstraction) than of axioms.

42. *Novum organum,* I, 105. Cf. *Filum labyrinthi,* sec. 10.

43. Bacon's sympathy with the views of Democritus is everywhere apparent, but he is careful to distinguish the particles of which he speaks from the atoms of Democritus (*Novum organum,* II, 8).

44. *Ibid.,* II, 5.

45. *De augmentis scientiarum,* Book III, chap. iv.

46. Ellis probably means the "efficient" causes.

47. Thus the laws that Bacon thinks of are laws of the causation *of our sensations* by certain states or motions of matter, while the laws that really have the practical value that Bacon thought of are laws of the causation *of states or changes of matter* by other states or changes of it.

48. *Valerius terminus,* chap. xi: "The fulness of direction to work and produce any effect consisteth in two conditions, certainty and liberty." From the explanation subjoined, it appears that, by speaking of a direction as certain, he means that it is *sufficient* to the effect desired, while by speaking of it as having to be also free, he means that it must be *necessary* to the effect as well. See also *Novum organum,* II, 4.

49. H. W. Blunt, "Bacon's Method of Science," *Proceedings of the Aristotelian Society,* N.S. IV (1904), 26.

50. *Novum organum,* II, 4. Cf. Herschel, *Discourse on the Study of Natural Philosophy,* secs. 79 ff.

51. Cf. Spinoza's "infinite modes" (*Ethics,* Part I, propositions 22, 23).

52. Blunt, "Bacon's Method of Science," p. 22. See also W. S. Jevons, *Principles of Science* (2nd ed.; London: Macmillan and Co., 1877), p. 506.

53. Whewell, *Philosophy of Discovery,* p. 152.

54. Cf. Blunt, "Bacon's Method of Science," p. 30.

55. *Novum organum,* II, 11, 12, 13, 15, 18, 20; also among the prerogative instances, aphorism 22.

56. *Ibid.,* I, 30: "For Interpretation is the true and natural work of the mind when freed from impediments"; also *Valerius terminus,* chap. xxii, and *De augmentis scientiarum,* Book V, chap. ii. Cf. J. F. Fries, *Die Geschichte der Philosophie* (2 vols.; Halle: Waisenhauses, 1837–40), Vol. II. A good deal of the criticism that has been directed against Bacon's method of tables and exclusion turns out to be much rather a criticism, by implication, of the *material* with which he thought science had to deal than of the *method* itself, which, given the virtually psychological character of that material, is as sound as any other application to a problem of empirical analysis of the principles of agreement and difference.

57. Christoph von Sigwart, *Logic,* trans. Helen Dendy (New York: Macmillan and Co., 1895).

58. Blunt, "Bacon's Method of Science," p. 18.

59. *Ibid.,* p. 19.

60. *Novum organum,* I, 61.

61. Cf. *ibid.,* I, 82; and *De interpretatione naturae sententiae duodecim,* sec. 8.

62. Cf. Jevons, *Principles of Science,* p. 507.

63. The translation adds that this was done only after consultation with the whole body.

64. Blunt, "Bacon's Method of Science," pp. 21, 29, 30.

65. *Ibid.,* p. 19.

CHAPTER FOUR

The Role of Experience in Descartes' Theory of Method

1. *Discours de la méthode,* Part I, *Œuvres de Descartes,* ed. C. Adam and P. Tannery (12 vols.; Paris: Léopold Cerf, 1897–1910), VI, 4 ff. All future references to the works of Descartes are to this edition.

2. *Regulae ad directionem ingenii,* Regula II, X, 362–63. Translation from *The Philosophical Works of Descartes* by Elisabeth S. Haldane and G. T. R. Ross (2 vols.; Cambridge, Eng.: Cambridge University Press, 1911), I, 3; hereafter cited as Haldane and Ross, *Descartes.*

3. Cf. Christopher Clavius, S.J., *Operum mathematicorum,* I, Prolegomena, 5 (cited by E. Gilson, *René Descartes' Discours de la méthode: Texte et commentaire* [Paris: J. Vrin, 1925], p. 128, hereafter cited as Gilson, *Texte et commentaire*): "Mathematicae disciplinae sic demonstrant omnia, de quibus suscipiunt disputationem, firmissimis rationibus, confirmantque, ita ut vere scientiam in auditoris animo gignant, omnemque prorsus dubitationem tollant: id quod aliis scientiis vix tribuere possumus, cum in eis saepenumero intellectus multitudine opinionum ac sententiarum varietate in veritate conclusionum judicanda suspensus haereat atque incertus. . . . Cum igitur disciplinae mathematicae veritatem adeo expetant, adamant, excolantque, ut non solum nihil quod sit falsum, verum etiam nihil quod tantum probabile existat, nihil denique admittant quod certissimis demonstrationibus non confirment, corroborentque, dubium esse non potest quin eis primus locus inter alias scientias omnes sit concedendus." Cf. G. Milhaud, *Descartes savant* (Paris: Alcan, 1921), p. 235: "Si Descartes est mécontent de l'enseignement de l'École, ce mécontentement lui-même et son désir de savoir autrement et davantage ne sont-ils pas dus en partie à ce qu'il avait appris?"

4. Cf. *Regulae ad directionem ingenii,* regula ii, X, 362: "Omnis scientia est cognitio certa et evidens." Cf. *Discours de la méthode,* VI, 12–13.

5. *Letter to Mersenne,* March 11, 1649, III, 39.

6. *Discours de la méthode,* Part V, VI, 41; Haldane and Ross, *Descartes,* I, 106–7.

7. *Ibid.*

8. *Letter to Mersenne,* November 13, 1629, I, 70.

9. *Le monde,* XI, 47.

10. *Discours de la méthode,* Part VI, VI, 68; Haldane and Ross, *Descartes,* I, 123.

11. *Letter to Mersenne,* May 27, 1638, II, 141.

12. *Discours de la méthode*, Part VI, VI, 63–64; Haldane and Ross, *Descartes*, I, 121.

13. For the motives of this reticence, cf. Gilson, *Texte et commentaire*, pp. 471–74.

14. Cf. *Letter to Vatier*, February 22, 1638, I, 563: "As to the hypothesis that I make at the beginning of the *Météores*, I could not demonstrate it *a priori*, without giving the whole of my Physics." Cf. *Discours de la méthode*, Part VI, VI, 76: "I consider myself able to deduce them from the primary truths which I explained above" (Haldane and Ross, *Descartes*, I, 129).

15. Cf. *Letter to Vatier*, February 22, 1638, I, 564: "les croyant déduire par ordre des premiers principes de ma Métaphysique."

16. *Regulae ad directionem ingenii*, regula iii, X, 368.

17. *Ibid.*, regula ii, X, 364–65; Haldane and Ross, *Descartes*, I, 4–5.

18. *Novum organum*, Book I, aphorism 50.

19. *Ibid.*, aphorism 82.

20. *The Great Instauration*, author's Preface.

21. *Novum organum*, Book I, aphorism 95.

22. *Ibid.*

23. *The Great Instauration*, author's Preface.

24. *Météores*, Discours VIII, VI, 340.

25. *Ibid.*, VI, 334.

26. *Letter to Plempius*, October 30, 1637, I, 420–21.

27. Cf. Gilson, *Texte et commentaire*, pp. 269–72, 397; Milhaud, *Descartes savant*, chap. ix, "Descartes expérimentateur," especially pp. 197 ff.

28. *Discours de la méthode*, VI, 21–22.

29. *Ibid.*, VI, 22; Haldane and Ross, *Descartes*, I, 94.

30. *Discours de la méthode*, Part III, VI, 29; Haldane and Ross, *Descartes*, I, 99. Cf. the Latin version "multa experimenta colligebam."

31. *Discours de la méthode*, Part VI, VI, 63–64; Haldane and Ross, *Descartes*, I, 121.

32. *Discours de la méthode*, Part V, VI, 50; Haldane and Ross, *Descartes*, I, 112.

33. *Discours de la méthode*, Part V, VI, 47; Haldane and Ross, *Descartes*, I, 110.

34. *Discours de la méthode*, Part V, VI, 50; Haldane and Ross, *Descartes*, I, 112.

35. *Regulae ad directionem ingenii*, regula xii, X, 427; Haldane and Ross, *Descartes*, I, 47.

36. *Discours de la méthode*, Part II, VI, 19–20.

37. Cf. *Regulae ad directionem ingenii*, regula iv, Vol. XIV.

38. Cf. Gilson, *Texte et commentaire*, pp. 216–17.

39. Cf. *Regulae ad directionem ingenii*, regula iv, X, 377; *Discours de la méthode*, Part II, VI, 21: "je me promettais de l'appliquer aussi utilement aux difficultés des autres sciences que j'avais fait à celles de l'Algèbre."

40. *Discours de la méthode*, Part II, VI, 22.

41. *Ibid.*, Part III, VI, 29; Haldane and Ross, *Descartes*, I, 99. Cf. especially "quasi semblables à celles des Mathématiques."

42. Gilson, *Texte et commentaire*, p. 272.

43. *Letter to Mersenne*, July 27, 1638, II, 268.

44. *Ibid.*, May 17, 1638, II, 141–42. Cf. also the sharp line Descartes draws between mathematics and physics in discussing the problem of the anaclastic line (regula viii).

45. *Letter to Mersenne*, July 27, 1638, II, 268.

46. Louis Liard, *Descartes* (Paris: G. Baillière et Cie., 1882), p. 74. We have not been able to verify the reference to Descartes.

47. *Regulae ad directionem ingenii*, regula v, X, 380; Haldane and Ross, *Descartes*, I, 15.

48. Bacon calls such accumulations by the name of "Natural History" (cf. *Novum organum*, I, 98; *Distributio operis; Parasceve; Sylva sylvarum*, etc.). Descartes refers to Bacon more than once in this connection. Cf. *Letter to Mersenne*, May 10, 1632, I, 251–52, where he desires to have compiled "l'histoire des apparences célestes, selon la méthode de Verulamius et que, sans y mettre aucunes raisons ni hypothèses, il nous décrivit exactement le Ciel, tel qu'il paraît maintenant, quelle situation à chaque étoile fixe au respect de ses voisines, quelle différence, ou de grosseur, ou de couleur, ou de clarté, ou d'être plus ou moins étincelantes, etc. . . ." Cf. *Letter to Mersenne*, December 23, 1630, I, 195–96. On the curious general accord between Descartes and Bacon and the basis upon which it rests cf. Milhaud, *Descartes savant*, chap. x, "Descartes et Bacon," pp. 213 ff. Also A. Lalande, "Quelques textes de Bacon et de Descartes," *Revue de Métaphysique et de Morale*, XIX (1911), 296–311.

49. Cf. *Discours de la méthode*, Part VI, VI, 63, 64–65, 75.

50. *Principiorum philosophiae*, Part III, chap. iv, VIII, 81–82.

51. *Discours de la méthode*, Part VI, VI, 63; Haldane and Ross, *Descartes*, I, 120.

52. *The Great Instauration*, Plan of the Work.

53. *Letter to Mersenne*, December 23, 1630, I, 195–96.

54. *Le recherche de la vérité*, X, 503; Haldane and Ross, *Descartes*, I, 309–10.

55. Cf. *Letter to Mersenne*, December 18, 1629, I, 99–100: "Pour le

quantum, je l'ignore, et encore qu'il se pût faire milles expériences pour le trouver à plus près, toutes fois pour ce qu'elles ne se peuvent justifier par raison, au moins que je puisse encore atteindre, je ne crois pas qu'on doive prendre la peine de les faire."

56. *Discours de la méthode*, Part VI, VI, 64; Haldane and Ross, *Descartes*, I, 121.

57. *Ibid.*

58. Gilson (*Texte et commentaire*, pp. 456–57) notes one difference between the Baconian *instantia crucis* and the procedure of Descartes at this point. Bacon's crucial instance has for its object rather to discriminate the cause amid the complexity of the given facts, while Descartes' recourse to experience is rather, in the present case, to decide between alternative explanations of a given phenomenon, each equally probable a priori, but only one of which accords with the actual facts of nature.

59. I.e., the possibilities of combination of material particles, in accordance with the laws of motion God has imposed upon them.

60. *Discours de la méthode*, Part VI, VI, 64–65; Haldane and Ross, *Descartes*, I, 121.

61. *Discours de la méthode*, Part V, VI, 52 ff. Cf. Gilson, *Texte et commentaire*, pp. 409 ff.

62. *La description du corps humain*, chap. xviii, XI, 242.

63. *Les principes de la philosophie*, Preface, IX, 17; Haldane and Ross, *Descartes*, I, 213. Cf. *La description du corps humain*, chap. xxvii, XI, 252–53: "Et bien que je n'aie pas voulu jusqu'ici entreprendre d'écrire mon sentiment touchant cette matière, à cause que je n'ai pu encore faire assez d'expériences, pour vérifier par leur moyen toutes les pensées que j'en ai eu: je ne puis néanmoins refuser d'en mettre ici en passant quelque chose de ce qui est le plus général, et dont j'espère que je ferai le moins en hasard ci-après de me dédire, lorsque nouvelles expériences me donneront davantage de lumière."

64. *Letter to Morin*, September 12, 1638, II, 366.

65. *Letter to Plempius*, October 30, 1637, I, 420–21.

66. *Letter to Huygens*, June, 1645, IV, 224–25.

67. *Principiorum philosophiae*, Part IV, chap. cc, VIII, 323; Haldane and Ross, *Descartes*, I, 296.

68. "Contra ego, si quae talis mora sensu perciperetur, totam meam Philosophiam funditus eversam fore inquiebam."

69. Liard, *Descartes*, pp. 121–24. Cf. Descartes, *Principiorum philosophiae*, Part II, chaps. xxvii–xli, xlvi ff., VIII, 55–65, 68 ff.

70. *Regulae ad directionem ingenii*, regula xii, X, 412–13: "mihi

sufficiet quam brevissime potero explicare quisnam modus concipiendi illud omne, quod in nobis est ad res cognoscendas, sit maxime utilis ad meum institutum. Neque reditis, nisi lubet, rem ita se habere: sed quid impediet quominus easdem suppositiones sequamini, si appareat nihil illas ex rerum varietate minuere sed tantum reddere omnia longe clariora? etc."

71. *Discours de la méthode*, Part V, VI, 42 ff.

72. *Les principes de la philosophie*, Part III, chap. xlv, IX, 123–24. So, too, in *Le monde*, XI, 31, 36.

73. *Letter to Mersenne*, May 17, 1638, II, 142.

74. *Ibid.*, pp. 143–44.

75. *Principiorum philosophiae*, Part IV, chap. cciv, VIII, 327 ff. Haldane and Ross, *Descartes*, I, 300.

76. *Ibid.*

77. *Discours de la méthode*, Part VI, VI, 76; Haldane and Ross, *Descartes*, I, 129.

78. *Letter to Vatier*, February 22, 1638, I, 563–64.

79. *Discours de la méthode*, Part VI, VI, 76; Haldane and Ross, *Descartes*, I, 128–29.

80. *Ibid.*

81. *Letter to Morin*, July 13, 1638, II, 198.

82. Cited above in *Letter to Vatier*.

83. *Les principes de la philosophie*, Part IV, chap. cciii, IX, 321.

84. *Letter to Morin*, July 13, 1638, II, 199. Cf. also *Letter to Plempius*, October 3, 1637, I, 423–24, especially, "Quae quamvis singula sejunctim considerata non nisi *probabiliter* persuadeant, omnia tamen simul spectata *demonstrant*."

85. *Letter to Morin*, July 13, 1638, II, 199.

86. *Ibid.*

87. *Principiorum philosophiae*, Part IV, chap. xlviii, VIII, 99.

88. *Ibid.*, chap. ccv, VIII, 328; Haldane and Ross, *Descartes*, I, 301.

89. *Letter to Morin*, July 13, 1638, II, 200.

90. *Letter to Mersenne*, October 28, 1640, III, 212.

91. Gilson (*Texte et commentaire*, p. 454) sums up Descartes' doctrine concerning the a posteriori proof of true causes as follows: "Le caractère auquel on reconnaît la vérité d'un principe scientifique est, en effet, son aptitude à expliquer un grand nombre de phénomènes sans recourir à des hypothèses supplémentaires inventées pour la circonstance. Les principes de l'École sont sans valeur parce que, les formes substantielles une fois admises, il faut admetter une forme spéciale pour expliquer les propriétés de chaque espèce de corps; les principes cartésiens sont vrais parce que, une

fois admis, ils expliquent à eux seuls les phénomènes les plus évidents de la nature, sans que jamais aucune hypothèse supplémentaire ne doive être invoquée." Cf. Newton's conception of "true causes."

92. *Letter to Mersenne,* October 11, 1638, II, 380.

93. *Principiorum philosophiae,* Part IV, chap. cci, VIII, 324–25; *Les principes de la philosophie,* Part IV, chap. cci, IX, 319–20; Haldane and Ross, *Descartes,* I, 297–98. Cf. *Cogitationes privatae,* X, 218–19: "Cognitio hominis de rebus naturalibus, tantum per similitudinem eorum quae sub sensum cadunt; et quidem eum verius philosophatum arbitramur, qui res quaesitas felicius assimilare poterit sensu cognitis."

94. *Principiorum philosophiae,* Part IV, chap. cc, VIII, 323; Haldane and Ross, *Descartes,* I, 296.

95. *Discours de la méthode,* Part VI, VI, 77; Haldane and Ross, *Descartes,* I, 129.

96. *Principiorum philosophiae,* Part IV, chap. cc, VIII, 323–24; Haldane and Ross, *Descartes,* I, 296–97.

97. *Discours de la méthode,* Part VI, VI, 77; Haldane and Ross, *Descartes,* I, 130. Cf. *Regulae ad directionem ingenii,* regula iv, X, 371: "quod etiam experientia comprobatur, cum saepissime videamus illos, qui litteris operam nunquam navarunt, longe solius et clarius de obviis rebus judicare, quam qui perpetuo in scholis sunt versati." Cf. *Les principes de la philosophie,* Preface, VIII, 8–9: "From this we must conclude that those who have learnt least about all that which has hitherto been named philosophy, are the most capable of understanding the truth" (Haldane and Ross, *Descartes,* I, 208).

98. *Les principes de la philosophie,* Part IV, chap. ccv, IX, 323.

99. *Principiorum philosophiae,* Part IV, chap. ccvi, VIII, 328–29; *Les principes de la philosophie,* chap. ccvi, IX, 324–25.

100. *Sur les V^es^ objections,* IX, 205–6.

101. *Regulae ad directionem ingenii,* regula xii, X, 424: "quoties ex re particulari vel contingenti aliquid generale et necessarium deduci posse judicamus" (Haldane and Ross, *Descartes,* I, 43).

102. *Responsio ad secundas objectiones,* VII, 140–41.

103. *Regulae ad directionem ingenii,* regula xii, X, 425; Haldane and Ross, *Descartes,* I, 45. Cf. regula vi, X, 383: "paucas esse dumtaxat naturas puras et simplices, quas primo et per se, non dependenter ab aliis ullis, sed vel in ipsis experimentis, vel lumine quodam in nobis insito, licet intueri"; also regula viii, X, 394: "de rebus tantum puris simplicibus et absolutis experientiam certam haberi posse." Cf. Heinz Heimsoeth, *Die Methode des Erkenntnis bei Descartes und Leibniz* (Giessen: Töpelmann, 1912–14), p. 71; also J. Berthet, "La méthode des Descartes avant le Discours,"

Revue de Metaphysique et de Morale, IV (1896), 399–415; J. L. Mursell, "The Function of Intuition in Descartes' Philosophy of Science," *Philosophical Review*, XXVIII (1919), 391–409: "when Descartes speaks of intuition, he is dealing with the actual practice and procedure of the expert investigator. The expert will develop and possess a power of immediately perceiving the essential factors of a complex situation."

104. *Regulae ad directionem ingenii*, regula xiv, X, 441; Haldane and Ross, *Descartes*, I, 55–56.

105. *Ibid.* Cf. Descartes, *Météores*, discours viii, Vi, 325 ff.; Liard, *Descartes*, p. 30 and n.

106. Cf. *Regulae ad directionem ingenii*, regula viii, X, 395: "atque si statim in secundo gradu illuminationis naturam non possit agnoscere, enumerabit . . . alias omnes potentias naturales, ut ex alicujus alterius cognitione saltem per imitationem . . . hanc etiam intelligat."

107. *Regulae ad directionem ingenii*, regula vii, X, 388.

108. *Ibid.*, regula xi, X, 409. Cf. regula vii.

109. *Ibid.*, p. 408.

110. *Ibid.*, regula vii, X, 389.

111. *Ibid.*, p. 390.

112. *Ibid.*

113. *Ibid.*

114. *Ibid.*, p. 391.

115. Cf. Berthet, "La méthode de Descartes avant le Discours," p. 404, n. 1.

116. *Regulae ad directionem ingenii*, regula xi, X, 407.

117. *Ibid.*, p. 408.

118. *Ibid.*, regula ix, X, 400; regula iii, X, 366, 369.

119. *Ibid.*, regula iii, X, 368.

CHAPTER FIVE

Thomas Hobbes and the Rationalistic Ideal

1. Cf. R. Blackbourne, *Vitae Hobbianae auctarium* in Thomas Hobbes, *Opera philosophica quae Latine scripsit*, ed. W. Molesworth (5 vols.; London: J. Bohn, 1839–45), I, xxv; hereafter cited as *Opera*. Cf. John Aubrey, *Brief Lives*, ed. Andrew Clark (2 vols.; Oxford: Clarendon Press, 1898), I, 70, 83, 331.

2. Max Köhler, "Die Naturphilosophie des Thomas Hobbes in ihrer Abhängigkeit von Bacon," *Archiv für Geschichte der Philosophie*, XV (1902), 370–99.

3. Cf. the assertion in *Human Nature* (*The English Works of Thomas Hobbes of Malmesbury*, ed. W. Molesworth [11 vols.; London: John Bohn, 1839], IV, 27; hereafter cited as *English Works*) that all knowledge consists of experience and remembrance; the teaching in *Leviathan*, *English Works*, III, 1, that all our concepts are derived from sensation; and the doctrine in *Human Nature*, *English Works*, IV, 3 ff., and *Leviathan*, pp. 1–3, that all the data of sense are subjective, which in *De corpore* is extended even to space and time. Hobbes, however, never pushes his phenomenalism to a denial of the objective reality of *motion*.

4. G. C. Robertson, s.v. "Hobbes" in *Encyclopaedia Britannica*, 11th ed., XIII, 546, n. 1.

5. Hobbes, *Vita*, *Opera*, I, xiv: "In peregrinatione illa inspicere coepit in Elementa Euclidis; et delectatus methodo illius, non tam ob theoremata illa, quam ob artem ratiocinandi, diligentissime perlegit." Cf. the longer account of Blackbourne, *Vitae Hobbianae auctarium*, *Opera*, I, xxvi. Also Aubrey, *Brief Lives*, I, 332.

6. Cf. *De principiis et ratiocinatione geometrarum*, Introduction, *Opera*, IV, 390: "Certitudo scientiarum omnium aequalis est, alioqui enim scientiae non essent: cum *scire* non suscipiat magis et minus. Physica, ethica, politica, si bene demonstratae essent, non minus certae essent quam pronunciata mathematica. . . ."

7. Cf. *De cive*, Epistola dedicatoria, *Opera*, II, 137.

8. *Leviathan*, pp. 23–24.

9. *Concerning Body*, *English Works*, I, 75.

10. *Ibid.*, p. 68.

11. D. G. James, *The Life of Reason: Hobbes, Locke, and Bolingbroke* (London: Longmans, Green and Co., 1949), p. 15.

12. *Concerning Body*, pp. 70 ff.

13. *Ibid.*

14. *Ibid.*, pp. 71–72.

15. *Ibid.*, p. 87.

16. *Human Nature*, pp. 17–18. Cf. *Leviathan*, pp. 35–38.

17. *Leviathan*, p. 71.

18. *Ibid.*, p. 6.

19. *Ibid.*, p. 16.

20. *Ibid.*, p. 14.

21. *Ibid.*, p. 15.

22. *Ibid.*

23. *Ibid.*, p. 16. Cf. also *Praefatio in Mersenni ballisticam*, *Opera*, V, 313 ff.; cf. *Leviathan*, p. 664; chap. ix, p. 71. Cf. *De homine*, *Opera*, II, 92: "*Scientia* intelligitur de theorematum, id est, de propositionum ge-

neralium veritate, id est, de veritate consequentiarum. Quando vero de veritate facti agitur, non proprie *scientia*, sed simpliciter *cognitio* dicitur."

24. Cf. "Hobbes" in *Encyclopaedia Britannica*, 11th ed., XIII, 546, n. 2.

25. *Examinatio et emendatio mathematicae hodiernae*, Dialogus V, *Opera*, IV, 179.

26. Leslie Stephen, *Hobbes* (New York: Macmillan Co., 1904), p. 54.

27. *Examinatio* . . . , Dialogus VI, pp. 228–29.

28. *De corpore*, *Opera*, I, 59: "Principia itaque scientiae omnium prima, sunt phantasmata sensus et imaginationis, quae quidem cognoscimus naturaliter quod sunt; quare autem sunt, seu a quibus proficiscuntur causis cognoscere ratiocinatione opus est. . . ." That it is analysis that leads to the discovery of first principles is the teaching also of *Praefatio* . . . , p. 312; *De corpore*, pp. 61, 65–66, and 251 ff. That analysis is a form of ratiocination is maintained in *ibid.*, pp. 59, 61, and 252 ff.

29. G. C. Robertson, *Hobbes* (Philadelphia: J. B. Lippincott, n.d.), p. 91. Cf. Hobbes, *De corpore*, pp. 59–62. It would be interesting to know how Hobbes would have reconciled these pages with his general nominalistic teaching that there is "nothing in the world universal but names; for the things named are every one of them individual and singular" (*Leviathan*, p. 21). Cf. *De corpore*, pp. 17–18: "Est ergo nomen hoc *universale*, non rei alicujus existentis in rerum natura, neque ideae, sive phantasmatis alicujus in animo formati, sed alicujus semper vocis sive nominis nomen."

30. Robertson, *Hobbes*, pp. 91–92.

31. Hobbes constantly insists that the indemonstrable first principles of science are all *definitions*. Cf. *De corpore*, p. 71: "Principia autem illa, solae definitiones sunt." Cf. *Examinatio* . . . , Dialogus II, pp. 26, 86; *De corpore*, pp. 33, 72; *Principia et problemata aliquot geometrica*, *Opera*, V, 157, 203–4.

32. ". . . conceptuum nostrorum simplicissimorum explicationes," *De corpore*, p. 62.

33. *Human Nature*, p. 20.

34. *Ibid.*, p. 18. Cf. *De corpore*, p. 14: "*Nomen est vox humana arbitratu hominis adhibita*," etc. Also *ibid.*, p. 32: "Deduci hinc quoque potest, veritates omnium primas, ortas esse ab arbitrio eorum qui nomina rebus primi imposuerunt, vel ab aliis posita acceperunt. Nam exempli causa verum est *hominem esse animal*, ideo quia eidem rei duo illa nomina imponi placuit." Also *ibid.*, p. 316: "Ratiocinationis principia prima, nempe definitiones, vera esse facimus nosmet ipsi, per consensionem circa rerum appellationes." Finally, *Principia* . . . , p. 157: "De veritate quidem definitionis legitimae, quoniam habent veritatem suam a consensu et arbitrio hominum rebus explicatis nomina suo libitu imponentium, dubitari non potest." It will be noted

that Hobbes is really involved in an inconsistency at this point. Definitions are to be the basis of all certain knowledge, inasmuch as such knowledge cannot be derived from experience. Yet Hobbes also holds that it is from experience of how men arbitrarily use words that we derive their definitions. Having first laid it down that "*experience concludeth nothing universally*" (*Human Nature*, p. 18), he continues on the same page as follows: "As in conjecture concerning things past and future, it is prudence to conclude from experience, what is likely to come to pass, or to have passed already; so it is an error to conclude from it, that *it is* so or so *called;* that is to say, we cannot from experience conclude, that anything is to be called *just* or *unjust*, *true* or *false*, or any proposition *universal* whatsoever, except it be from the remembrance of the use of names imposed arbitrarily by men: for example to have heard a sentence given in the like case, the like sentence a thousand times is not enough to conclude that the sentence is just; though most men have no other means to conclude by: but it is *necessary*, for the drawing of such conclusion, to *trace* and *find out*, by many experiences, what men do mean by calling things just and unjust." From which it appears that, although it is an error to conclude from experience that anything is called just, yet in order that we may validly conclude that anything *is* just we must nevertheless first conclude from experience as to what is called just! On a somewhat later page (*ibid.*, p. 27) we are told that, although our knowledge of "how things are called" is itself "but *experience*," i.e., "experience men have from the proper use of *names* in language," and therefore, like all experience, "but remembrance," yet such knowledge of "how things are called" differs radically from the knowledge that consists of the "experience of the effects of things that work upon us from *without*." This latter, inasmuch as "experience concludeth nothing universally," can at best be called *history* and is more or less conjectural (cf. *ibid.*, p. 18); but experience and remembrance of "how things are called" constitutes "knowledge of the *truth of propositions*," is somehow derived not from sense but from "understanding," and partakes of the certainty of "science"!

35. *Leviathan*, p. 30. Cf. *ibid.*, pp. 52–53; *Human Nature*, p. 27; *Praefatio . . . ,*" pp. 314–16. Note especially the nominalistic interpretation of the whole procedure on the page last cited: "Quod si dicamur universalia intelligere, nihilque sit universale praeter nomen, intellectio non erit ipsarum rerum, sed nominum, et orationis ex nominibus compositae. . . . Hinc ratio dicitur facultas *syllogisandi*, cum ratiocinatio sit continua propositionum in unam summam collectio, vel calculus nominum. . . . Ubi supponenda recta ratiocinatio, quae sumens initium ab accurata nominum explicatione procedit per syllogismum, seu continuam verarum propositionum connexionem," etc.

36. *Leviathan*, p. 23.

37. *Ibid.*, p. 32.

38. Robertson, *Hobbes*, pp. 87–88. Cf. Hobbes, *Opera*, V, 257–58.

39. *Concerning Body*, p. 36.

40. *Ibid.*, p. 31.

41. James, *The Life of Reason*, pp. 26–27. James apparently limits this judgment—or, at any rate, he should limit it—to Hobbes's philosophy of nature. His political philosophy is another story. Most recent commentators agree with James's judgment of Hobbes's philosophy of science and are at pains to show that his political philosophy in no way really depends upon it. F. J. E. Woodbridge writes: "Hobbes's physics is the weakest part of his system. It contains little in the way of a contribution to the sciences of his own day or since. It is interesting chiefly because of its attempt to align the psychology and the politics with that temper of mind which has produced our natural sciences. And it is interesting as a part of Hobbes's system. He would keep a unity in nature from the movements of bodies in space to the movements of thought in a sovereign's mind. There is often novelty in his explanations of physical occurrences, but there is little novelty in his general conception of what the physical world is like. In this he shared the views of those with whom he associated. He shared their views in general, but with the method by which they supported them—mathematical theory and experimentation—he had so little sympathy, and of it so little knowledge, that he wasted many years and many words in attacking those—like Boyle, for example—whom his professions should have led him to support, and in whom he might have found friends instead of enemies. In the matter of the new physics, he was less an aid than a hindrance. . . . The *Leviathan* really needs neither an antecedent physics nor metaphysics to support it. For any genuine appreciation of its value and power, the reader will do well to forget the larger setting in which Hobbes would place it. The worth of the book lies in the picture of man in his social and political relations which its author draws" (*Hobbes Selections*, ed. F. J. E. Woodbridge [London: Charles Scribner's Sons, 1930], pp. xxiii-xxv). A. D. Lindsay writes: "The *De Corpore*, the exposition of his scientific materialism, was published in 1655. Unfortunately it contained a rash mathematical adventure, Hobbes' claim to have squared the circle, which drew him into a long and fierce controversy with the Savilian professor of mathematics at Oxford—Wallis. . . . His love for geometry was greater than his knowledge" (*Leviathan*, ed. A. D. Lindsay [New York: E. P. Dutton and Company, 1950], p. xi). According to Leo Strauss: "He was a moralist rather than a scientist: the book which he sought to read was the book of man rather than the book of nature; and he brought no 'naturalistic' or 'mechanistic' preconceptions to his reading of

the book of man" (*The Political Philosophy of Thomas Hobbes*, trans. E. M. Sinclair [Oxford: Clarendon Press, 1936], p. viii). Cf. also Sterling Lamprecht's introduction to Hobbes's *De cive* (New York: Appleton-Century-Crofts, 1949). Cf. Hobbes, *Concerning Body*, pp. 73–74.

42. *De corpore*, p. 62.

43. *Ibid.*, pp. 111–12.

44. *Ibid.*, p. 75.

45. *Opera*, V, 157.

46. *De corpore*, p. 71.

47. Cf. *ibid.*, pp. 33, 72; *Principia* . . . , pp. 157, 203–4.

48. In *Concerning Body*, pp. 91 ff.

49. *Ibid.*, p. 94.

50. *Ibid.*, p. 105.

51. *Ibid.*, p. 102.

52. James, *The Life of Reason*, p. 18.

53. F. Tönnies, "Anmerkungen über die Philosophie des Hobbes," *Vierteljahrschrift für wissenschaftliche Philosophie*, IV (1880), 63.

54. *Principia* . . . , p. 157.

55. Tönnies, "Anmerkungen über die Philosophie des Hobbes," pp. 63–64.

56. *Human Nature*, p. 29. Hobbes is here thinking mainly of the *reductio ad absurdum*, however. The only test he mentions for the falsity of a supposition is that it will lead to some "*absurd* or impossible conclusion." Discrepancy with observed facts is not mentioned.

57. *Leviathan*, p. 664.

58. Chaps. v, ix.

59. Cf. p. 35.

60. Cf. p. 37. But even in chapter xlvi (p. 664) he still says that "nothing is produced by reasoning aright, but general, eternal, and immutable truth."

61. *Epistola dedicatoria:* "Confido enim . . . in tribus libelli hujus partibus prioribus ex definitionibus; in quarta ex hypothesibus non absurdis omnia esse legitime demonstrata." Also *Opera*, I, 315–16, 430–31, where it is admitted that hypotheses other than those employed by the author may do just as well or even better.

62. *De homine*, p. 93; *Six Lessons to the Professors of the Mathematics*, *Epistle Dedicatory*, *English Works*, VII, 183–84.

63. *Seven Philosophical Problems*, *English Works*, VII, 11.

64. *Ibid.*, *Epistle Dedicatory*, pp. 3–4. Hobbes's comment, "for there is no effect which the power of God cannot produce by many several ways," reminds one of a similar remark of Descartes (*Principiorum philosophiae*, Part III, VIII, 100–1).

65. *De homine,* p. 93: "fieri non potest ut non aliqua etiam a physico demonstratione a priore demonstranda sint. Itaque physica, vera inquam physica, quae geometriae innititur, inter mathematicas mixtas numerari solet."

66. Tönnies, "Anmerkungen über die Philosophie des Hobbes," p. 70.

67. *De natura aeris, Opera,* IV, 247: "omnis hypotheseos lex haec est, ut quae supponuntur omnia debeant esse sua natura possibilia, id est, cogitabilia."

68. *Ibid.,* p. 254: "Hypothesim legitimam faciunt duae res; quarum prima est, ut sit conceptibilis, id est, non absurda; altera, ut ab ea concessa inferri possit phaenomeni necessitas."

69. *De corpore,* p. 68. The translation is that of the *Elements of Philosophy, the First Section, Concerning Body, English Works,* I, 77. This translation was not made by Hobbes and must be used with caution.

70. *Human Nature,* p. 30; *Of Liberty and Necessity, English Works,* IV, 274–75.

71. *Of Liberty and Necessity,* p. 276.

72. *De corpore,* p. 108.

73. *Ibid.,* p. 107.

74. *Of Liberty and Necessity,* pp. 274–77.

75. *Ibid.,* pp. 266–67.

76. *Ibid.,* pp. 246–47.

77. Cf. *De corpore,* p. 68. Hobbes follows this account of the method with an example—the determination of the cause of light.

CHAPTER SIX

Isaac Newton and the Hypothetico-Deductive Method

1. Translation from *The Theory of Light,* ed. Henry Crew (New York: American Book Co., 1900).

2. L. Bloch, *La philosophie de Newton* (Paris: Alcan, 1908), p. 129: "The *Principia* from its first edition (1672) was written on the model of Euclid. The *Opticks* in its successive editions (1704, 1717) became more and more geometrical." Cf. W. E. Strong, "Newtonian Explications of Natural Philosophy," *Journal of the History of Ideas,* XVIII (1957), 49–83.

3. Bloch, *La philosophie de Newton,* p. 130.

4. *Principia,* Book II, sec. 7, Final Scholium, experiment 4 (*The Mathematical Principles of Natural Philosophy,* trans. Andrew Motte [1st American ed.; New York: D. Adee, 1846], p. 348; this edition is cited hereafter as *Principia,* trans. Motte): "I began the foregoing experiments to investi-

gate the resistances of fluids, before I was acquainted with the theory laid down in the Propositions immediately preceding." For a revision of Motte's translation—in which antiquated phraseology is modernized and old mathematical terms are replaced with modern notations—cf. *Sir Isaac Newton's Mathematical Principles of Natural Philosophy and His System of the World,* Motte's translation revised and supplied with a historical and explanatory appendix by Florian Cajori (Berkeley: University of California Press, 1947). This new edition is excellent, but the changes in translation make no difference in the interpretation of Newton's views on the logic of science.

5. Cf. Bloch, *La philosophie de Newton,* p. 130.

6. *Principia,* Book III, trans. Motte, p. 383.

7. Cf. Bloch, *La philosophie de Newton,* p. 129.

8. *Ibid.,* p. 131.

9. *Principia,* Motte's translation, new edition revised by William Davis (3 vols.; London: Sherwood, Neely, and Jones, 1819), II, 314; this edition is cited hereafter as *Principia,* trans. Motte-Davis. Cf. *Opticks: or, a Treatise of the Reflections, Refractions, Inflections and Colours of Light* (2nd ed. with additions; London, 1718), query 31, p. 380: "For Hypotheses are not to be regarded in experimental philosophy." Also *Responsio ad Pardies,* Philosophical Transactions, No. 84 (Isaac Newton, *Opuscula,* ed. J. Castillioneus [Lausanne and Geneva, 1744], II, 322), where Newton says that his theory of light is not properly to be called a hypothesis, inasmuch as it seems "to contain nothing else than certain properties of light, which I have discovered and regard it not difficult to prove; and if I had not perceived them to be true I would have preferred to reject them as futile and inane speculation, rather than to acknowledge them as my hypothesis" (translation from E. A. Burtt, *The Metaphysical Foundations of Modern Science* [New York: Harcourt, Brace and Co., 1925], p. 212).

10. Trans. Motte-Davis, I, xxii.

11. *Ibid.,* p. xiv.

12. Bloch, *La philosophie de Newton,* pp. 465–66.

13. *Principia,* Book III, trans. Motte, p. 385.

14. *Opticks,* query 31, p. 380.

15. *Letter to Oldenburg,* July, 1672 (Isaac Newton, *Opera quae exstant omnia commentariis illustrabat* Samuel Horsley [London: J. Nichols, 1779–85], IV, 342; cited hereafter as *Opera*).

16. *Opticks,* query 31, p. 364: "Even the Rays of Light seem to be hard Bodies; for otherwise they would not retain different properties in their different Sides."

17. *Letter to Oldenburg,* July, 1672, *Opera,* IV, 324: "It is true that

from my theory I argue the corporeity of light, but I do it without any absolute positiveness, as the word *perhaps* intimates, and make it at most but a very plausible consequence of the doctrine, and not a fundamental presupposition, nor so much as any part of it, which was wholly comprehended in the precedent propositions."

18. *Letter to Oldenburg*, January 25, 1675–76, printed in D. Brewster, *Memoirs of the Life, Writings, and Discoveries of Sir Isaac Newton* (Edinburgh: Thomas Constable and Co., 1855), I, 391–92.

19. Burtt, *The Metaphysical Foundations . . .*, p. 264. Cf. *Opera*, IV, 380.

20. Cf. *A Letter to the Hon. Mr. Boyle on the Cause of Gravitation*, February, 1678–79, *Opera*, IV, 384: "by what has been said, you will readily discern whether, in these conjectures, there be any degree of probability; which is all I aim at. For my own part, I have so little fancy to things of this nature, that, had not your encouragement moved me to it, I should never, I think, have thus far put pen to paper about them."

21. Cf. Burtt, *The Metaphysical Foundations . . .*, p. 264.

22. *Ibid.*, p. 275.

23. *Ibid.*

24. Cf. *Opticks*, p. 369. "There are therefore Agents in nature able to make the Particles of Bodies stick together by very strong Attractions. And it is the business of experimental philosophy to find them out."

25. Cf. *Principia*, Book II, sec. 1, prop. iv, Scholium, trans. Motte, p. 257: "But, yet, that the resistance of bodies is in the ratio of the velocity, is more a mathematical hypothesis than a physical one." Cf. also the title of section 9, *Principia*, Book II.

26. *Ibid.*, Book II, sec. 9, Scholium, n. 2.

27. Cf. Bloch, *La philosophie de Newton*, pp. 441, 475.

28. The first edition contained only queries 1 to 7, part of queries 8, 9, 10, 11, queries 12 to 15, and the beginning of query 16. The edition of 1706 adds the end of queries 8, 10, and 11, and queries 25 to 31. The edition of 1717 gives for the first time queries 17 to 24 and the latter part of query 31. Cf. Bloch, *La philosophie de Newton*, p. 477.

29. *Letter to Oldenburg*, July, 1672, *Opera*, IV, 320.

30. Cf. *Secunda responsio ad Pardies*, Philosophical Transactions, No. 85, p. 5014, *Opuscula*, II, 329: "If any one offers conjectures about the truth of things from the mere possibility of hypotheses, I do not see how anything certain can be determined in any science; since one can always think up other and yet other hypotheses which will seem to furnish new difficulties. Wherefore I judged that one should abstain from considering hypotheses, as from a fallacious argument."

31. Cf. also his reaction to the criticism that he has by no means proved that universal gravitation is the *sole* force at work in producing celestial movements. Even though the law of gravitation may be sufficient to account for these motions so far as they are known to us, there may nevertheless be at work also other forces, producing, for example, translations of the whole system, of which the Newtonian theory takes no account. Newton's reply is that hypothetical forces producing merely hypothetical effects may safely be ignored. Cf. *Principia, The System of the World*, trans. Motte, p. 518: "It may be alleged that the sun and planets are impelled by some other force equally and in the direction of parallel lines; but by such a force no change would happen in the situation of the planets to one another, nor any sensible effect follow: but our business is with the causes of sensible effects. Let us, therefore, neglect every such force as imaginary and precarious, and of no use in the phaenomena of the heavens." Cf. also Newton's First Rule of Philosophizing (quoted later in this chapter).

32. Cf. *Opticks*, query 31, p. 380 (quoted above).

33. Cf. *Letter to Oldenburg*, July, 1672, *Opera*, IV, 321: "And therefore I could wish all objections were suspended taken from hypotheses, or any other heads than these two: of shewing the insufficiency of experiments to determine these queries, or prove any other parts of my theory, by assigning the flaws and defects in my conclusions drawn from them; or of producing other experiments, which directly contradict me, if any such may seem to occur. For if the experiments, which I urge, be defective, it cannot be difficult to shew these defects; but if valid, then by proving the theory, they must render all objections invalid."

34. Cf. Roger Cotes, Preface to the second edition of the *Principia*, trans. Motte-Davis, I, xiv. Experimental philosophers "frame no hypotheses, nor receive them into philosophy otherwise than as questions whose truth may be disputed."

35. Cf. *Secunda responsio ad Pardies*, Philosophical Transactions, No. 85, p. 5014, *Opuscula*, II, 329: "For hypotheses ought to be fitted merely to explain the properties of things and not attempt to predetermine them except so far as they can be an aid to experiments" (translation from Burtt, *The Metaphysical Foundations* . . . , p. 211 n.).

36. Cf. preceding note and *Opticks*, query 31, p. 351: "For we must learn from the Phaenomena of Nature what Bodies attract one another, and what are the Laws and Properties of the Attraction, before we enquire the Cause by which the Attraction is perform'd." Also *Letter to Oldenburg*, July, 1672, *Opera*, IV, 320: "And this I would have done in a due method; the laws of refraction being thoroughly enquired into and determined, before the nature of colours be taken into consideration."

37. Cf. *Secunda responsio ad Pardies,* Philosophical Transactions, No. 85, p. 5014, *Opuscula,* II, 233: "Postquam proprietates Lucis, his, et similibus Experimentis satis exploratae fuerint . . . Hypotheses exinde dijudicandae sunt, et quae non possint conciliari rejiciendae."

38. *Ibid.,* p. 329: "For the best and safest method of philosophizing seems to be, first diligently to investigate the properties of things and establish them by experiments, and then later seek hypotheses to explain them." "For hypotheses ought to be fitted merely to explain the properties of things and not attempt to predetermine them . . ." (translation from Burtt, *The Metaphysical Foundations* . . . , p. 211 n.).

39. *Principia,* Book I, definition viii, trans. Motte-Davis, I, 5.

40. *Ibid.,* p. 6.

41. *Principia,* Book I, sec. 11, trans. Motte, I, 144.

42. *Principia,* Book I, sec. 11, prop. lxix, theorem xxix, Scholium, trans. Motte-Davis, I, 174.

43. *Principia, The System of the World,* trans. Motte, p. 526.

44. Cf. *Opticks,* Book II, Part III, prop. viii, p. 243: "The Rays of Light whether they be very small Bodies projected, or only Motion or Force propagated, are moved in right Lines."

45. *Ibid.,* p. 255: "What kind of action or disposition this is; whether it consists in a circulating or vibratory motion of the Ray, or of the Medium, or something else, I do not here enquire." *Ibid.,* p. 256: "I content myself with the bare Discovery, that the Rays of Light are by some cause or other alternately disposed to be reflected or refracted for many vicissitudes." *Letter to Oldenburg,* January, 1672–73, *Opera,* IV, 350: "But to examine how colours may be thus explained hypothetically, is beside my purpose." (Horsley's text reads, "is, besides, my purpose," but this is inconsistent, not only with the context, but also with the Latin of *Opuscula,* II, 362: *"extra propositum meum est."*) "I never intended to shew wherein consists the nature and difference of colours, but only to shew that *de facto* they are original and immutable qualities of the rays which exhibit them."

46. Cf. *Principia,* Book I, sec. 14, prop. xcvi, Scholium, trans. Motte, p. 246: "Therefore because of the analogy there is between the propagation of the rays of light and the motion of bodies, I thought it not amiss to add the following Propositions for optical uses; not at all considering the nature of the rays of Light, or inquiring whether they are bodies or not; but only determining the trajectories of bodies which are extremely like the trajectories of the rays." Cf. *Letter to Oldenburg,* July, 1672, *Opera,* IV, 324–25: "I chose to decline them all [i.e., all hypotheses] and speak of light abstractedly as something or other propagated everywhere in straight lines from luminous bodies, without determining what that thing is; whether a con-

fused mixture of difform qualities, or modes of bodies, or of bodies themselves; or of any virtues, powers or beings whatsoever." Cf. *Theory of Light and Colours, Opera,* IV, 305: "But to determine more absolutely what light is, after what manner refracted, and by what modes or actions it produceth in our minds the phantasms of colours, is not so easy: and I shall not mingle conjectures with certainties."

47. Cf. *Secunda responsio ad Pardies,* Philosophical Transactions, No. 85, p. 5014, *Opuscula,* II, 333: "Sed levissimi negotii est accommodare Hypotheses ad hanc Doctrinam." Cf. *Letter to Oldenburg,* January, 1672–73, *Opera,* IV, 350: Newton intended "to leave it to others to explicate, by mechanical hypotheses, the nature and difference of the qualities; which I take to be no very difficult matter."

48. Cf. *Letter to Oldenburg,* July, 1672, *Opera,* IV, 324: "But I knew that the properties, which I declared of light, were in some measure capable of being explicated not only by that, but by many other mathematical hypotheses; and therefore I chose to decline them all." Cf. *Secunda responsio ad Pardies,* Philosophical Translations, No. 85, *Opuscula,* II, 333.

49. Cf. *Opticks,* Book II, Part III, prop. xii, p. 155: "Those that are averse from assenting to any new Discoveries, but such as they can explain by an hypothesis, may for the present suppose," etc. Cf. *An Hypothesis Explaining the Properties of Light Discoursed of in My Several Papers, Letter to Oldenburg,* January 25, 1675–76, Brewster, *Memoirs . . . ,* I, 391–92: "and therefore because I have observed the heads of some great virtuosos to run much upon hypotheses, as if my discoveries wanted an hypothesis to explain them by, and found that some, when I could not make them take my meaning when I spake of the nature of light and colours abstractedly, have readily apprehended it when I illustrated my discourse by an hypothesis; for this reason I have here thought it fit to send you a description of the circumstances of this hypothesis, as much tending to the illustration of the papers I herewith send you." Cf. the alternative suggestions of *Principia,* Book I, sec. ii, prop. xlix, Scholium, trans. Motte, p. 217 (quoted above); also the suggestion (*System of the World,* trans. Motte, p. 526) that it is as if the planets were bodies held together by a rope: "Two bodies may be mutually attracted to each other by the contraction of a cord interposed."

50. First Rule of Philosophizing, *Principia,* Book III, trans. Motte, p. 384. Perhaps I go too far in making this latter point. Cf. Leon Brunschvicg, *L'expérience humaine et la causalité physique* (Paris: F. Alcan, 1922), p. 234: "Lorsque l'on pose le problème du point de vue historique, pour le XVIIe siècle, on ne voit nullement que, soit Newton, soit l'école newtonienne, ait jamais songé à exclure de la science la recherche des causes. La critique de la relation causale est tout entière l'œuvre de l'occasionalisme

cartésien, qui a directement inspiré sur ce point Berkeley et Hume . . . *sed causam Gravitatis nondum assignavi.* Newton n'a pas encore résolu le problème de la cause, et il juge antiscientifique de proposer des conjectures, alors qu'il n'est pas en état d'en établir la vérité. Cela ne veut pas du tout dire qu'à ses yeux cette vérité serait sans intérêt intrinsèque ou sans portée scientifique. Cela signifie seulement qu'ayant découvert la formule de la loi, Newton s'est trouvé, ainsi que l'avait Galilée, arrêté devant la question soulevée par cette découverte elle-même: Quelle est la cause de la loi?"

51. Cotes, Preface, I, xiii.

52. *Opticks*, query 31, pp. 376–77. Cf. also *ibid.*, p. 364: "others tell us that Bodies are glued together by rest, that is, by an occult Quality, or rather by nothing." Hence it is that "the moderns, laying aside substantial forms and occult qualities, have endeavored to subject the phaenomena of nature to the laws of mathematics" (*Principia*, Preface of 1686, trans. Motte-Davis, I, ix).

53. Cotes, Preface, I, xxii.

54. Nicholas Malebranche, *De la recherche de la vérité*, ed. F. Bouillier (2 vols.; Paris: Garnier frères, 1880), II, 51.

55. *Ibid.*, p. 56.

56. *Ibid.*, pp. 56–57.

57. Cf. Bloch, *La philosophie de Newton*, pp. 419 ff.

58. J. F. W. Herschel, *A Preliminary Discourse on the Study of Natural Philosophy* (London: Longman, Brown, Green, and Longmans, 1842), sec. 138.

59. William Whewell, *Philosophy of Discovery* (3rd ed.; London: John W. Parker and Son, 1860), p. 186.

60. *Ibid.*, p. 189.

61. *Ibid.*, p. 190.

62. *Ibid.*, p. 191.

63. *Ibid.*, p. 192. As I have pointed out elsewhere (chapter 4), this is the Cartesian conception of a "true cause." Cf. also J. S. Mill, *System of Logic* (8th ed.; New York: Harpers, 1900), Book III, chap. xiv, par. 4, p. 353: "What is true in the maxim is, that the cause, though not known previously, should be capable of being known thereafter: that its existence should be capable of being detected, and its connexion with the effect ascribed to it should be susceptible of being proved, by independent evidence. The hypothesis, by suggesting observations and experiments, puts us on the road to that independent evidence, if it be really attainable; and till it be attained, the hypothesis ought only to count for a more or less plausible conjecture."

64. *Principia*, trans. Motte-Davis, I, xxiv.

65. *Ibid.*, p. xxiii.

66. *Ibid.*, II, 314.

67. *Ibid.*, I, x.

68. Cf. *Opticks*, query 31, p. 364: "All bodies seem to be composed of hard particles." *Ibid.*, p. 375: "All these things being consider'd, it seems probable to me that God in the Beginning form'd Matter in solid, massy, hard, impenetrable, moveable Particles, of such Sizes and Figures, and with such other Properties, and in such Proportion to Space, as most conduced to the End for which he form'd them."

69. Cf. chapter 3.

70. For the contrast of Newton with Descartes cf. *Principia*, Rules of Reasoning in Philosophy, rule iii, trans. Motte, p. 384: "We no other way know the extension of bodies than by our senses." "That all bodies are impenetrable we gather not from reason, but from sensation."

71. Cf. *Opticks*, query 31, p. 364: "And therefore Hardness may be reckoned the Property of all uncompounded Matter. At least this seems to be as evident as the universal impenetrability of Matter. For all Bodies, so far as Experience reaches, are either hard, or may be harden'd; and we have no other Evidence of universal Impenetrability, besides a large Experience without experimental Exception." Cf. also Third Rule of Philosophizing, *Principia*, trans. Motte, p. 384.

72. Cf. *Opticks*, query 31, p. 364: "And for explaining how this may be, some have invented hooked Atoms, which is begging the Question."

73. *Ibid.*, query 28, pp. 343–44. Cf. *Letter to Oldenburg*, July, 1672, *Opera*, IV, 335: "And he that shall explicate this case mechanically must conquer a double impossibility." Cf. Bloch, *La philosophie de Newton*, p. 399.

74. Cf. *Principia*, Book I, sec. 14, prop. xcvi, Scholium, trans. Motte, p. 246. Also Book II, sec. 5, Scholium, pp. 392–93: "But whether elastic fluids do really consist of particles so repelling each other, is a physical question. We have here demonstrated mathematically the property of fluids consisting of particles of this kind, and hence philosophers may take occasion to discuss that question."

75. Cf. *Opticks*, query 31, p. 380, as quoted above; also: "By this way of Analysis we may proceed from Compounds to Ingredients, and from Motions to the forces producing them; and in general, from Effects to their Causes, and from particular Causes to more general ones, till the Argument end in the most general. This is the Method of Analysis."

76. Trans. Motte-Davis, I, x. Cf. Cotes, Preface, p. xiv: "There is left, then, the third class, which profess experimental philosophy. These, indeed, derive the causes of all things from the most simple principles possible; but,

then, they assume nothing as a principle that is not proved by phaenomena. . . . They proceed, therefore in a two-fold method, synthetical and analytical. From some select phaenomena they deduce by analysis the forces of nature, and the more simple laws of forces; and from thence by synthesis shew the constitution of the rest."

77. *Principia*, trans. Motte-Davis, I, ix.

78. Cf. chapter 4.

79. Cf. *Principia*, Book II, sec. 7, Scholium, experiment 4, trans. Motte, p. 348 (quoted above).

80. Cf. Fourth Rule of Philosophizing (quoted above). "Continually he called in experimental verification, even for the solution of questions whose answers would seem to be involved in the very meanings of his terms, such as the proportionality of resistance to density [*Opticks*, p. 340]. Having defined mass in terms of density and also in terms of resistance, such proportionality would seem to be involved in the very meaning of the words" (Burtt, *The Metaphysical Foundations* . . . , p. 209).

81. Cf. *Opticks*, query 31, pp. 380–81: "And the Synthesis consists in assuming the Causes discover'd and establish'd as Principles, and by them explaining the Phaenomena proceeding from them, and proving the explanations." *Ibid.*, Book II, Part IV, observation 12, p. 288: "all which [sc. phaenomena], so far as I have yet observed, follow from the Propositions at the end of the third part of this Book, and so conspire to confirm the truth of those Propositions." So much even Descartes is prepared to admit. Cf. the account of Descartes in chapter 4.

82. Cf. *Opticks*, Book II, Part II, p. 215: "These are the principal Phaenomena of thin Plates of Bubbles, whose Explications depend on the properties of Light, which I have heretofore deliver'd. And these you see do necessarily follow from them, and agree with them, even to their very least circumstances; and not only so, but do very much tend to their proof."

83. Burtt, *The Metaphysical Foundations* . . . , p. 216. *Opticks*, p. 66.

84. Burtt, *The Metaphysical Foundations* . . . , p. 216.

85. Bloch, *La philosophie de Newton*, p. 436. Cf. *Principia*, Book II, sec. 2, prop. x, trans. Motte, p. 271: "Then the density of the medium would come out as -*a*/*e*. But nature does not admit of a negative density, that is, a density which accelerates the motion of bodies; and therefore it cannot naturally come to pass that a body by ascending from P should describe the quadrant PF of a circle."

86. How frankly Newton can make use of a purely provisional principle may be illustrated from *Lectiones opticae*, Part I, sec. 2, par. xliii, *Opuscula*, p. 118: "Hujus quidem Theorematis certitudinem ab experimentis nondum

habeo depromptam; sed cum a veritate vix multum discrepare videatur, nihil veritus sum in praesentia gratis assumere. Post hac forte, vel experientia confirmabo, vel, si falsum invenero corrigam."

87. We find him employing such expressions as the following: "But since the description of this curve is difficult, a solution by approximation will be preferable" (*Principia*, Book I, sec. 6, prop. xxxi, problem xxiii, Scholium, trans. Motte-Davis, II, 100). *Ibid.*, p. 102: "It will be sufficient if that angle is found by a rude calculus in numbers near the truth." *Ibid.*: "And by repeating the computation the place [sc. of the body] may be found perpetually to greater and greater accuracy." *Ibid.*, p. 103: "This practice seems expeditious enough, because the angles . . . being very small, it will be sufficient to find two or three of their first figures. But it is likewise sufficiently accurate to answer to the theory of the planet's motions. For even in the orbit of Mars . . . the error will scarcely exceed one second."

88. Cf. Bloch, *La philosophie de Newton*, p. 282; *Principia*, Book II, sec. 6, General Scholium, trans. Motte, p. 318, where it is a case of formulating the mathematical laws governing the resistance of a fluid to the movement of a solid. Newton writes: "But the greatest of the globes I used in these experiments was not perfectly spherical, and therefore in this calculation I have, for brevity's sake, neglected some little niceties; being not very solicitous for an accurate calculus in an experiment that was not very accurate." Cf. also *The System of the World*, trans. Motte, p. 519: "In examining this proportion, we are to use the mean distances . . . and to neglect those little fractions, which, in defining the orbits, may have arisen from the insensible errors of observation, or may be ascribed to other causes which we shall afterwards explain."

89. Cf. *The System of the World*, passage just quoted above.

90. Cf. *ibid.*, p. 512: "But our purpose is only to trace out the quantity and properties of this force from phaenomena, and apply what we discover in some simple cases as principles, by which, in a mathematical way, we may estimate the effects thereof in more involved cases; for it would be endless and impossible to bring every particular to direct and immediate observation."

91. Book II, sec. 7, prop. xxxvii, trans. Motte, p. 340.

92. Book I, sec. 11, trans. Motte, p. 144: "I have hitherto been treating of the attractions of bodies toward an immovable centre; though very probably there is no such thing existing in nature." Cf. Bloch, *La philosophie de Newton*, pp. 447–49. Cf. also *Principia*, Book II, sec. 1, prop. iv, Scholium, trans. Motte, p. 257: "But, yet, that the resistance of bodies is in the ratio of the velocity, is more a mathematical hypothesis than a physical one."

93. Brewster, *Memoirs* . . . , II, 403.

94. W. E. H. Lecky, *History of the Rise and Influence of the Spirit of*

Rationalism in Europe (2 vols.; rev. ed.; New York: D. Appleton and Co., 1884), I, 292 n.

95. It is very strange, however, to find Bloch (*La philosophie de Newton,* p. 423, n. 2) stating that Newton likewise never refers to Descartes. On the contrary he frequently does so.

96. Cf. Bloch, *La philosophie de Newton,* p. 423.

97. *Novum organum,* Book II, aphorism 45.

98. *Ibid.,* aphorism 36.

99. *Ibid.,* aphorisms 36, 48.

100. *Ibid.,* aphorism 22. For further discussion of the point cf. Bloch, *La philosophie de Newton,* p. 422.

101. Cf., e.g., *Isaaci Newtoni scala graduum coloris et frigoris,* Philosophical Transactions, No. 270, *Opuscula,* II, 417; *Principia,* Book II, sec. 7, prop. xl, Scholium; *Opticks,* Book II, Part I, observations 7, 19; *ibid.,* Part III, prop. x; *ibid.,* Book III, Part I, observations 3, 9.

102. Cf. Bacon, *Novum organum,* Book II, aphorism 36; Newton, *Theory of Light and Colours, Opera,* IV, 298: "The gradual removal of these suspicions at length led me to the *experimentum crucis* which was this." Cf. Bloch, *La philosophie de Newton,* p. 440.

103. *Letter to Oldenburg, Opera,* IV, 370.

104. *Opticks,* query 31, p. 380.

105. Compare also Newton's "for it would be endless and impossible to bring every particular to direct and immediate observation" (*The System of the World,* trans. Motte, p. 512) or his "in examining this proportion, we are to . . . neglect those little fractions . . ." (*ibid.,* p. 519) with Bacon's "inquisitionem rerum justam et plenam, demptis individuis et gradibus rerum et variationibus minutis (id quod ad scientias satis est)" (*Partis instaurationis secundae delineatio et argumentum,* in *The Works of Francis Bacon,* ed. James Spedding, Robert Leslie Ellis, and Douglas Denon Heath [14 vols.; London: Longman and Co., 1857–59], VII, 45).

106. Cf. *Novum organum,* Book II, aphorism 8: "And inquiries into nature have the best result, when they begin with physics and end with mathematics. Again, let no one be afraid of high numbers or minute fractions."

107. Cf. Bloch, *La philosophie de Newton,* pp. 429 ff.

108. Compare, e.g., Bacon's "not pretty and probable conjectures, but certain and demonstrable knowledge" (*Novum organum,* Preface, last paragraph), or "a form of induction which shall . . . lead to an inevitable conclusion" (*The Great Instauration,* Plan of the Work, *The Works of Bacon,* VIII, 42) with Newton's "although the arguing from Experiments and Observations by Induction be no Demonstration of general Conclusions" (*Opticks,* query 31, p. 380). Cf. chapter 3.

109. *Principia,* Book III, trans. Motte, pp. 304–5.

110. Burtt, *The Metaphysical Foundations* . . . , p. 214, comments as follows: "The later more mathematical expression of this principle is that where different events are expressed by the same equations, they must be regarded as produced by the same forces." So also Bloch, *La philosophie de Newton,* p. 456.

111. The restriction of the application of this rule to "qualities which admit neither intension nor remission of degree" is curious. Bloch thinks that Newton thereby intended to exclude from the application of the rule such properties as viscosity or friction. These properties are susceptible of more or less, and what can decrease indefinitely may in the end disappear altogether. It would therefore be rash to attribute these properties to the minute particles of matter. So, too, resistance decreases as rarefaction increases. It would therefore be rash to attribute this property to such a rarefied medium as the ether. Bloch, *La philosophie de Newton,* pp. 462–63.

112. This expression is frequently repeated by Newton. Cf. *Opticks,* Book I, Part I, prop. vi, pp. 65–66; query 31, pp. 351–72. We omit the illustrations of the application of this rule, which Newton here adds.

113. Cotes, Preface, pp. xxi–xxii, states the substance of the second and third rules as follows: "This axiom . . . is received by all philosophers, namely, that effects of the same kind, that is, whose known properties are the same, take their rise from the same causes, and have the same unknown properties also. . . . All philosophy is founded on this rule; for if that be taken away we can affirm nothing of universals. The constitution of particular things is known by observation and experiments; and when that is done, it is by this rule that we judge universally of the nature of such things in general."

114. Cf. Whewell, *Philosophy of Discovery,* pp. 185–86: "Thus the first Rule is designed to strengthen the inference of gravitation from the celestial phenomena, by describing it as a *vera causa,* a true cause; the second Rule countenances the doctrine that the planetary motions are governed by mechanical forces, as terrestrial motions are; the third rule appears intended to justify the assertion of gravitation as a *universal* quality of bodies; and the fourth contains, along with a general declaration of the authority of induction, the author's usual protest against hypotheses, levelled at the Cartesian hypotheses especially."

115. Upon this point Burtt comments as follows (*The Metaphysical Foundations* . . . , p. 215): "Are these apriorisms speculative assumptions about the structure of the universe which make it always possible to reduce its phenomena to laws, especially mathematical laws; or were they to Newton a matter of method merely, to be used tentatively as a principle of further

inquiry? It is perhaps impossible to answer this question with absolute confidence. At those times when the theological basis of Newton's science was uppermost in his mind, it is probable that he would have answered substantially as Galileo and Descartes did. But in his strictly scientific paragraphs the emphasis is overwhelmingly in favour of their tentative, positivistic character, hence the fourth rule of reasoning in philosophy . . . must be regarded as imposing definite limits on all of the other three." Cf. Burtt's comment on the Fourth Rule (quoted above).

CHAPTER SEVEN

David Hume on Causation

1. David Hume, *A Treatise of Human Nature*, ed. L. A. Selby-Bigge (Oxford: Clarendon Press, 1946), p. 165.
2. *Ibid.*, p. 165.
3. *Ibid.*, p. 211.
4. *Ibid.*, p. 218.
5. *Ibid.*, p. 29.
6. *Ibid.*, p. 165.
7. *Ibid.*, p. 173.
8. *Ibid.*, pp. 104–5.
9. *Ibid.*, pp. 173 ff.
10. See *ibid.*, pp. 90–91, for Hume's own objections to it.
11. *Ibid.*, p. 87.
12. *Ibid.*, pp. 88–89.
13. *Ibid.*, p. 90.
14. Also *ibid.*, p. 105.
15. *Ibid.*, p. 173.
16. *Ibid.*, pp. 413, 463.
17. *Ibid.*, p. 165.

CHAPTER EIGHT

John F. W. Herschel's Methods of Experimental Inquiry

1. William Minto, *Logic, Inductive and Deductive* (New York: Charles Scribner's Sons, 1904), p. 257.
2. John F. W. Herschel, *Preliminary Discourse on the Study of Natural Philosophy* (Cabinet Cyclopaedia ed.; London: Longman, Brown, Green, and Longmans, 1842), sec. 65.

3. Herschel is one of the few English philosophers of science to hold such a high opinion of Bacon.

4. Herschel, *Preliminary Discourse* . . . , sec. 96.

5. *Ibid.*, sec. 105.

6. The relevant passages in the *Opticks* are in query 31 and in the *Principia* in the section entitled "Regulae Philosophandi" and, at the end of the book, in the General Scholium where occurs his famous "*hypotheses non fingo.*"

7. Cf. Alexander Bain, *Logic* (2nd ed.; 2 vols.; London: Longmans, Green and Co., 1899), II, 408. W. S. Jevons, *The Principles of Science* (2nd ed. corr.; New York: Macmillan Co., 1905), pp. 581 ff.

8. Herschel, *Preliminary Discourse* . . . , sec. 66.

9. *Ibid.*, sec. 67.

10. *Ibid.*

11. *Ibid.*, sec. 109.

12. *Ibid.*, secs. 227 ff.

13. *Ibid.*, sec. 130.

14. *Ibid.*, sec. 132.

15. *Ibid.*, sec. 110.

16. *Ibid.*

17. *Ibid.*, sec. 76.

18. *Ibid.*, sec. 74.

19. *Ibid.*

20. *Ibid.*, sec. 109.

21. Cf. Bacon, *Novum organum*, II, 4.

22. Herschel, *Preliminary Discourse* . . . , sec. 80.

23. *Ibid.*

24. The example of Bacon in using the word "axiom" to designate the laws of nature is followed by Herschel—also often by Newton. See Dugald Stewart, *Collected Works* (7 vols.; Cambridge, Mass.: Hilliard and Brown, 1829), Vol. II, chap. iv.

25. Herschel, *Preliminary Discourse* . . . , sec. 78.

26. *Ibid.*, sec. 77.

27. *Ibid.*

28. *Ibid.*

29. *Ibid.*, sec. 78. The "cause" of sensation is declared by Herschel to be "much more obscure" still than that of motion (sec. 82). As thus obscure, the "cause of sensation," e.g., of auditory sensation, cannot be taken to refer to the vibration of such objects as bells, or of the air, of course. Herschel, indeed, makes it clear in section 82 that by the "cause of sensation" he in-

tends something that has to sensation a relation analogous to that of an "effort of memory or imagination" to the images which that effort causes to appear in our minds!

30. *Ibid.*, sec. 77.

31. For example, as some psychologists have suggested, in the muscles of the breathing apparatus.

32. Herschel, *Preliminary Discourse* . . . , sec. 81.

33. *Ibid.*, sec. 141.

34. Cf. chapters 3 and 10 in this volume.

35. Herschel, *Preliminary Discourse* . . . , sec. 91.

36. *Ibid.*, sec. 93.

37. *Ibid.*, sec. 94.

38. *Ibid.*, sec. 26.

39. *Ibid.*, sec. 92. Cf., however, sec. 162.

40. *Ibid.*, sec. 187.

41. Morris R. Cohen, *The Meaning of Human History* (La Salle, Ill.: Open Court, 1947), p. 103.

42. Cf. Bacon, *New Atlantis*, the end.

43. Herschel, *Preliminary Discourse* . . . , sec. 144.

44. *Ibid.*, sec. 145.

45. *Ibid.*, secs. 146–62.

46. *Ibid.*, sec. 170.

47. *Ibid.*, sec. 176.

48. *Ibid.*, sec. 184.

49. *Ibid.*, sec. 177.

50. *Ibid.*, sec. 180.

51. *Ibid.*, sec. 187.

52. *Ibid.*, sec. 191.

53. In view of Augustus De Morgan's opinion (expressed in his *A Budget of Paradoxes* [2 vols.; Chicago: Open Court Publishing Co., 1915]) that Bacon's writings neither influenced Newton nor could have been of any possible value to him, and even that Newton had probably not read them, it is interesting to note Herschel's statement, with regard to one of Bacon's "travelling instances," that, "in reading this, and many other instances in the *Novum Organum*, one would almost suppose (had it been written) that its author had taken them from Newton's *Optics*"! The correspondence Herschel notes—though it obviously cannot be construed as evidence either that Newton "borrowed" from Bacon or even that he had read his work—at least shows that Bacon's observations were not all as intrinsically worthless as the expressions of some of his critics would lead one to

believe. Cf. McVey Napier, "Remarks Illustrative of the Scope and Influence of the Philosophical Writings of Lord Bacon," *Transactions Royal Society of Edinburgh*, VIII, 384.

54. Herschel, *Preliminary Discourse* . . . , sec. 201.

55. *Ibid.*, sec. 202.

56. *Ibid.*, sec. 208.

57. *Ibid.*, sec. 209.

58. *Ibid.*, sec. 210.

59. *Ibid.*, sec. 214.

60. *Ibid.*, sec. 216.

61. *Ibid.*, sec. 219.

62. *Ibid.*, sec. 218.

63. Herschel even says (*ibid.*, sec. 86) that "the axioms of geometry themselves may be regarded as in some sort an appeal to experience, not corporeal, but mental . . . these axioms, however self-evident, are still general propositions so far of the inductive kind, that, independently of experience, they would not present themselves to the mind. The only difference between these and axioms obtained from extensive induction is this, that in raising the axioms of geometry, the instances offer themselves spontaneously . . . and are few and simple; in raising those of nature, they are infinitely numerous, complicated and remote."

64. In the thirteenth century Roger Bacon exhibited a more genuine empiricism and a greater practical mastery of scientific method than did Francis Bacon in the seventeenth. But Roger's example did not influence his contemporaries or his successors as it would have done had he lived three hundred years later.

65. See for instance Norman Campbell's *Physics: The Elements* (Cambrdge, Eng.: Cambridge University Press, 1920); and more recently J. H. Woodger's *The Technique of Theory Construction* (Chicago: University of Chicago Press, 1939). The various judgments above as to errors and confusions in Herschel's *Discourse* have had to be expressed dogmatically here since any attempt to justify them would have required far too much space. The grounds on which the writer bases most of them may, however, be found in C. J. Ducasse, *Causation and the Types of Necessity* (Seattle: University of Washington Press, 1924) and in a paper entitled "The Nature and Function of Theory in Ethics," *Ethics*, LI (1940), 22–37.

66. J. S. Mill, *System of Logic* (8th ed.; New York: Harper and Bros., 1874), Book III, chap. ix, sec. 3.

67. W. S. Jevons, *Pure Logic and Other Minor Works*, ed. Robert Adamson and Harriet A. Jevons (London: Macmillan Co., 1890), p. 251 n.

68. *Novum organum,* II, aphorisms 11, 12, 13, 15, 18, 20; also among the Prerogative Instances.

69. Book I, Part III, sec. 15, aphorism 22: "Rules by which to judge causes and effects."

CHAPTER NINE

William Whewell's Philosophy of Scientific Discovery

1. William Whewell, *On the Philosophy of Discovery* (3rd ed.; London: John W. Parker and Son, 1860), p. 223.
2. H. L. Mansel, *Prologomena logica* (2nd ed.; Oxford: W. Graham, 1851), Appendix, note A.
3. G. H. Lewes, *The Biographical History of Philosophy* (New York: D. Appleton and Co., 1885), pp. 661–74.
4. Whewell, *On the Philosophy of Discovery,* p. 334.
5. *Ibid.,* p. 312.
6. *Ibid.,* pp. 336, 343–44; Whewell, *History of Scientific Ideas* (3rd ed.; 2 vols.; London: John W. Parker and Son, 1858), I, 87.
7. Whewell, *History of Scientific Ideas,* I, 25.
8. *Ibid.,* p. 27.
9. *Ibid.*
10. *Ibid.,* pp. 29, 30.
11. *Ibid.,* pp. 30, 31.
12. *Ibid.,* p. 34.
13. *Ibid.,* p. 75.
14. *Ibid.,* p. 74.
15. *Ibid.,* p. 91.
16. *Ibid.,* p. 183.
17. Whewell, *Novum organon renovatum* (3rd ed.; London: John W. Parker and Son, 1858), p. 187.
18. Whewell, *History of Scientific Ideas,* I, 82, 83.
19. *Ibid.,* p. 38.
20. *Ibid.,* p. 40.
21. *Ibid.,* pp. 41–43.
22. *Ibid.,* p. 44.
23. *Ibid.,* p. 49.
24. *Ibid.,* p. 59.
25. *Ibid.,* p. 61.
26. *Ibid.,* p. 63.
27. J. S. Mill, *System of Logic* (8th ed.; New York: Harper and Bros., 1874), Book II, chap. v.

28. Whewell, *History of Scientific Ideas*, I, 67–68.
29. Mill, *System of Logic*, Book II, chap. v, sec. 4.
30. Whewell, *On the Philosophy of Discovery*, pp. 345–46.
31. Mill, *System of Logic*, Book II, chap. v, sec. 6.
32. Whewell, *On the Philosophy of Discovery*, p. 336.
33. *Ibid.*, pp. 338–39.
34. *Ibid.*, p. 339.
35. *Ibid.*, p. 340.
36. *Ibid.*, p. 344.
37. Whewell, *History of Scientific Ideas*, I, 78.
38. Mill's contention, it should be noted, was that experience furnishes us ample ground for *believing* the axioms. But the point at issue is not whether our experience contains facts psychologically adequate to cause us to believe the axioms necessarily and universally true, but whether it furnishes logically adequate evidence of their being necessarily and universally true. Belief is often brought about otherwise than by logically adequate evidence; and on the other hand the latter does not always bring about belief. The two are thus distinct matters.
39. Whewell, *History of Scientific Ideas*, I, 267.
40. *Ibid.*, pp. 266, 267.
41. Mill, *System of Logic*, Book II, chap. v, sec. 6.
42. Whewell, *History of Scientific Ideas*, II, 26.
43. Whewell, *On the Philosophy of Discovery*, p. 340.
44. J. F. W. Herschel and W. Hamilton both wrote reviews of Whewell's *Philosophy of the Inductive Sciences*, in which objections to his doctrine of necessary truths, and to other views of his, are raised. Herschel's review, which on the whole was favorable, appeared in the *Quarterly Review*, LXVIII (1841), 177–238; Whewell's reply to some of the criticisms in it is printed as Appendix F to *On the Philosophy of Discovery*. W. Hamilton reviewed in the *Edinburgh Review* both the *History of the Inductive Sciences* (LXVI [1837], 110–51) and *Philosophy of the Inductive Sciences* (LXXIV [1842], 265–306). Both reviews, in which Hamilton shows off his vast erudition, are not only adverse but sarcastically contemptuous. See also Hamilton's *Discussions on Philosophy, Literature, Education, and University Reform* (New York: Harper and Bros., 1853), pp. 322–23.
45. Mansel, *Prologomena logica*, Appendix, note A.
46. Friedrich Paulsen, *Immanuel Kant, His Life and Doctrine*, trans. J. E. Creighton and Albert Lefevre from rev. German ed. (New York: Charles Scribner's Sons, 1902), pp. 178, 179.

47. Whewell, *Novum organon renovatum*, p. 29.
48. *Ibid.*, p. 71.
49. *Ibid.*, p. 31.
50. *Ibid.*, p. 49.
51. *Ibid.*, p. 42.
52. *Ibid.*
53. *Ibid.*
54. *Ibid.*, p. 36.
55. *Ibid.*, p. 37.
56. *Ibid.*, pp. 39–40.
57. *Ibid.*, p. 40.
58. *Ibid.*, p. 43.
59. *Ibid.*, pp. 44, 46.
60. *Ibid.*, p. 143.
61. *Ibid.*, pp. 180, 181.
62. *Ibid.*, p. 51.
63. *Ibid.*, p. 53.
64. *Ibid.*, p. 54.
65. *Ibid.*
66. *Ibid.*, p. 56.
67. *Ibid.*, p. 60.
68. *Ibid.*, p. 70.
69. *Ibid.*, p. 71.
70. *Ibid.*, p. 73.
71. *Ibid.*, p. 74.
72. *Ibid.*, p. 110.
73. *Ibid.*, p. 75.
74. *Ibid.*, pp. 110, 111.
75. *Ibid.*, pp. 76–78.
76. *Ibid.*, p. 110.
77. *Ibid.*, p. 111.
78. *Ibid.*, p. 115.
79. *Ibid.*, p. 192.
80. *Ibid.*, p. 78.
81. *Ibid.*, p. 79.
82. *Ibid.*, p. 81.
83. *Ibid.*, p. 187.
84. *Ibid.*, p. 193.
85. *Ibid.*
86. *Ibid.*, p. 202.

87. *Ibid.*, p. 206.

88. *Ibid.*, p. 207.

89. This, then, may be said to constitute the postulate upon which is dependent such validity as the processes of extrapolation and interpolation possess.

90. Whewell, *Novum organon renovatum*, p. 208.

91. *Ibid.*, p. 202.

92. *Ibid.*, p. 221.

93. *Ibid.*, p. 224.

94. The advent of the quantum theory would then throw considerable doubt upon the reliability of the intuitive evidence upon which Whewell alleges the law of continuity to be based.

95. Whewell, *Novum organon renovatum*, p. 224.

96. *Ibid.*, p. 228.

97. *Ibid.*, p. 229.

98. *Ibid.*, p. 232.

99. *Ibid.*, p. 228.

100. Whewell, *On the Philosophy of Discovery*, p. 263.

101. For example, Alexander Bain, *Logic* (2nd ed.; 2 vols.; London: Longmans, Green and Co., 1899), Vol. II, Appendix ii.

102. Compare Louis Liard, *Les logiciens anglais contemporains* (3rd ed.; Paris: F. Alcan, 1890), p. 13.

103. Whewell, *History of Scientific Ideas*, I, 173.

104. *Ibid.*, p. 184.

105. *Ibid.*, pp. 173–74.

106. Whewell, *Novum organon renovatum*, p. 125.

107. Whewell, *History of Scientific Ideas*, I, 185 ff.

108. *Ibid.*, p. 184.

109. Whewell, *Novum organon renovatum*, p. 83.

110. *Ibid.*, p. 85.

111. *Ibid.*, p. 86.

112. *Ibid.*, p. 87.

113. Hamilton, *Discussions on Philosophy* . . . , p. 580. Cf. W. M. Thorburn, "The Myth of Occam's Razor," *Mind*, N.S. XXVII (1918), 345–53.

114. Whewell, *Novum organon renovatum*, p. 91.

115. John Venn, *The Principles of Empirical or Inductive Logic* (2nd ed.; London: Macmillan Co., 1907), p. 354 n.

116. Mill, *System of Logic*, p. 222.

117. *Ibid.*, p. 208.

CHAPTER TEN

John Stuart Mill's System of Logic

1. *System of Logic* (8th ed.; New York: Harper and Bros., 1874), p. 208.
2. *Ibid.*, p. 210.
3. *Ibid.*
4. *Ibid.*, p. 221.
5. *Ibid.*, p. 223.
6. *Ibid.*, p. 142.
7. *Ibid.*, p. 153.
8. *Ibid.*, pp. 153–54.
9. *Ibid.*, p. 208.
10. *Ibid.*, p. 209.
11. *Ibid.*, p. 452.
12. *Ibid.*, p. 209.
13. William Whewell, *On the Philosophy of Discovery* (3rd ed.; London: John W. Parker and Son, 1860), pp. 247–48.
14. Mill, *System of Logic*, p. 453.
15. For a discussion of some of the other inconsistencies in the *System of Logic*, see W. S. Jevons' *Pure Logic and Other Minor Works*, ed. Robert Adamson and Harriet A. Jevons (London: Macmillan Co., 1890), Part II; cf. Carveth Read's comments on Jevons' criticisms in *Mind*, XVI (1891), 106–10. Cf. also Morris R. Cohen and Ernest Nagel, *An Introduction to Logic and Scientific Method* (New York: Harcourt, Brace and Co., 1934), pp. 249 ff.
16. Mill, *System of Logic*, p. 223.
17. *Ibid.*, p. 224.
18. Cf. Whewell, *On the Philosophy of Discovery*, pp. 240 ff.
19. See Mill, *System of Logic*, end of note at bottom of p. 225.
20. *Ibid.*, p. 401.
21. *Ibid.*, p. 402.
22. *Ibid.*, p. 403.
23. *Ibid.*, p. 271.
24. *Ibid.*
25. *Ibid.*, p. 236.
26. *Ibid.*
27. *Ibid.*
28. *Ibid.*, p. 241.
29. *Ibid.*, p. 244.

30. *Ibid.*, p. 245.
31. *Ibid.*
32. *Ibid.*, p. 246.
33. *Ibid.*
34. Cf. Jevons, *Pure Logic and Other Minor Works*, p. 251.
35. Mill, *System of Logic*, p. 325.
36. *Ibid.*
37. *Ibid.*, p. 328.
38. *Ibid.*, p. 329.
39. *Ibid.*, p. 330.
40. *Ibid.*, p. 350.
41. Whewell, *On the Philosophy of Discovery*, pp. 282 ff.
42. Mill, *System of Logic*, "Induction Is Proof," Book III, chap. ii, sec. 5.

CHAPTER ELEVEN

W. S. Jevons on Induction and Probability

1. W. S. Jevons, *The Principles of Science* (2nd ed.; London: Macmillan Co., 1924), p. 11.
2. *Ibid.*, pp. 11–12.
3. John F. W. Herschel, *Preliminary Discourse on the Study of Natural Philosophy* (London: Longman, Brown, Green, and Longmans, 1842), sec. 184.
4. Jevons, *The Principles of Science*, pp. 88–90, 91 ff.
5. *Ibid.*, p. 125.
6. *Ibid.*, pp. 122–23.
7. *Ibid.*, pp. 124–25.
8. *Ibid.*, p. 146.
9. *Ibid.*
10. *Ibid.*, p. 148.
11. *Ibid.*, pp. 146–47.
12. *Ibid.*, p. 149.
13. *Ibid.*, pp. 149–50.
14. *Ibid.*, p. 150.
15. *Ibid.*, p. 222.
16. *Ibid.*, p. 226.
17. *Ibid.*, p. 198.
18. *Ibid.*, p. 222.
19. *Ibid.*, p. 221.

20. *Ibid.*, p. 197.
21. *Ibid.*, p. 203.
22. *Ibid.*, p. 204.
23. *Ibid.*, p. 200.
24. *Ibid.*, p. 210. Cf. p. 212.
25. *Ibid.*, p. 198.
26. *Ibid.*
27. *Ibid.*, pp. 198–99.
28. *Ibid.*, pp. 242–43.
29. *Ibid.*, p. 243.
30. *Ibid.*, p. 251.
31. *Ibid.*, pp. 253–55.
32. *Ibid.*, pp. 257–58.
33. *Ibid.*, p. 255.
34. Charles S. Peirce, *Collected Papers*, ed. Charles Hartshorne and Paul Weiss (8 vols.; Cambridge, Mass.: Harvard University Press, 1931–58), Vol. II, par. 747.
35. Ernest Nagel, *Principles of the Theory of Probability* (Vol. I, No. 6, of *International Encyclopedia of Unified Science*) (Chicago: University of Chicago Press, 1939), p. 45.
36. Jevons, *The Principles of Science*, pp. 258–59.
37. *Ibid.*, p. 207.
38. *Ibid.*
39. *Ibid.*, p. 12.
40. *Ibid.*, p. 265.
41. *Ibid.*, p. 267.
42. *Ibid.*
43. *Ibid.*, p. 504.
44. *Ibid.*, p. 511.
45. *Ibid.*, p. 581.
46. *Ibid.*, chapters xxii and xxiii.
47. *Ibid.*, pp. 490–92.
48. Cf. J. O. Urmson, "Two of the Senses of 'Probable,'" in *Philosophy and Analysis*, ed. Margaret Macdonald (New York: Philosophical Library, 1954), pp. 191–99.

CHAPTER TWELVE

Charles Sanders Peirce's Search for a Method

1. Thomas A. Goudge, *The Thought of C. S. Peirce* (Toronto: University of Toronto Press, 1950), p. 157. The next most complete com-

mentary on Peirce's inductive thought will be found in Justus Buchler, *Charles Peirce's Empiricism* (London: Kegan Paul, Trench, Trubner and Co., 1939).

2. Charles S. Peirce, *Collected Papers*, ed. Charles Hartshorne and Paul Weiss (8 vols.; Cambridge, Mass.: Harvard University Press, 1931–58), Vol. II, par. 702. All references to *Collected Papers* are made by noting number of volume and paragraph.

3. *Ibid.*, II, 756 ff.

4. *Ibid.*, II, 758.

5. *Ibid.*, II, 756.

6. *Ibid.*, II, 703; VI, 40.

7. *Ibid.*, II, 725.

8. *Ibid.*, II, 696.

9. *Ibid.*, II, 725; cf. II, 696.

10. *Ibid.*, II, 738.

11. *Ibid.*, II, 740.

12. Cf. Max Black, *Problems of Analysis* (Ithaca, N.Y.: Cornell University Press, 1954), p. 169. Cf. pp. 157 ff.

13. Goudge, *The Thought of C. S. Peirce*, pp. 180–81.

14. Peirce, *Collected Papers*, II, 747.

15. *Ibid.*, II, 746.

16. *Ibid.*, II, 745.

17. *Ibid.*, II, 746; cf. II, 684.

18. *Ibid.*, II, 761.

19. *Ibid.*, VI, 100.

20. *Ibid.*, II, 766.

21. *Ibid.*

22. *Ibid.*

23. *Ibid.*, II, 703.

24. *Ibid.*

25. *Ibid.*, II, 769; cf. V, 575 ff.

26. *Ibid.*, II, 784.

27. *Ibid.*, VI, 40.

28. *Ibid.*

29. *Ibid.*, II, 726.

30. *Ibid.*, VI, 41.

31. *Ibid.*, II, 693.

32. *Ibid.*, II, 624, 636, 640.

33. *Ibid.*, II, 625.

34. *Ibid.*, II, 712.

35. Cf. Goudge, *The Thought of C. S. Peirce*, p. 197.

36. Peirce, *Collected Papers*, II, 776; I, 68.
37. *Ibid.*, II, 96.
38. *Ibid.*, II, 786.
39. *Ibid.*, II, 634.
40. *Ibid.*
41. *Ibid.*, V, 591, 592.
42. *Ibid.*, II, 642.
43. *Ibid.*, II, 657, 658.
44. *Ibid.*, II, 650.
45. *Ibid.*, II, 675.
46. *Ibid.*, II, 758.
47. *Ibid.*, II, 661.
48. *Ibid.*, II, 664.
49. *Ibid.*, II, 665.
50. *Ibid.*, II, 666.
51. *Ibid.*, II, 652, 669.
52. *Ibid.*, II, 657.
53. *Ibid.*, II, 780.
54. *Ibid.*, II, 96.
55. Goudge, *The Thought of C. S. Peirce*, p. 167.
56. The bulk of the commentary on Peirce is not concerned primarily with those concepts of his which have been of particular interest to us. In addition to the commentaries already cited, the following items should be examined: Thomas A. Goudge, "Peirce's Treatment of Induction," *Philosophy of Science*, VII (1940), 56–68, and Philip P. Wiener, chapter on Peirce in *Evolution and the Founders of Pragmatism* (Cambridge, Mass.: Harvard University Press, 1949). Cf. Edward H. Madden, "Charles Peirce e la ricerca di un metodo," *Rivista di Filosofia*, XLIX (1958), 1–20.

CHAPTER THIRTEEN

Chauncey Wright and the American Functionalists

1. Charles S. Peirce, "The Doctrine of Necessity," in *Collected Papers*, ed. Charles Hartshorne and Paul Weiss (8 vols.; Cambridge, Mass.: Harvard University Press, 1931–58), Vol. VI, pars. 35–65. I have broken the chronological order here (Wright was nine years older than Peirce) for three reasons: (1) The Jevons and Peirce chapters form a closely knit unit; (2) Wright was a strong influence on James and, via him, on Dewey and the American functionalists; and (3) Wright leads us to many contemporary discussions of semantics and methodology.

2. Chauncey Wright, *Philosophical Discussions*, ed. C. E. Norton (New York: Henry Holt and Co., 1877), pp. 4 ff., 9 ff., 17 ff., 130 ff., 137–38, 141, 143–44, 173 ff., 177–79, 190 ff., 199–205, 244 ff.
3. *Ibid.*, pp. 137–38, 179.
4. *Ibid.*, pp. 173–74.
5. *Ibid.*, p. 131.
6. *Ibid.*, pp. 131–32.
7. *Ibid.*, p. 141.
8. G. J. Warnock, " 'Every Event Has a Cause,' " *Logic and Language, Second Series*, ed. A. G. N. Flew (Oxford: Basil Blackwell, 1953), p. 96.
9. *Ibid.*, p. 98.
10. Peirce, *Collected Papers*, Vol. VI, par. 60.
11. Wright, *Philosophical Discussions*, p. 4.
12. *Ibid.*, pp. 199–205.
13. *Ibid.*, pp. 199–200.
14. *Ibid.*, pp. 200–1.
15. *Ibid.*, pp. 201–2.
16. *Ibid.*, p. 202.
17. A. W. Burks in his Introduction to the chapter on Peirce in *Classic American Philosophers*, ed. Max Fisch (New York: Appleton-Century-Crofts, 1951), pp. 41–53.
18. Cf. R. B. Braithwaite's discussion of counterfactuals and lawfulness in *Scientific Explanation* (Cambridge, Eng.: Cambridge University Press, 1953), particularly pp. 295–304.
19. Cf. Gustav Bergmann, *The Metaphysics of Logical Positivism* (New York: Longmans, Green and Co., 1954), pp. 262 ff.
20. Cf. C. G. Hempel and Paul Oppenheim, "Studies in the Logic of Explanation," *Philosophy of Science*, XV (1948), 135–75; Nelson Goodman, *Fact, Fiction, and Forecast* (Cambridge, Mass.: Harvard University Press, 1955), pp. 87 ff.; Braithwaite, *Scientific Explanation*, pp. 295 ff.
21. Braithwaite, *Scientific Explanation*, pp. 295 ff.; Goodman, *Fact, Fiction, and Forecast*, pp. 87 ff.
22. Burks, Introduction, pp. 41–53.
23. Philip P. Wiener, *Evolution and the Founders of Pragmatism* (Cambridge, Mass.: Harvard University Press, 1949), pp. 48 ff., and "Chauncey Wright's Defense of Darwin and the Neutrality of Science," *Journal of the History of Ideas*, VI (1945), 19–45.
24. *The Letters of Chauncey Wright*, ed. James B. Thayer (Cambridge, Mass.: John Wilson and Son, 1878), p. 132.
25. Wright, *Philosophical Discussions*, p. 248.
26. Philip P. Wiener, in his reply to Herbert Schneider's review of his

book, *Evolution and the Founders of Pragmatism*, in *Journal of the History of Ideas*, XI (1950), 246–47.

27. E. H. Madden, "Wright, James, and Radical Empiricism," *Journal of Philosophy*, LI (1954), 868–74. Cf. Gail Kennedy's illuminating article, "The Pragmatic Naturalism of Chauncey Wright," *Studies in the History of Ideas*, ed. Department of Philosophy, Columbia University (New York: Columbia University Press, 1935), III, 477–503.

28. Wright, *Philosophical Discussions*, pp. 5 ff., 9.

29. *Ibid.*, p. 9.

30. *Ibid.*

31. This view permeates his "A Physical Theory of the Universe," *ibid.*, pp. 1–34.

32. *Ibid.*, pp. 17–34.

33. *Ibid.*, pp. 43–96.

34. *Ibid.*, p. 87.

35. *Ibid.*, pp. 43–96, 201.

36. Herbert Schneider, *A History of American Philosophy* (New York: Columbia University Press, 1946), p. 348.

37. Wright, *Philosophical Discussions*, p. 206.

38. *Ibid.*, p. 210.

39. *Ibid.*, p. 223.

40. *Ibid.*, p. 229.

41. *Ibid.*, p. 231.

42. *Ibid.*, p. 234.

43. *Ibid.*, p. 206.

44. James R. Angell, "The Province of Functional Psychology," *Psychological Review*, XIV (1907), 61–91.

45. John Dewey, "The Reflex Arc Concept in Psychology," *Psychological Review*, III (1896), 357–70.

46. Cf. also Angell, "The Province of Functional Psychology," pp. 83 ff.

47. Dewey, "The Reflex Arc Concept in Psychology," p. 364.

48. Wright, *Philosophical Discussions*, p. 234.

49. *Ibid.*, p. 219.

50. Ralph Barton Perry, *Thought and Character of William James* (2 vols.; Boston: Little, Brown and Co., 1935), I, 520 ff.

51. Wright, *Philosophical Discussions*, p. 291.

52. James Mill, *Analysis of the Phenomena of the Human Mind*, ed. J. S. Mill *et al.* (2 vols.; London: Longmans, Green, Reader and Dyer, 1869), I, 111 ff., text and footnotes.

53. William James, *Principles of Psychology* (2 vols.; New York: Henry Holt and Co., 1896), II, 359.

54. *Studies in the History of Ideas,* ed. Department of Philosophy, Columbia University (New York: Columbia University Press, 1925), II, 368–69. I have dealt with the subject matter of this chapter in the following articles and have drawn on them for the bulk of this chapter: "Chauncey Wright and the Logic of Psychology" (with Marian C. Madden), *Philosophy of Science,* XIX (1952), 325–32; "Pragmatism, Positivism, and Chauncey Wright," *Philosophy and Phenomenological Research,* XIV (1953), 62–71; "Chance and Counterfacts in Wright and Peirce," *The Review of Metaphysics,* IX (1956), 420–32; "Chauncey Wright's Life and Work: Some New Material," *Journal of the History of Ideas,* XV (1954), 445–55. For a full bibliography on all phases of Wright's work, see these articles and the notes to the Wright chapter in Philip P. Wiener's *Evolution and the Founders of Pragmatism.*

Index